Vernal Pools

Vernal Pools

Natural History and Conservation

Elizabeth A. Colburn, Ph.D.

Harvard Forest
Petersham, Massachusetts

The McDonald & Woodward Publishing Company
Blacksburg, Virginia

The McDonald & Woodward Publishing Company
Blacksburg, Virginia, and Granville, Ohio
www.mwpubco.com

Vernal Pools: Natural History and Conservation

Printed in the United States of America
by McNaughton & Gunn, Inc., Saline, Michigan

15 14 13 12 11 10 9 8   10 9 8 7 6 5 4 3 2

First Printing October 2004
Second Printing April 2008

Library of Congress Cataloging-in-Publication Data

Colburn, Elizabeth A., 1952-
Vernal pools : natural history and conservation / Elizabeth A. Colburn.
p. cm.
Includes bibliographical references (p.).
ISBN 0-939923-92-0 (hardcover : alk. paper) -- ISBN 0-939923-91-2 (softcover : alk. paper)
1. Vernal pool ecology. 2. Vernal pools. I. Title.
QH541.5.P63C64 2004
577.63'6--dc22

2004008637

Contents

Preface

Every year, I know spring has arrived when I first hear the wood frogs and spring peepers calling from the vernal pool behind my neighbor's house. Until recently, few people besides herpetologists and invertebrate zoologists paid attention to these small, seasonally flooded, fishless waters. Since the early 1980s, however, the picture has changed. Research around the world has made people aware of global declines in amphibian populations. Local efforts to document the status of amphibians across the United States and southern Canada have raised general appreciation of frogs, treefrogs, toads, and salamanders as interesting animals, and of vernal pools as important breeding habitats. With awareness has come a new interest in and appreciation of fairy shrimp, water beetles, and other invertebrates. From town conservation boards to elementary school classrooms to federal agencies, vernal pools have become the focus of education and protection. Although the emphasis has been primarily on amphibians, there has been a heightened desire to understand all aspects of the biology of vernal pools, and conservation biologists have become more aware of the broad contributions of vernal pools to regional biodiversity. Increasingly, conservation efforts focus on protecting both vernal pools and adjacent upland habitats, with emphasis on their combined significance for amphibians, turtles, and invertebrates.

During the past twenty-five years I have had the good fortune to study, teach about, and work for the protection of vernal pools and other small wetlands in the formerly glaciated northeastern United States. I have heard repeatedly that conservation efforts are being hampered by gaps in the understanding of the physical characteristics of vernal pools and the ecology of vernal pool wildlife. Again and again, I have been asked by teachers, government regulators, environmental consultants, journalists, naturalists, and researchers to provide information on sources of data and citations of scientific publications. It has become clear that there is a need for an in-depth summary of the scientific and technical literature on vernal pools, and a demand for detailed information about the ecology and biology of vernal pools — beyond that found in field guides and general overviews.

Vernal Pools is my effort to meet this need and satisfy this demand for a better understanding of pools that occur in the formerly glaciated portions of southeastern Canada and the northeastern United States. My goal has been to provide, through a comprehensive review of published literature and of unpublished data from a variety of studies, including my own research, an authoritative synthesis of the current understanding of the habitat characteristics of vernal pools, the plants and animals that live in them, and the factors that govern the interactions among pool organisms and their environment. In addition, I hope that the information presented here, together with the discussion of policy questions associated with vernal pool conservation, will contribute to the long-term protection of vernal-pool-dependent wildlife.

Organization

This book has five major parts: it provides a general overview of vernal pool ecology, an assessment of physical habitat factors, an in-depth summary of the biology and ecology of vernal pool wildlife species and of vernal pools as ecosystems, a discussion of issues and options relevant to the conservation of vernal pools, and a presentation of supplementary material that documents and supports the text.

Chapter 1 introduces vernal pools of northeastern North America. It summarizes the geographic distribution of pools covered in this book, provides a comprehensive definition of vernal pools, and describes each of the characteristics that define pools. This overview is expanded upon and documented in the rest of the book.

Chapters 2 and 3 summarize information on the physical habitat characteristics of vernal pools. Chapter 2 focuses on hydrology, including water sources and flooding regimes. Chapter 3 reviews the origins and positions in the landscape, substrate, surface area and depth, water chemistry, and temperature of pools. Collectively, these chapters provide guidance for conducting inventories of vernal pool habitats, present data that are essential to engineering designs for the restoration or creation of pools, and provide background information for the discussions of species' distributions and life history strategies that follow.

The ten chapters that make up the bulk of the book present the natural history and ecology of vernal pools. Topics include overviews of bacteria, algae, fungi, protists, and higher plants associated with vernal pools; life history strategies of animals that inhabit vernal pools; detailed species accounts of invertebrates and vertebrates that are most commonly encountered in vernal pools of the northeast (other animals found in vernal pools but not diagnostic of them are also discussed, but in less detail); and energy flows, seasonal cycles, and changes in community composition over time.

The final chapter discusses conservation. Topics include relationships among vernal pools; the connections between pools, other freshwater habitats, and uplands; major threats to vernal pools and pool-dependent wildlife; policy questions relating to pool conservation; existing conservation tools; and recommendations for additional conservation efforts.

Following the main text are a glossary of technical terms; an appendix listing all animals reported from northeastern vernal pools, with annotations of their habitat, distribution, and life history; a bibliography; and an index.

Technical Information

Because this book will be used by a variety of readers for many different purposes, the text presents both general summaries of, and technical information about, our knowledge of northeastern vernal pools. Measurements are provided in both metric and standard units. Scientific citations are presented first as superscripts in the main text, then as author and date of record at the end of each chapter, and finally as a full citation in the bibliography. All technical terms are defined in the glossary.

Future Research

Each chapter concludes with a list of suggestions for additional research, including both scientific studies and natural history observations by school groups and amateur naturalists. It is worth keeping in mind that there is considerable overlap among topics covered in different chapters — for example, evaluation of the productivity of vernal pools and of the transfer of energy from pools to upland forests involves hydrology, energetics, population studies on amphibians and insects, and relationships between upland areas and pools. It is hoped that the brief summaries of research questions provided here will stimulate interest among academic and governmental researchers, as well as naturalists, classroom teachers, and others, and will serve as an impetus for funding to support additional studies on vernal pools.

Acknowledgments

Funding for the collection of some of the original data presented in this book, and for the preparation of the manuscript, was provided by Sweet Water Trust, the National Fish and Wildlife Foundation, Dr. Josie Murray, the Fund for Protection of Wildlife and Natural Areas, the National Park Service, The Fuller Foundation, and the Massachusetts Audubon Society's Center for Biological Conservation. A Charles Bullard Fellowship from the Harvard Forest, Harvard University, provided additional support for final writing and editing.

I want to thank the following individuals for insights provided in conversations about temporary waters and their inhabitants at various times over the past fifteen years: Len Ambergy, Kyla Bennett, Bob Brooks, Beth Bullock, Matt Burne, Aram Calhoun, Joe Choiniere, Bob Cook, Danielle DiMauro, Len Ferrington, Marilyn Flor, David Foster, Steve Golladay, Judy Helgen, Mike Higgins, Gail Howe, Pat Huckery, Scott Jackson, Leo Kenney, Elissa Landre, Stafford Madison, Wende Mahaney, Harold Manners, Mark McCollough, Peter Paton, Steve Pelletier, Chris

Phillips, Sigrid Pickering, John Portnoy, Bob Prescott, Alan Quackenbush, Heidi Ricci, Steve Roble, Donald Schall, Daniel Schneider, Matt Schweisberg, David Small, Douglas Smith, Janice Stone, Robert Sobczak, Bob Speare, Tom Tyning, Doug Williams, and Bryan Windmiller. Special acknowledgement and thanks are due to Rosemary J. Mackay, who thirty years ago shared her enthusiasm for vernal pools and their caddisflies with a fledgling graduate student, who served subsequently as a role model and inspiration extraordinaire for scientific excellence and high editorial standards, and whose comprehensive review and helpful comments on the manuscript of this book contributed to a much better product.

The following people helped me collect some of the original data included here: Maria Aliberti, Carol Batdorf, Emily Brunkhurst, Pat Buckley, Matt Burne, Lachelle Campbell, Ed Faison, Fran Garretson, Ricky Holt, Nele Janssen, Holly Jensen-Herrin, Dave King, Claudia Lipschitz, Joan Milam, Kate Musgrove, Jarol Olarte, Polly Patterson, Katy Rolih, Fred Saint-Ours, Peter Severance, Jackie Sones, Bruce Wenning, and Gina Yazzie. Their hard days in the field and long hours at the microscope are greatly appreciated.

Chris Leahy, Joe Larson, Joan Milam, and John Mitchell provided much-appreciated assistance and advice during my search for a publisher. I feel very fortunate to have discovered McDonald and Woodward, and I owe a great debt of thanks to Jerry McDonald for his enthusiasm for the project, his patience, his editorial guidance, and his support.

I would like to thank the following people for helping me in my efforts to check facts on obscure species, verify taxonomy, or clarify details of their research: Rebecca Newcomb Homan, Eugen K. Kempf, Cathy Johnson, Koen Martens, Doug Smith, Stephen Weeks, and Grace Wyngaard. Thanks, too, to Ian M. Smith, who contributed substantially to my understanding of water mites and clarified my presentation of this group.

Special acknowledgement and thanks are due those colleagues whose prowess with the camera produces wonderful images of small animals and vernal pools, and who agreed to have their photos included in this book. They include Aram Calhoun, Scott Cooper, David Foster, Fran Garretson Clapp, Scott Jackson, Holly Jensen-Herrin, Leo Kenney, Gary Meszaros, Daniel Schneider, and Steve Weeks. Thanks, also, to Doug Smith for permission to include his illustrations of fairy shrimp. I am grateful, as well, to the Massachusetts Audubon Society for permission to use the watercolor of a vernal pool in early spring by Barry van Dusen. Thank you to Chris Leahy and Kristin Eldridge for tracking down the original artwork!

I am indebted to the family and estate of Ann Haven Morgan for permission to reproduce illustrations from *Field Book of Ponds and Streams*, originally published by George Putnam and Sons in 1930 and long out of print. Dr. Morgan's book introduced me to freshwater biology, and I am delighted to be able to share some of the pictures that have fascinated me since I was a young girl.

Preface

This book would not have been possible without the efforts of the naturalists and biologists who have studied vernal pools and their wildlife over the past 250 years. I am inexpressibly thankful for all of those who have collected plants and animals, worked to identify and classify them, spent time studying their natural history and ecology, and written down the results of the research for the benefit of others.

Over time, this book evolved from an abbreviated summary to a comprehensive review. Many people helped me over the years of research and writing. Kyla Bennett, Aram Calhoun, and Wende Mahaney kindly reviewed the first draft manuscript and provided invaluable comments and suggestions. Joseph Larson gave me helpful advice on formatting and text flow. Chris Leahy, Joan Milam, Thomas J. Rawinski, and Ian M. Smith reviewed the sections on odonates, turtles, vegetation, and water mites, respectively. The late John Post contributed to the text on conservation restrictions, and Aaron Ellison, Ed Faison, David Foster, Holly Jensen-Herrin, Brooks Mathewson, and John O'Keefe provided additional feedback on the conservation chapter. My husband, Lee Mirkovic — who has good-naturedly lived in a chaos of nets, waders, specimens, keys, journal articles, piles of draft manuscripts, and with a distracted wife, for most of our marriage — applied his editor's eye to the final draft and rescued some of the more obscure sentences. Editorial reviews and suggestions from Rosemary Mackay, Judy Moore, and Jerry McDonald contributed immensely to improvements in the manuscript. My mother, Alice B. Robinson, and brother, Ted Colburn, shared the critical job of proofreading at the final stage of production. I am grateful for the time and care all of them gave to this task. Any errors and flaws are my responsibility alone.

Finally, I want to acknowledge and thank all those people who head out to vernal pools each spring to celebrate the migration of mole salamanders and wood frogs, who glory in the sight of fairy shrimp and the sound of spring peepers, who are working to make others aware of the fascinating biology of these tiny wetlands, and who are involved in the fight at the local, state, provincial, regional, or national level to preserve vernal pools and their associated woodlands for future generations of wildlife and people. This book is dedicated to them.

Spring Pool

These pools that, though in forests, still reflect
The total sky almost without defect,
And like the flowers beside them, chill and shiver,
Will like the flowers beside them, soon be gone,
And yet not out by any brook or river,
But up by roots to bring dark foliage on.

The trees that have it in their pent-up buds
To darken nature and be summer woods—
Let them think twice before they use their powers
To blot out and drink up and sweep away
These flowery waters and these watery flowers
From snow that melted only yesterday.

— Robert Frost

Chapter 1

Introducing Vernal Pools

Most people who have spent much time in the woods are aware of the small ponds that appear in low areas in early spring and often disappear by mid to late summer. These small, fishless water bodies are known popularly as vernal pools. They are homes for remarkable animals including the aptly named fairy shrimp — so-called both because of their tendency to appear in certain pools unpredictably from one year to the next and because, like spring ephemeral wildflowers, they complete their life cycle early and quickly, and are gone well before summer begins. Vernal pools are best known for their role in the lives of woodland amphibians, including wood frogs and mole salamanders (especially spotted, blue-spotted, Jefferson's, marbled, and small-mouthed salamanders), and a variety of invertebrates, including clam shrimp and fairy shrimp (Plate 1). All of these animals depend on vernal pools as breeding sites.

There is something almost magical about visiting a flooded woodland hollow in early spring and seeing hundreds of salmon-colored fairy shrimps swimming lazily above the submerged leaves, watching caddisflies in their miniature log-cabin houses lumbering along the bottom, and catching sight of a spotted salamander as it journeys to the surface for a gulp of air during its brief sojourn in the pool. Hearing the short-lived concert provided by quacking wood frogs is a pleasure eagerly anticipated during long late-winter days when it seems as though spring will never come. Each year, the burgeoning of life in tiny pools across the landscape carries a message of renewal and hope. It is an annual miracle, a uniquely miniature aquatic world, distinct from the surrounding woodlands, yet, as we shall see, dependent upon them.

In some cases, the pool whose crystal-clear water seethed with activity in early spring is an apparently stagnant mudhole when visited in midsummer. Look more closely — the basin is teeming with salamander larvae, snails, paddle-legged water bugs of various kinds and colors, an infinite variety of water beetles, worms, midge larvae, recently transformed wood frogs and spring peepers, and toad tadpoles with tiny legs and rapidly disappearing tails. In other instances, all that remains of the spring pool is a low area covered by ferns or stained leaves. If you have the right search image, you can find empty caddis cases and perhaps a few scattered snail shells or the valves of fingernail clams littering the bottom. Under the leaves are eggs, cysts, and other resting stages of animals ready to emerge and start the cycle anew. All around the pool, the woods shelter frogs, toads, and salamanders who will return to the pool again next spring to mate and lay their eggs.

Vernal pools are interesting for many reasons, including their unique contributions to biodiversity. These waters differ from other freshwater habitats in their alternation between extended flooding and periodic drying. This distinguishing hydrologic cycle gives aquatic animals enough time to complete their life cycles during the flooded phase, while drawdown keeps out fish and allows for the production of nutritious food from the damp, decaying leaves. Lacking predatory fish and offering abundant food, vernal pools provide specialized habitats for a remarkable variety of wildlife species that do not occur in permanently flooded waters that do contain fish.

A particularly attractive feature of vernal pools is their accessibility. They are small and shallow, and many of their most characteristic animal inhabitants are readily identified and highly visible in early spring. Anyone with an interest in natural history — whether a third-grade class recording the date when salamander egg masses first appear in a pool or a serious student of water beetles investigating the different species making use of pools with different flooding regimes — can readily locate and learn about vernal pools and their wildlife.

This chapter introduces vernal pools of glaciated northeastern North America. After defining the geographic area covered by this book and providing some insights into the long-standing debate about the term "vernal pool," it provides a five-part working definition of vernal pools and presents an overview of each of the definition's components. More detailed information about each defining factor is provided in later chapters.

Geographic Distribution of Vernal Pools

Temporary pools are found worldwide. Seasonally flooded habitats and their faunas share many common features, whether they occur in deserts, prairies, large river floodplains, alpine meadows, or temperate forests. In this book I concentrate on woodland vernal pools of the glaciated northeastern third of North America — "the glaciated northeast" (Figure 1). The area covered extends well beyond the region commonly called the Northeast in the United States — it includes areas in Manitoba, Ontario, Quebec, and the Canadian Maritime Provinces; the New England states; New York; northern New Jersey; northern Pennsylvania; Michigan; Wisconsin; eastern and northern Minnesota; wooded areas of Illinois, Iowa, northern Missouri, eastern Kansas, and eastern Nebraska; and much of Ohio and Indiana. These pools occur in landscapes that have been glaciated relatively recently in geologic time, that were predominantly forested before European settlement, and that tend to revert to forest when left alone unless they have been severely altered.

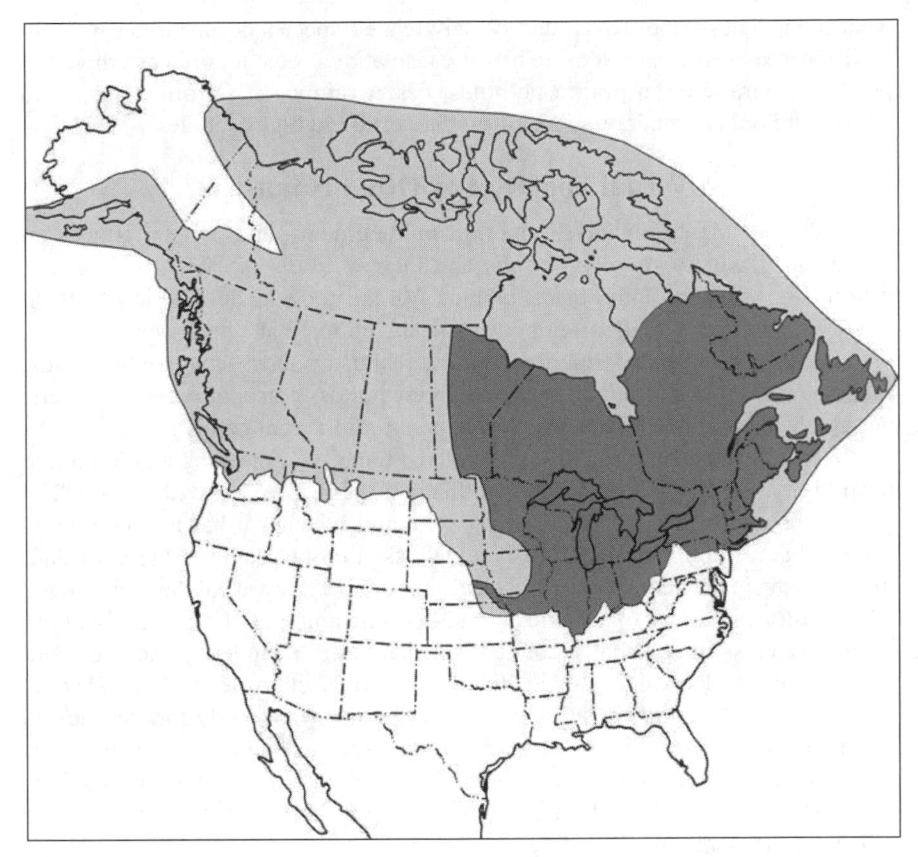

Figure 1. The lighter gray shading on this map shows the part of North America that was covered by large continental and montane glaciers during the Pleistocene — the most recent period of extensive glaciation. The darker gray identifies the area covered in this book and which, for convenience, is referred to as "the glaciated northeast." This area corresponds with the eastern portion of the formerly glaciated area that supported primarily forests or woodlands on the eve of the European settlement of North America. At the peak of glaciation, the ice extended farther to the northeast, and about AD 1600 the forests extended farther to the northwest, than is shown here.

While the focus is on the glaciated northeast, much of the information on habitat characteristics and biota of northeastern vernal pools applies to temporary waters elsewhere. Physical features of temporary pools are similar along the Central Atlantic Coastal Plain[1], in northwestern North America[2], in the southeastern United States[3], and in Europe[4]. Many of the species of amphibians and invertebrates found here range south through the Mid-Atlantic states to northern Florida and west to the Mississippi River, and closely related species occur farther west. In addition, seasonal cycles and life history adaptations seen in northeastern vernal pools are paralleled in prairie potholes, desert rainpools, Carolina Bays, and snowmelt pools in tundra regions of the far north and high altitudes.

A Vernal Pool by Any Other Name . . . .

Books, scientific papers, government documents, and popular articles on seasonally flooded woodland ponds and their wildlife use a wide variety of terminology for describing these habitats. Vernal pools as defined in this book have been variously called temporary ponds, ephemeral ponds, spring pools, woodland pools, semi-permanent ponds, fishless ponds, salamander ponds, intermittent woodland pools, seasonal forest ponds, seasonally astatic waters, geographically isolated wetlands, vernal pools, and vernal ponds[5].

There has been heated discussion about the proper meaning and use of the term "vernal pool," and the issue is unlikely to be resolved any time soon. The debate centers on popular usage of the term and whether it has a scientifically accepted definition. "Vernal pool" is widely used in the scientific literature and, increasingly, in the general print and electronic media. A search of the World-Wide Web illustrates the lack of a single, agreed-upon meaning[6]. In California, the term refers to a class of seasonally flooded waters that dry completely each year and support unique plants and animals[7]. In northeastern North America, "vernal pool" commonly refers to any small, shallow, fluctuating water body that reaches its maximum size in spring, lacks fish, and provides breeding habitat for certain species of woodland amphibians and/or characteristic invertebrate species[8]. This interpretation of vernal pools includes water bodies that occur over a broad range of timing and duration of flooding.

Some people feel very strongly that "vernal pool" should be used narrowly only to define pools that fill in spring ("vernal" is derived from the Latin word "vernalis," which means "pertaining to the spring") and dry sometime later the same year. However, even in the scientific literature the term is used to refer to pools that fill and dry in spring, pools that fill in spring and dry any time during the year, and pools that dry in spring regardless of when they fill[9]. All of these types of pools, as well as semi-permanent ponds that lack fish and meet the other physical and biological criteria of vernal pools listed below, fall under the umbrella definition of vernal pools in this book.

Unless vernal pools are being visited regularly and do not become buried under deep snow in winter, I have found in practice that it is difficult to determine precisely when many of them start to fill with water. Some pools almost never dry completely; instead they hold some water in the deepest points of their basins during all but the severest droughts. Some pools dry in summer and begin to refill in fall as groundwater tables rise and deciduous trees drop their leaves. But, all vernal pools reach their maximum depths and surface areas in the spring. Further,

springtime is when we
conservation is at the
sympathetic to the argu
that fill in spring, using

In this book, vernal p
relating to woodland contex
as follows.

- *Woodland context*:
areas.
- *Isolation*: vernal poo
inlet or outlet, and they ha
permanently flooded water bo
- *Size*: vernal pools are
- *Hydrology*: vernal poo ... ximum water levels and volumes in the sprin ... ually or every few years — standing water either disappear ... mpletely, or water levels drop substantially during the summer, exposing most of the pool bottom and retaining only a fraction of the peak volume; and they contain water for a minimum of two months in most years.
- *Biological community*: vernal pools lack established fish populations, and they provide habitat for animals that are adapted to seasonal drawdown and require the absence of fish predation for successful reproduction.

Each of these defining characteristics is discussed briefly, below. Later chapters provide additional information.

Woodland Context

In their natural state, most northeastern vernal pools are found in or adjacent to wooded areas. Temporary pools also occur in grassland regions and in deforested portions of the glaciated northeast, but the composition of the biological community in grassland pools differs from that in woodland pools[10]. Grassland temporary pools are not discussed in detail in this book.

Trees contribute to a pool's food web, help maintain cool water temperatures, and affect pool hydrology. Food, temperature, and the duration of flooding all affect the development of aquatic animals and can influence which species are present. Thus, trees and other plants surrounding or growing in a vernal pool can have a great influence on the animals that live in the pool.

The annual inputs of leaves from the trees around vernal pools support a detritus-based food web. The feeding behavior of vernal pool animals reflects the importance of leaves in the pool's economy. Animals such as caddisflies and isopods, common in many vernal pools, are described as "shredders" because

...er pieces. "Collectors" as varied as filter-
...np, and sediment-ingesting worms and midge
...ns of decomposing leaf litter and its community of
...es of different kinds of trees vary in their nutritional
... leach out of them into water, and how much aquatic
... as food. Nutrients released by the decaying leaves support
...ae, the extent of which is also affected by the amount of canopy
...ed by nearby trees — the more open the canopy, the greater the
... algal growth.

...hading by trees also helps moderate water temperatures. The eggs of many ...uatic species hatch in response to thermal cues, and the growth rates of aquatic animals are often closely tied to water temperatures. Shading can also help limit evaporation. On the other hand, during the growing season, trees in and around vernal pools take up water and contribute to pool drawdown. Because different species of trees differ in their water use and, if they are species that lose their leaves annually, in the timing of springtime leaf-out and autumn leaf-fall, differences in the composition of the plant community surrounding vernal pools can have a great influence on the pools' flooding patterns and temperature regimes.

Upland forests, as well as plants immediately adjacent to a pool, affect the wildlife of vernal pools. The amphibians that breed in the pools are forest animals that spend most of their lives in the upland woods and play an important role in energy flow on the forest floor. Juvenile amphibians are recruited from vernal pools into surrounding forests in astounding numbers, but the ecological roles of these amphibians in temperate forests, unlike the roles of their tropical cousins, are not well documented.

Recent literature on the conservation of forest biodiversity emphasizes the importance of vernal pools as part of the forest ecosystem[11]. In a survey of the wildlife habitat value of wetlands in New York, vernal pools ranked highest of all wetlands in total number of amphibian species, number of species per site, and total number of amphibian larvae[12]. Amphibians were associated only with wooded sites or with sites wooded at the edges. Various studies have noted the relationship between the contiguity of wooded upland and amphibian breeding in vernal pools[13]. In fact, some vernal-pool-dependent amphibians will not cross open areas to reach breeding pools.

Isolation

Vernal pools occur in isolated depressions or basins with defined boundaries. The completeness of physical isolation varies — some vernal pools are completely surrounded by upland woods, some are fed and drained by intermittent streams, some receive spillover flooding from adjacent water bodies, some are hydrologically connected to other waters by groundwater, and some occur within larger freshwater wetlands and are part of the overall wetland system — but pools do not have continuous surface-water connections to other waters during most of their biologically active season. The lack of a year-round surface-water connection with permanently flooded water bodies helps vernal pools remain fishless. All vernal pools are biologically connected to other pools, wetlands, and permanent waters by the dispersal of pool animals.

Depressions that contain pools include former stream channels and floodplain ponds, glacial kettle holes, swales, perched basins on relatively impermeable soils or bedrock, woodland hollows, and pit-and-mound topography within wooded swamps. Vernal pools also have developed over time in excavations created by human activity. Sometimes small numbers of vernal-pool-dependent amphibians may breed in depressions created by wind-throw of forest trees, but such small basins often fail to retain water long enough to support the biological communities typical of vernal pools. Vernal pools as defined here do not include streams and roadside ditches.

Pool Size

Vernal pools span a range of surface area and depth, but they generally are small and shallow (see Chapter 3) — even the largest pools are small compared to most permanently flooded ponds and lakes. Small size has a number of physical and biological effects. It is one reason why most pools dry annually. The shallow water is readily mixed by the wind, which helps to maintain oxygenation throughout the water column. Because vernal pools are shallow, water temperatures increase rapidly as spring progresses toward summer, stimulating growth rates of animals in the pools. Shallow water depths allow sunlight to reach the substrate, contributing to photosynthesis by algae — at least early in the season before canopy closure and increased water color limit light penetration. During the summer, the surrounding trees shed dense shade on the smallest and shallowest pools, limiting plant growth. Progressively larger pools with more of their basin open to the sky are apt to have more of the bottom covered by plants.

Hydrology

Hydrology involves all aspects of the presence and movement of water, including water sources, patterns of change in water levels, and the timing and duration of flooding (see Chapter 2). For vernal pools, a fluctuating water regime with alternating periods of flooding and drying is a fundamental characteristic (plates 2, 3). Although regular drying is important, vernal pools stay flooded for substantial periods of time — two to three months at a minimum, and often much longer. Due to this relatively extended period of flooding, as opposed to the one to two weeks of inundation typical of ephemeral waters such as rainwater pools, a great variety of animals can complete their development in vernal pools. The duration of flooding ("hydroperiod") affects the composition of the aquatic community. The shorter the period of inundation, the faster animals living in the pool need to complete their life cycles, and the greater must be their ability to avoid pool drying or to withstand desiccation.

The Importance of Drying

The periodic drying of vernal pools means that most aquatic animals cannot live in them because relatively few such species are equipped to withstand the disappearance of water. Pool drying prevents fish populations from becoming established. It limits the distribution of large, invertebrate predators, including some water beetles and dragonflies. It excludes some amphibians that typically need to overwinter in water as tadpoles, including green frogs and bullfrogs. The periodic exposure of the substrate to air affects the decomposition of organic detritus and increases the quality of food. The high nutritional value of detritus that is exposed to terrestrial bacteria and fungi may help the fauna of vernal pools compensate for the limited growing season that is imposed by seasonal drawdown (see Chapter 13). Seasonal drawdown thus allows those species with adaptations for withstanding or avoiding desiccation to take advantage of the abundant food and reduced predation in pools that are flooded for only a few months each year.

Flooding Patterns

Fed by melting snow and spring rainfall, northeastern vernal pools reach their largest surface area and contain their greatest volume of water in spring. This seasonal maximum in water levels, as well as the springtime burst in biological activity, is the reason that the term "vernal pool" is applied to these habitats. Although they are deepest and largest in the spring, many vernal pools start to fill in fall or during mid-winter thaws.

Hydrologically, vernal pools lie within a continuum of flooding frequency and duration, bounded at the dry end by ephemeral pools that flood briefly in response to rainfall but dry in only a few days or weeks, and at the wet end by permanent ponds. Ephemeral pools support their own unique assemblages of species, some of which are also able to use vernal pools, but they do not support the biological communities that characterize vernal pools. Vernal pools, therefore, range from relatively short-lived, spring-filling pools that normally are dry by early summer, to waters that dry only occasionally, sometimes as rarely as once every ten years.

"Annual vernal pools" are those pools that usually fill and dry every year. Depending on the timing of filling and drawdown, the duration of flooding in annual pools varies from as short as two to three months to as long as eleven months. Only species that can withstand seasonal drying are able to complete their life cycles in these pools.

A small proportion of vernal pools remains continuously flooded for several years at a time, in some cases drying only every five to ten years. Although these "semi-permanent vernal pools" do not become dry each year, they are shallow, they lack fish, and they include species that require fishless conditions and can survive the periodic drying. Semi-permanent pools commonly experience substantial annual drawdown, but they differ from annual vernal pools by retaining a significant volume of water in most years. They thus continue to provide aquatic habitat for plants, invertebrates, and vertebrates throughout the summer and fall when most annual vernal pools are dry. As a result, during their periods of continuous flooding, semi-permanent pools are also likely to support some species that cannot withstand drawdown. Some of these species probably experience periodic local extinction when the pools do dry, and they become reestablished by colonizers from other habitats when the pool floods again. Semi-permanent pools are usually larger than the average vernal pool, and they may be particularly important for the viability of some amphibian populations.

Sometimes, wildlife species that are typical of vernal pools occur in water bodies that are permanently flooded, often as a result of human alteration. Unless they are shallow enough that they freeze completely in winter, presumably, such ponds have the potential to support fish populations; indeed, the literature contains a number of references to vernal pools that were dammed, excavated, or otherwise altered expressly for the purpose of establishing fish. Once fish are introduced into a pool, however, most vernal-pool-dependent species can no longer successfully complete their development there. Such artificially modified water bodies will not be discussed in any detail in this book.

Within-Pool Hydrologic Variability

Over the long term, individual vernal pools fill and dry according to a relatively predictable annual or multi-year cycle. However, the frequency and duration of flooding in any pool shifts depending on the weather conditions in a given year. Accordingly, the animal species that successfully exploit these variable habitats possess great flexibility in the details of their life cycles and habitat requirements.

The Biological Community

For vernal pools in the glaciated northeast, the two defining biological characteristics are the absence of fish predation, which is considered to be a key factor behind the presence of many vernal-pool-dependent animal species[14], and the presence of species that can survive pool drying. Many of the animals that occur in vernal pools are not found in other waters. In a given pool, the species present depend on the site's physical characteristics and land-use history, geography, the presence of other species, and chance.

Plants of Vernal Pools

Although northeastern vernal pools are defined by their faunas, they also support a variety of plants. Like animals, plants in vernal pools must contend with seasonal inundation followed by drawdown. Their distributions are affected by the depth and duration of flooding, the nature of the substrate, and the availability of nutrients and sunlight. In contrast to animal species, the plants found in vernal pools in the glaciated northeast are typical wetland species found locally in other habitats, including marshes and swamps (Chapter 5). This situation is different from that in temporary waters in areas that have not been glaciated in the recent past, such as California vernal pools and Carolina Bays, which are characterized by unique plant communities with many endemic species[15].

Vernal Pool Animals

Northeastern vernal pools support distinctive faunas of invertebrates and vertebrates. From a biodiversity perspective, the number of invertebrate species that has been reported from vernal pools is particularly impressive. The importance of vernal pools as breeding habitat for amphibians such as mole salamanders and wood frogs has received particular attention over the past decade, and recent research has demonstrated that pools are important feeding areas for reptiles such as spotted and Blanding's turtles.

The number of taxa found consistently in annually drying vernal pools is large. These taxa are listed in Table 1 and the Appendix, and they are emphasized in chapters 7 to 12. Some of these species are restricted to habitats that dry seasonally and/or breed preferentially in habitats that lack predatory fish. Their reproductive success, and hence the maintenance of their populations, depends on the physical, biological, and hydrologic conditions in vernal pools. These species, sometimes described as obligate vernal pool species, or indicator species, are identified with an asterisk (*) in Table 1, and they best typify the vernal pool fauna. As suggested in the definition presented earlier in this chapter, animal communities that include a number of these species can be considered to be indicative of vernal pools.

More than 550 species of multicellular animals have been reported from northeastern vernal pools. If animals that have been identified only to the family or generic level are included, the number exceeds 700. It is beyond the scope of this book to discuss all the animals found in vernal pools; instead the focus is on those groups and species that are found most commonly in, and that are typical of, northeastern pools. A detailed, annotated list of all species reported in the literature from vernal pools is presented in the Appendix. This list includes species from semi-permanent pools as well as from spring- and fall-filling annual vernal pools, provides information on pool habitat characteristics, and summarizes information on life history strategies. The life history strategies referenced in the Appendix are discussed in greater detail in Chapter 6.

Some of the species found consistently in vernal pools occur in a wide variety of aquatic habitats, ranging from temporary pools to permanent waters. They are commonly described as facultative vernal pool species because they do not require the unique habitat conditions of vernal pools — seasonal drawdown and lack of fish predation — to maintain viable populations. However, in some geographic areas, vernal pools may provide the primary breeding habitat for some of these species, and these localized populations are indeed dependent on vernal pools.

Certain features of Table 1 and the Appendix are particularly noteworthy. First, species richness in northeastern vernal pools is centered in the microcrustaceans and the aquatic insects, particularly the water beetles (Coleoptera), true flies (Diptera), water bugs (Hemiptera), and caddisflies (Trichoptera). In temporary pools in central Italy and Australia, and in Carolina Bays in the southeastern United States, microcrustacean diversity, particularly in the cladocerans and copepods, is also high, and some of the genera — and even some of the species — are the same as in northeastern pools[16]. Second, relatively few species are restricted to a single class of vernal pools, such as short-hydroperiod, spring-filling pools. Rather, most taxa have been reported from a range of temporary habitats. It is unclear whether this apparent breadth in species' distributions along the vernal pool hydrologic gradient represents actual plasticity in habitat preferences or whether it reflects inadequate descriptions of the hydrologic conditions in pools. Third, even in a single family, vernal pool wildlife species display a remarkable diversity of strategies for surviving the period when the pool is dry. These themes will be explored further in the chapters on the biology of vernal pool wildlife.

Local and Regional Biodiversity

The contributions of vernal pools to local and regional biodiversity have received little attention. Biodiversity can be considered in several ways. One is in terms of the total number of species in a given location — known as species richness or alpha (α) diversity. Another is in terms of the similarities or differences in species composition between sites, or along gradients of habitat characteristics. These between-habitat differences are known as beta (β) diversity. A third way is in terms of overall regional or landscape richness — gamma (γ) diversity.

Taxonomic richness and species' distributions vary widely among vernal pools. Some differences are geographic. Comparisons of the overall species composition of vernal pools in different locations in Ontario, Michigan, and Massachusetts show both overlap in species and marked differences, with the distributions of some taxa apparently restricted geographically[17].

Species richness may differ dramatically between adjacent pools. Further, two pools that are close to each other and with similar numbers of species — pools with comparable alpha diversity — may show very little overlap in the composition of their biota. In other words, vernal pools show high levels of beta diversity. Significant environmental factors contributing to this beta diversity include water chemistry, vegetation, and hydroperiod (see chapters 6 to 13 for further discussion).

Within a geographic area, the presence of complexes of vernal pools contributes to greater overall regional biodiversity, or gamma diversity, as a result

Table 1. Species of Animals that are Characteristic of Northeastern Vernal Pools

Vertebrates
Amphibians
Ambystomatidae, mole salamanders
Ambystoma laterale, blue-spotted salamander
Ambystoma jeffersonianum, Jefferson's salamander*
Ambystoma laterale x jeffersonianum, diploid, triploid, and tetraploid hybrids*
Ambystoma laterale x jeffersonianum x tigrinum x texanum, various hybrids
Ambystoma maculatum, spotted salamander*
Ambystoma opacum, marbled salamander*
Ambystoma texanum, small-mouthed salamander*
Ambystoma tigrinum, tiger salamander
Ranidae, true frogs
Rana sylvatica, wood frog*
Hylidae, treefrogs
Pseudacris crucifer, spring peeper
Pseudacris triseriata triseriata, western chorus frog*
Reptiles
Emydidae, turtles
Clemmys guttata, spotted turtle
Emydoidea blandingii, Blanding's turtle

Crustaceans
Branchiopods
Anomola, water fleas or cladocerans
Chydoridae
Alona guttata, A. rectangula
Chydorus sphaericus
Daphniidae
Ceriodaphnia quadrangula, C. reticulata
Daphnia ephemeralis, D. obtusa, D. pulex*
Scapholeberis kingi
Simocephalus exspinosus, S. serratulus, S. vetulus
Anostraca, fairy shrimp*
Branchinectidae*
*Branchinecta paludosa**
Chirocephalidae*
Eubranchipus spp.*, including *E. bundyi*, E. holmani*, E. intricatus*, E. neglectus*, E. ornatus*, E. serratus,* E. vernalis**
Streptocephalidae*
Streptocephalus paludosus, S. seali**

(continued)

Table 1. (continued)

Laevicaudata, clam shrimp*
  Lynceidae*
    *Lynceus brachyurus**
  Copepods
  Calanoida, Diaptomidae
    Diaptomus stagnalis
  Cyclopoida, Cyclopidae
    Cyclops haueri, C. vernalis*
    *Diacyclops bicuspidatus, D. navus**
    *Megacyclops latipes**
    Microcyclops varicans rubellus
  Harpacticoida, Canthocamptidae
    Attheyella americana
    Bryocamptus spp., *minutus* complex
    Canthocamptus staphylinoides
  Ostracods or seed shrimp
  Candonidae
    Candona spp., including *C. candida*, C. decora*, C. distincta, C. hartwigi*, C. inopinata**
    Pseudocandona albicans, P. hartwigi**
  Cyclocyprididae
    *Cyclocypris laevis**
    Cypria spp., including *C. exculpta, C. ophthalmica*, C. palustera**
  Cyprididae
    *Bradleycypris tincta**
    Bradleystrandesia fuscata, B. splendida**
    Cypridopsis vidua
    Cypris pubera, C. subglobosa*
    *Eucypris crassa, E.hystrix**
    *Sarscypridopsis aculeata**
    Spirocypris horridus
  Notodromadidae
    *Cyprois marginata**

Insects

  Coleoptera, beetles
  Dytiscidae, predaceous diving beetles
    Acilius spp., including *A. mediatus, A. semisulcatus*, A. sylvanus*
    *Agabetes acuductus**
    Agabus anthracinus, A. bifarius, A. confinis, A. erichsoni*, A. opacus*, A. phaeopterus*, A. semipunctatus**
    *Colymbetes sculptilis**
    *Dytiscus fasciventris**

(continued)

Table 1. (continued)

Hydroporus spp.*, especially *H. despectus, H. fuscipennis, H. niger, H. notabilis, H. tenebrosus*
*Hygrotus impressopunctatus**, *H. laccophilinus**, *H. sylvanus,** *H. turbidus**
*Ilybius biguttulus, I. discedens, I. ignarus**
*Laccophilus biguttatus, L. fasciatus**, *L. maculosus, L. proximus, L. undatus**
*Laccornis difformis**
*Liodessus affinis, L. fuscatus**
*Matus bicarinatus**
Neoporus lobatus, N. undulatus
*Rhantus binotatus**, *R. consimilis*
*Uvarus suburbanus**
Gyrinidae
Gyrinus affinis, G. lecontei
Haliplidae, crawling water beetles
*Haliplus blanchardi, H. borealis, H. immaculicollis**, *H. longulus, H. ohioensis**, *H. subguttatus*
*Peltodytes edentulus**, *P. muticus**, *P. pedunculatus, P. tortulosus*
Helophoridae
*Helophorus aquaticus**, *H. grandis**, *H. lacustris**, *H. lineatus**, *H. orientalis**
Hydrophilidae, water scavenger beetles
Anacaena limbata
Cymbiodyta sp.*
*Enochrus cinctus, E. hamiltoni**, *E. ochraceous*
Hydrobius fuscipes
*Hydrochara obtusatus**
Tropisternus blatchleyi modestus, T. mixtus, T. natator
Diptera, true flies
Ceratopogonidae, no-see-ums, biting midges, punkies, or sandflies. Common genera include:
Alluaudomyia, Atrichopogon, Bezzia, Palpomyia
Chaoboridae, phantom midges
Chaoborus americanus, C. flavicans, C. punctipennis
Eucorethra underwoodi
*Mochlonyx cinctipes**, *M. velutinus**
Chironomidae, non-biting midges. Many genera and species; very little work has been done on chironomids in vernal pools. Common members of this group include:
Acricotopus sp., *Chironomus* spp., *Corynoneura* sp., *Cricotopus* sp., *Einfeldia dorsalis, Endochironomus nigricans,*

(continued)

Table 1. (continued)

Eukiefferella sp., *Guttipelopia* sp., *Hydrobaenus* spp., *Limnohyphes* spp., *Micropsectra* spp., *Paraphenocladius* spp., *Paratanytarsus* sp., *Prodiamesa* sp., *Psectrocladius* sp.,
Psectrotanypus daryi, *Smittia* sp., *Trissocladius* sp.
Culicidae, mosquitoes
Aedes spp., especially *A. canadensis**, *A. communis**, *A. diantaeus**, *A. euedes*, *A. excrucians*, *A. fitchii**, *A. grossbecki**, *A. intrudens**, *A. provocans*, *A. punctor**, *A. stimulans**, *A. vexans*
Psorophora sp.
Sciomyzidae, marsh flies
Antichaeta melanosoma
Hedria mixta
Renocera spp.
Sepedon fuscipennis
Tabanidae, horseflies and deerflies
Chrysops sp.
Tabanus sp.
Tipulidae, craneflies
Tipula spp.
Ephemeroptera, mayflies
Arthropleidae
Arthroplea bipunctata
Baetidae
Callibaetis spp., including *C. ferrugineus*
Cloeon sp.
Hemiptera, water bugs
Corixidae, water boatmen
Callicorixa alaskensis, *C. audeni*.
Hesperocorixa spp., including *H. atopodonta*, *H. michiganensis*, *H. vulgaris*
Sigara spp., including *S. alternata*, *S. decoratella*, *S. grossolineata*, *S. knighti*, *S. modesta*, *S. solensis*
Gerridae, water striders
Gerris buenoi, *G. marginatus*, *G. remigis*
Limnoporus dissortis
Notonectidae, backswimmers
Notonecta undulata
Odonata, Anisoptera, dragonflies
Aeshnidae, darners
*Epiaeshna heros**
Libellulidae, skimmers
*Sympetrum obtrusum**, *S. rubicundulum**

(continued)

Table 1. (continued)

Odonata, Zygoptera, damselflies
  Lestidae, spreadwing damselflies
    *Lestes congener, L. dryas**, L. eurinus*, L. forcipatus*, L. rectangularis, L. unguiculatus**
Trichoptera, caddisflies
  Leptoceridae, swimming caddisflies
    Oecetis sp.
    Triaenodes aba, T. nox
  Limnephilidae, northern (or "log-cabin") case makers
    Anabolia bimaculatus, A. sordida**
    *Asynarchus batchawanus**
    Ironoquia parvula, I. punctatissima**
    *Lenarchus crassus**
    Limnephilus indivisus, L. sericeus*, L. submonilifer**
    Nemotaulius hostilis
    *Platycentropus radiatus**
  Phryganeidae, giant tube case makers
    Banksiola crotchii, B. dossuaria
    Ptilostimis ocellifera, P. postica**
  Polycentropodidae, free-living retreat makers
    Plectrocnemia aureola, P. crassicornis*, Holocentropus flavus**

Water Mites
  Arrenuridae
    Arrenurus (Arrenurus) planus, A. (Megaluracurus) neobirgei*, A. (Megaluracurus) neomamillanus*, A. (Megaluracurus) rotundus, A. (Micruracurus) pseudosetiger, A. (Truncaturus) acuminatus*, A. (Truncaturus) angustilimbatus*, A. (Truncaturus) danbyensis, A. (Truncaturus) kenki, A. (Truncaturus) lacrimatus, A. (Truncaturus) palustris, A. (Truncaturus) ringwoodi*, A. (Truncaturus) rubropyriformis*
  Eylaidae
    Eylais spp., including *E. harmani**
  Hydrachnidae*
    Hydrachna baculoscutata, H. conjecta*, H. crenulata*, H. magniscutata*, H. militaria*, H. rotunda*, H. stipata**
  Hydryphantidae*
    *Euthyas truncata**
    Hydryphantes spp.*, *H. ruber**
    Thyas spp.*, including *T. barbigera**, *T. stolli**
    *Thyasides sphagnorum**
    Thyopsis spp.*, *T. cancellata**
    *Zschokkea bruzelii**

(continued)

Table 1. (continued)

Limnesiidae
Limnesia spp., including *L. maculata*
Limnochares sp.
Mideopsidae
Xystonotus robustus
Oxidae
Oxus connatus
Pionidae
Piona spp., including *P. carnea, P. clavicornis, P. constricta, P. mitchelli, P. napio, P. neumani*
Pionopsis lutescens paludis
Tiphys spp.*, including *T. americanus*, T. brevipes*, T. simulans*, T. vernalis**

Molluscs
Gastropoda, snails
Lymnaeidae, pond snails
Fossaria parva, F. modicella*
Stagnicola caperata, S. elodes*, S. reflexa*
Physidae, tadpole snails
*Aplexa elongata**
*Physa gyrina, P. heterostropha, P. vernalis**
Physella ancillaria
Planorbidae, rams-horn snails
Gyraulus circumstriatus, G. crista*, G. parvus**
*Planorbula armigera**
*Promenetus exacuous**
Pelecypoda, bivalves
Sphaeriidae, fingernail clams
*Musculium partumeium**, *M. securis**
*Pisidium casertanum**
*Sphaerium occidentale**

Segmented Worms
Hirudinea, leeches
Erpobdellidae
*Erpobdella punctata**
*Mooreobdella bucera**
Glossiphoniidae
*Desserobdella picta**
Helobdella fusca, H. stagnalis**
*Marvinmeyeria lucida**
Placobdella ornata, P. parasitica*

(continued)

Table 1. (continued)

Oligochaeta, segmented worms: this is a group with large numbers of representatives in vernal pools. Unfortunately, few species are recorded from the literature

Flatworms
- Tricladida
  - Planariidae
    - *Hymanella retenuova**
    - *Phagocata velata**
- Neorhabdocoela
  - Dalyellidae
    - *Dalyellia viridis**

Rotifers
- Bdelloidea, Bdelloida
  - Philodinidae
    - *Philodina roseola*
    - *Rotaria citrina*
- Monogononta, Ploimida
  - Brachionidae
    - *Epiphanes senta*

Notes: Species marked with an asterisk (*) usually are found in annual vernal pools. For more detailed information on the animals listed in this table, see Appendix and chapters 7–12.

of the between-pool differences in community composition. For example, as part of an ongoing study of fourteen vernal pools within a 260-ha (1-mi^2) area in Cape Cod National Seashore, Massachusetts, my colleagues and I have identified more than 200 kinds of macroscopic crustaceans, water mites, and aquatic insects (not including the dipterans — the true flies). Pool E9, a semi-permanent pool with the greatest taxonomic richness of all the study sites, accounts for fewer than half of the total species present in all the pools. In the next richest pool, 40% of the species present are animals that do not occur in pool E9. Nearby, a small, spring-filling, long-hydroperiod vernal pool supports sixty-two taxa of which twenty-two, or more than 35%, have not been found in pool E9. It is evident that the biodiversity supported by this cluster of pools with different basin profiles, hydrologic regimes, and vegetation is far greater than that supported by any one pool (Figure 2).

In addition to overall differences in species composition between pools and between geographic areas, the biological community of individual vernal pools changes from one year to the next in response to weather-related changes in hydroperiod. Species typical of semi-permanent pools become established in annual vernal pools when water levels remain unusually high during unusually wet years, and the same species disappear even from semi-permanent pools during droughts.

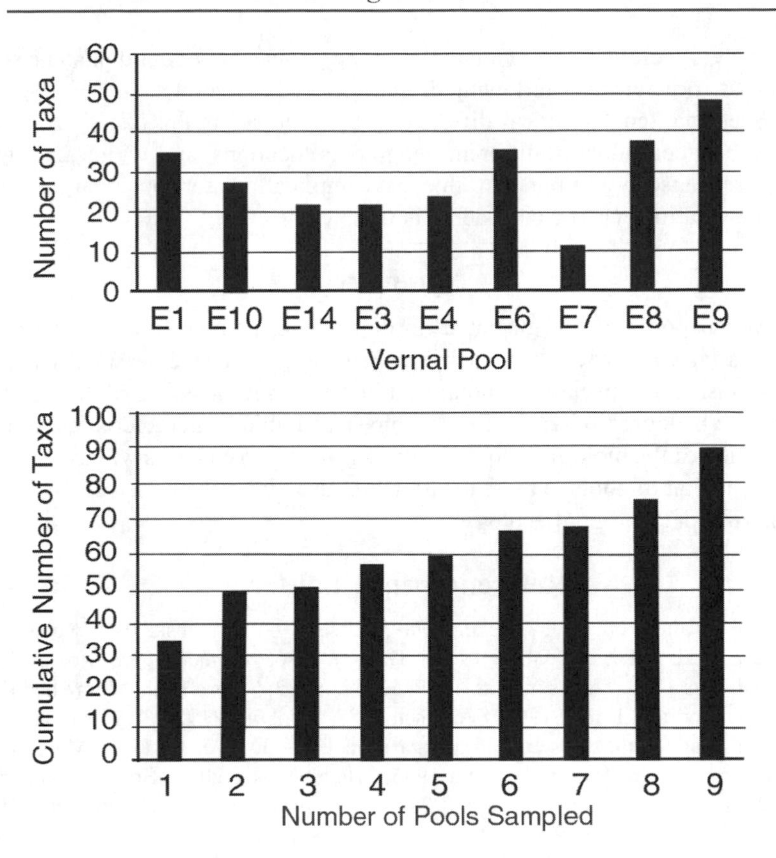

Figure 2. Individual vernal pools located close to one another can vary widely in the number and kinds of taxa they support, and clusters of pools typically support more taxa, overall, than any one pool within the cluster. The top graph shows that the number of non-dipteran animal taxa collected in early 1995 from nine vernal pools in a cluster on Cape Cod, Massachusetts, ranged from eleven (pool E7) to forty-eight (pool E9). The bottom graph shows how the cumulative number of taxa found increased as more pools were sampled, starting with pool E1 at the left and adding each of the pools in the top graph in turn. With only pool E1 sampled, there were thirty-four taxa. The addition of pool E10 brought the total number of taxa collected to fifty, which means that sixteen of the twenty-eight taxa found in E10 were not found in E1. Pool E14 added one new taxon; pool E3, which contained the same number of taxa as E14, added six new taxa and brought the total to fifty-seven; and so on. The number of taxa found in the entire cluster of nine pools on this sampling date was eighty-nine, or forty-one more than in the most diverse pool, E9, alone. (Colburn unpublished data.)

Conversely, species typical of annually drying pools can be found in some semi-permanent pools in years following drought-related drawdown.

Such marked biological differences between pools that are close to one another, between pools in different geographic locations, and within individual pools in response to weather over time, have implications for conservation policy. These implications will be considered in Chapter 14.

Summary

Vernal pools of the glaciated northeast broadly fit a five-part definition encompassing woodland location, physical isolation, small size, seasonally varying water levels, and a specialized biota. Within the constraints of this definition they show a high degree of variability in physical habitat characteristics and the composition of the biological community. The next two chapters will examine the physical habitat of pools in more detail, with subsequent chapters focusing on the biological aspects of pool ecology.

References for Chapter 1

[1] Albers and Prouty 1987, Ireland 1989, Massey 1990, Rawinski 1997, Roble 1998, Thompson et al. 1980, Zappalorti 1993. [2] Daborn 1976a; Donald 1983; Mozley 1928, 1932; Williams 1983. [3] Jackson et al. 1989, Modlin 1982, Moore 1963, Moore and Faust 1972, Raymond and Hardy 1991, Semlitsch 1987, Stenhouse 1985, Taylor et al. 1990. [4] Bazzanti et al. 1996, Crosetti and Margaritora 1987, Eyre et al. 1992, Maier 1989, Mastrantuono 1994, Serrano and Toja 1995. [5] Biebighauser 2003; Brooks et al. 1998; Burne 2001; Dexter 1946 ff.; Downer 1988; Edman and Clark 1988; Higgins and Merritt 1997, 1998; James 1961, 1966a, 1966b, 1967; Jokinen and Morrison 1998; Kenk 1949; Kenney and Burne 2000; Klemens 1998; Leibowitz and Nadeau 2003; Massachusetts Department of Public Health 1940; Massachusetts Division of Fisheries and Wildlife 1988; Schneider and Frost 1996; Tiner 2003a, 2003b; Tiner et al. 2002; Wiggins et al. 1980; Williams 1983. [6] Search of the World-Wide Web conducted June 15, 2003: *http://search.netscape.com/nscp_results.adp?source=NSCPNetSearch&query=vernal+pool.* [7] Keeley and Zedler 1998, Zedler 1987. [8] Biebighauser 2003; Colburn 1996; Kenney 1994, 1996a; Kenney and Burne 2000; Massachusetts Division of Fisheries and Wildlife 1988; New Jersey Division of Fish and Wildlife 2003; Skelly 1998; University of Rhode Island 2001. [9] Clarke 1981, Edman and Clark 1988, Massachusetts Department of Public Health 1940, Wiggins 1973, Wiggins et al. 1980, Wissinger et al. 1999. [10] Larson 1985. [11] deMaynadier and Hunter 1995, 1998; Windmiller 1996. [12] Kiviat et al. 1994. [13] Kenney 1991, Stone 1992, Thompson et al. 1980, Whitlock et al. 1994, Windmiller 1996. [14] Bishop 1941, Dexter 1953, Dodson and Frey 1991, Pennak 1978, Wellborn et al. 1996. [15] Ross 1987. [16] Crosetti and Margaritora 1987, Lake et al. 1989, Mahoney et al. 1990, Taylor et al. 1990. [17] Colburn unpublished data, Kenk 1949, Wiggins et al. 1980, Williams 1983.

Chapter 2

Hydrology

The contrast between the flooded spring pool and the dry, late-summer basin is one of the most dramatic features of many vernal pools. Not surprisingly, for wildlife in vernal pools as well as in other wetlands[1], water — where it comes from, when it arrives, how much of it there is, and how long it persists — is the most important element of the physical habitat. An understanding of vernal pool hydrology is thus essential to an understanding of vernal pool ecology. Successional cycles in vernal pools, the life cycles of vernal pool animals, community dynamics within pools, and the population ecology of vernal pool amphibians are closely linked to the timing, duration, and maximum depths of flooding (see chapters 6 to 13). This chapter summarizes information on the water sources and flooding patterns of vernal pools. It also suggests an approach to classifying pool hydrology which may prove useful for research and conservation.

Water Sources

Depending on its origin and location within the landscape, a vernal pool may receive water from rainfall, surface runoff, intermittent streamflow, groundwater, and/or overbank flooding from nearby water bodies (Figure 3). There is general consensus that the chronology of filling of pools and the maximum depth and volume they attain in a given season are correlated with rainfall and snowmelt. However, few comprehensive hydrologic studies have been carried out on vernal pools and, as a result, the general relationships between precipitation, surface runoff, groundwater elevations, and vernal pool flooding remain unquantified. Field studies suggest that the source of water helps to determine whether certain taxa, such as stoneflies or members of some mayfly families, are likely to be present in a vernal pool (see chapters 6 to 13)[2]. Quantitative assessments of the relative contributions of precipitation, runoff, and groundwater to vernal pools are needed.

Precipitation and Surface Runoff

All northeastern vernal pools receive direct inputs of water from precipitation. Depending on the pool's surface area, the amount of snow and rainfall, and local

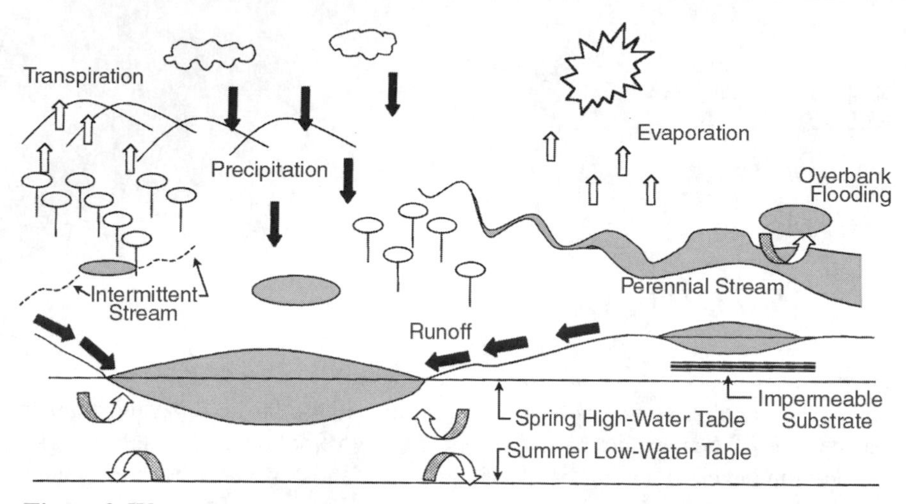

Figure 3. Water enters vernal pools from precipitation, surface runoff, intermittent streamflows, groundwater, and, in some cases, overbank flooding from rivers and other water bodies. Water leaves pools through evapotranspiration, intermittent streamflows, and groundwater recharge.

conditions of soil, vegetation, and groundwater, such inputs may be significant[3]. For example, water levels in vernal pools respond markedly to the direct input of precipitation in summer on Cape Cod, after rainstorms that do not produce surface runoff or cause the groundwater table to rise. Similar rapid increases in water levels after rainfall have been observed in the Connecticut Valley in Massachusetts.

Surface runoff is presumed to be a major source of water for vernal pools. Runoff from rainstorms and melting snow is often cited as the primary factor causing the springtime maximum in water levels and volumes. In perched situations, where a vernal pool sits on an impermeable substrate and is cut off from the groundwater table, water levels depend exclusively on precipitation and surface runoff. Anecdotal data from perched vernal pools in peat-lined depressions on bedrock mountain ridges, basins in poorly sorted till deposits, or low areas over clay lenses suggest that some of these pools may be quite "flashy" hydrologically, in that their depth and volume increase rapidly in response to storm events and fall quickly afterward. Such pools may experience several periods of filling and drying over a season and may support populations of summer-adapted species that do not occur in vernal pools with the more common pattern of spring maxima and gradual drawdown. Depending on the configuration of the basin and the volume of runoff, some perched pools may remain flooded for long periods of time, drying only in late summer or early fall.

By definition, vernal pools lack surface-water connections to other water bodies most of the time. However, many vernal pools have inlets and outlets that flow during the high-water periods of late winter and early spring. In some cases these intermittent streams may flow out of or into wetlands or other water bodies, and in other cases they just drain surrounding uplands. For example, three vernal pools in Duchess County, New York, appear to be hydrologically perched with intermittent inlet and outlet channels[4]. They are separated from other wetlands by uplands (although some lie within 100 m, or about 330 ft, of larger wetlands), and they generally contain water from late fall to early or mid summer. Their maximum water table is above that of other wetlands, including red maple swamps, fens, and clay meadows, and their summer drawdown is less than in fens and similar to red maple swamps.

Intermittent streamflow may contribute to regional biodiversity in vernal pool faunas. In Connecticut and Massachusetts, biological communities of vernal pools fed by intermittent streamflow differ from those of pools that are completely isolated[5]. For example, they sometimes support rolled-winged and nemourid stoneflies and prong-gilled mayflies, insects that rarely are found in standing waters.

Groundwater

Many vernal pool depressions intersect the groundwater table. A multi-year study on outer Cape Cod, involving monthly observations of surface water levels and more than one hundred monitoring wells, shows that water levels in fourteen vernal pools fluctuate with the groundwater table[6]. The pools lie along a hydrologic gradient in a glacial collapse valley in outwash sediments, and they differ in their annual drawdown. Even when the pools are without standing water, their sediments remain saturated at or near the surface, and wetland plants cover the bottom in summer. Interestingly, during summer drawdown, the water table in the vernal pool basins remains above that of the underlying groundwater, the result of hydrologic isolation of the pools by the extensive underlying deposits of organic material.

Available evidence suggests that many vernal pools in glacial deposits in other parts of the glaciated northeast also intersect the groundwater table[7]. In Mansfield, Connecticut, the discharge of groundwater from an esker feeds a vernal pool nearby. The water depth of an artificial vernal pool excavated in glacial till deposits in Cornwall, Connecticut, fluctuates dramatically as the area's groundwater table moves up and down. In evaluating proposals that would alter vernal pools in Massachusetts, I have found that standard hydrologic methods for calculating runoff from storms routinely predict that existing pools should either be dry or contain much less water than is observed in the field. Such calculations presume that surface runoff is the only source of water entering depressions and ignore the potential contribution of groundwater.

Floodwater

In floodplains, vernal pools receive water from a combination of precipitation, surface runoff, groundwater, and overbank flooding. Year-to-year and within-season variability in the hydrologic regime of these pools is great. Floodplain pools may periodically be populated by fish carried in by floods, and fairy shrimp and other vernal pool inhabitants with limited swimming abilities may be swept away by flood waters[8]. There has been little research into the hydrologic variations in floodplain pools and the responses of their biological communities to floods.

Hydrologic Classification: Timing and Duration of Flooding

We know that vernal pools lie on a hydrologic continuum between ephemeral pools and deep, permanent ponds. Numerous studies of vegetated wetlands have shown that differences in hydrology are associated with differences in the wetlands' landscape positions and water sources[9]. Less is known about vernal pools. Much of the information about pool hydrology is anecdotal, collected as part of studies of pool-dependent amphibians or invertebrates. Relatively few studies examine pools over several seasons, and most of the literature on vernal pools is imprecise about the pools' hydrologic regimes.

Available information on hydrology through the year for vernal pools in Ontario, Michigan, and Massachusetts, indicates that variability in maximum depth, volume, and hydroperiod, both between pools at a given time and within individual pools from one year to the next, is a common characteristic of northeastern vernal pools[10]. The predictability of the flooding regime appears to be related to the water sources and the local geologic conditions, as well as to the timing of storms and the overall volume of precipitation in a given season. Figure 4 shows the estimated fluctuations in water level over twenty years in the deepest part of a vernal pool on Cape Cod. The estimates, based on a model derived from two years of monitoring the relationship between water levels in the pool and a groundwater monitoring well, highlight the amount of short- and long-term variability that can occur in some vernal pools[11].

Despite high between-year variability, individual pools seem to follow a broadly predictable pattern of annual filling and drying, with the variation ranging around the average condition. The distributions of plants and animals, and the likelihood that a given species will be able to reproduce successfully in a pool in a given year, are closely tied to the pool's hydrologic regime. Therefore, if we can classify vernal pools based on their hydrology, it should contribute to our understanding of the biological community.

Often, the literature on the biota and ecology of vernal pools does not provide enough information about when the pool normally fills and dries, and the criteria used by various authors to determine whether and when a pool is "dry" are often undefined. A vernal pool may be described as dry but have saturated soils or

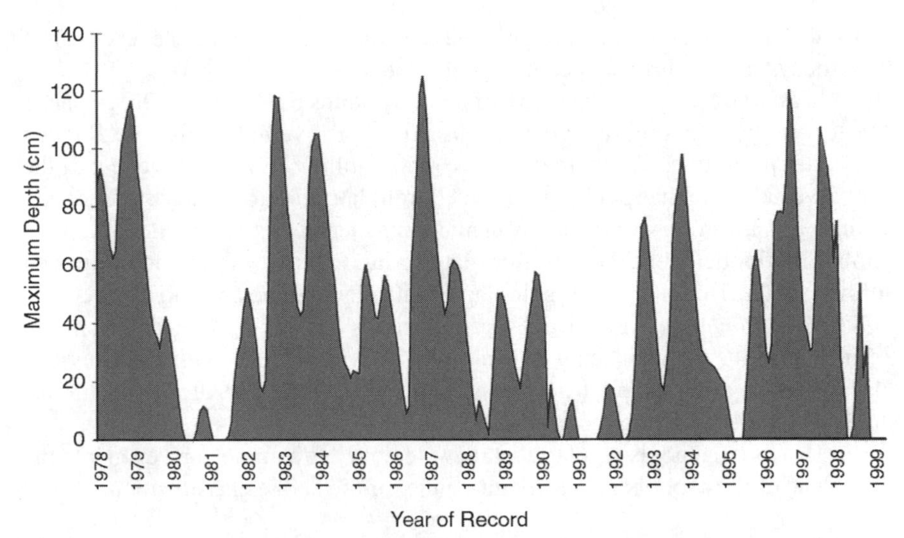

Figure 4. The water levels in vernal pools fluctuate dramatically within and between years. This illustration charts water-level fluctuations in a vernal pool in Eastham, Cape Cod, Massachusetts, from 1978 to 1999. Water levels for 1978 through 1995 were extrapolated from long-term monthly groundwater monitoring data by regression of the relationship between groundwater and surface-water elevations as monitored from 1996 to 1997. Depths for 1996 to 1999 were measured in the field. (Data for this model are from Sobczak 1999 and Sobczak et al. 2003.)

contain water or ice in a portion of its basin, or as continuously flooded when it draws down to two small "puddles"[12].

Previous Hydrologic Classifications

Two hydrologic classifications of vernal pools have previously been proposed[13]. Each has limitations, and neither fully encompasses the range of hydrologic variation found in northeastern vernal pools.

In their comprehensive assessment of the adaptation of animals to life in temporary ponds, published in 1980, Glenn Wiggins, Rosemary Mackay, and Ian Smith identify three hydrologic classes of ponds: "vernal temporary pools" fill in spring and dry by early summer, "autumnal temporary pools" fill in autumn and dry by early summer, and "permanent pools" retain water year-round. This classification is very useful in clarifying the hydrologic conditions in a particular pool in any given year. It is somewhat confusing, however, because a given pool may be a vernal temporary pool one year, an autumnal temporary pool the following year, and a permanent pool the year after. Two of the pools these authors studied showed all three patterns, a third pool was a vernal temporary pool every year of their study, and a fourth was continuously flooded.

In *Pond Life*, a popular guide to temporary ponds published in 1968, Charles Masters suggests a slightly different classification. "Vernal pools" are flooded from late spring to early summer and are dry or solidly frozen for the rest of the year. "Vernal-autumnal pools" are flooded from late spring to early summer and again in late fall and early winter, dry in mid summer, and frozen in winter. "Aestival pools" are flooded continuously from late spring to early winter and are frozen solid in winter. This latter class is similar to that of Paul S. Welch, who in his classic 1952 book, *Limnology*, identified "aestival ponds" as having free water only in spring through fall and being frozen solid in winter[14]. As with the previous classification, a vernal pool may experience more than one of these patterns, depending on the weather conditions in a given year.

Many vernal pools regularly retain water into late summer, and even into the fall. None of these pools fits easily into either of the above classifications.

Proposed Hydrologic Classification

Here, I suggest a revised hydrologic classification for use as a working model in vernal pool studies and conservation (Table 2). It specifically addresses how long a pool usually remains flooded as well as when it fills. The classification should be tested through consistent record keeping for a large number of pools over an extended period of time.

The proposed classes are derived from a synthesis of information in the literature and unpublished data. They encompass the hydrologic continuum, extending from ephemeral pools that generally do not meet the criteria for vernal pools to permanent ponds that rarely provide suitable conditions for vernal pool wildlife.

Two hydrologic characteristics are particularly important to vernal pool wildlife. First is the timing of drawdown and drying. How long water persists into the summer is an important factor controlling the composition of the biological community, and it is particularly critical for amphibians and some aquatic invertebrates with long larval periods. The usual timing of annual filling, whether in the fall or spring, is the second important consideration, because it affects the length of the dry period and the suitability of a pool for fall-breeding species, such as marbled salamanders.

Many vernal pools do not dry completely but retain an area of saturated sediment or shallow water within a small part of thc basin. Such pools may provide habitat for species that lack mechanisms for withstanding or avoiding desiccation and would be excluded from pools that dry completely. For the purposes of the proposed hydrologic classification, I suggest that pools that retain an area of standing water, even if that area represents only 1% of the pool's maximum area, be classified as "incompletely dry" as distinct from "dry" or "continuously flooded." This distinction would separate pools that dry completely, pools that remain

Table 2. Proposed Hydrologic Classes of Vernal Pools and Average Duration of Flooding

Hydrologic Class	Average Duration of Flooding (Months)
Short-cycle, spring-filling pools	3 – 4
Long-cycle, spring-filling pools	5 – 8
Short-cycle, fall-filling pools	7 – 9
Long-cycle, fall-filling pools	9 – 11
Semi-permanent pools	36 – 120

continuously flooded with a significant amount of water, and pools that become mostly dry but may provide refugia for drought-intolerant species. Maintaining this distinction should help efforts to determine the degree of drought-tolerance in different pool inhabitants.

The proposed hydrologic classes are as follows.

Ephemeral or Rainwater Pools

"Ephemeral pools," also known as "rainwater pools," fall at the dry end of the hydrologic continuum. They often occur in perched situations. They fill in response to rainfall and, often, do not retain water for more than a few weeks. Ephemeral pools can exhibit the "short-cycle" pattern of vernal pool activity (see below), but they are also more likely to refill as a result of summer rainstorms than are typical short-cycle pools. They generally fail to retain water for sufficient periods of time to meet our definition of vernal pools.

Ephemeral pools support distinct communities of protists, rotifers, crustaceans, diatoms, algae, and insects. They are important habitats for warmwater crustaceans such as clam shrimp, and some serve as breeding pools for spadefoot toads. They are thus of importance for biodiversity and of considerable conservation interest. However, the short duration of flooding in ephemeral pools prevents the establishment of biological communities typical of vernal pools.

Short-Cycle Pools

"Short-cycle pools" are annual vernal pools that, having reached their spring maximum in depth and volume, shrink rapidly once snowmelt and spring rains are complete and plant growth, with associated evapotranspiration, begins. They remain flooded long enough to meet the definition of a vernal pool. In most years short-cycle pools are dry by late June or early July. During the dry period, if canopy cover is not too dense, the basin of short-cycle pools may be vegetated with ferns, mosses, and herbaceous annual plants. Wetland shrubs commonly occur along the pool's edges and in the basin.

Short-cycle pools provide breeding habitat for wood frogs and spring peepers, but in many years they do not remain flooded long enough to allow the successful development of the larvae of spring-breeding mole salamanders. These pools often contain fairy shrimp and a variety of other vernal pool indicator invertebrates, some of which are uncommon in pools with longer hydroperiods (see chapters 6 to 13). In particularly wet years, short-cycle pools remain flooded longer than usual and may provide breeding habitat for salamanders, green frogs, and long-lived invertebrates. During unusually dry years, these pools may fail to fill.

Short-cycle pools exhibit two patterns of flooding. Both spring- and fall-filling pools reach their maximum surface area in spring, following snowmelt and runoff from spring rains. A given pool may exhibit both patterns, depending on local variations in groundwater elevations and rainfall from one year to the next.

"Short-cycle, spring-filling pools" are usually dry or incompletely dry in winter, and they fill in spring. These pools generally contain water for three to four months of the year. Those that dry completely correspond broadly to the vernal temporary pools of Wiggins et al. and to the vernal pools of Masters.

"Short-cycle, fall-filling pools," like short-cycle, spring-filling pools, dry in late spring or early summer, but they start to fill in late fall or early to mid winter, gaining 50% or more of their peak volume and depth at this time. They are typically flooded for seven to nine months. These fall-filling pools correspond broadly to the autumnal temporary pools of Wiggins et al. and the vernal-autumnal pools of Masters. They support aquatic life during the winter if the water is deep enough to prevent freezing to the bottom, and they can serve as breeding pools for marbled salamanders.

Long-Cycle Pools

"Long-cycle pools" are annual vernal pools that remain flooded longer than short-cycle pools. They hold water into the summer and typically dry in mid to late summer or early fall. Long-cycle pools usually stay flooded long enough to support salamander breeding. Their faunas include a greater variety of invertebrate and vertebrate species than short-cycle pools. Long-cycle pools are not included in previous classifications.

In many long-cycle pools, the sediments remain saturated at or near the surface after standing water has disappeared. If canopy cover is not too dense, such pools often support extensive growths of shrubs, ferns, grasses, sedges, rushes, and other perennial wetland plants. In association with the long hydroperiod of these pools, organic materials often accumulate to substantial depths in the sediments.

Occasionally, during unusually wet conditions, long-cycle pools will remain partly flooded for an entire year. In such years they may successfully support green frogs, large skimmer dragonflies, and other species that overwinter as aquatic larvae. Provided that a long-cycle pool does not remain continuously flooded more often than once in five years, I would still consider it to be an annual vernal pool rather than a semi-permanent pool.

As with short-cycle pools, two patterns of flooding are seen in long-cycle pools. Both types of pools attain their maximum depths and surface areas in the spring.

"Long-cycle, spring-filling pools" fill in spring from snowmelt, storm runoff, overbank flooding from streams, and/or groundwater, and they retain water well into the summer. Such pools typically contain a substantial volume of standing water for five to eight months.

"Long-cycle, fall-filling pools" usually gain 50% or more of their volume and depth in autumn or early to mid winter, become fully flooded in spring, and remain flooded from late fall through mid summer. They often contain water for nine to eleven months of the year, although in summer the area and volume of standing water are greatly reduced from the peak levels of spring.

Partially Drying Pools

Some vernal pools draw down but consistently retain a relatively small area of standing water or saturated substrate. They may occur under a flooding regime that is short- or long-cycle and either spring- or fall-filling. These "partially drying pools" are expected to differ from pools that dry completely in supporting some species that cannot resist or avoid drawdown. They are expected to differ from semi-permanent pools in lacking substantial populations of species whose life cycles span multiple years, such as bullfrogs and green frogs and some large dragonflies.

Semi-Permanent Vernal Pools

Some pools generally remain continuously flooded for several consecutive years. They occasionally dry completely, but this may occur as rarely as only once every five to ten years. Although their water levels reach a seasonal maximum in spring and drop somewhat through the summer, these "semi-permanent pools" retain substantial volumes and depths of water throughout the year, and their mean water depths rarely fall below 20 cm (8 in). They provide aquatic habitat for a variety of amphibian larvae and freshwater invertebrates. Such pools may sustain introduced fish for periods of several years, but the occasional drying prevents the permanent establishment of fish populations. Wiggins and his colleagues define such pools as permanent pools in years that they are flooded continuously.

As noted above, semi-permanent pools are often larger than short- or long-cycle pools. Some of them have been artificially deepened by impoundment or dredging. Fish, if introduced, may survive for several seasons. Other vertebrates, such as green frogs and bullfrogs, and invertebrates such as some aeshnid and large libellulid dragonflies, all of which require more than one year to complete their life cycles, may be present. In such pools, predation and other inter-specific interactions may be very important in determining which species are present and able to survive.

Permanent Fishless Ponds

Permanent ponds are perennially flooded and do not dry or draw down significantly, except during droughts. They provide potential habitat for fish, bullfrogs, large dragonflies, and other species typical of permanent waters. Fish, however, are generally excluded from shallow permanent ponds that freeze to the bottom or that become anoxic during winter ice cover. Such "permanent fishless ponds" support invertebrate communities that differ from those in waters that contain fish. Some amphibian species typical of vernal pools, especially spotted salamanders, breed in some permanent ponds without fish. In some cases, permanent ponds supporting vernal pool wildlife represent former vernal pools that have been altered by human activity.

Ranking of Pools Along the Hydrologic Continuum

Based on this proposed hydrologic classification, vernal pools can be ranked in order of increasing hydroperiod (Table 2, Figure 5). This classification should be considered a preliminary working model, and it should be evaluated as additional information becomes available on the hydrologic cycle in a number of vernal pools through time. The identification of the average hydrologic regime of individual pools will need to be based on observations taken over a period of several years, because, as noted earlier, the timing and duration of flooding in vernal pools vary with year-to-year variations in weather. It may prove useful to use this classification variably to describe different conditions in a given pool from year to year, in the way that Wiggins and his colleagues applied their classification[15]. Certainly, in addition to information on the average hydrologic conditions found in a given vernal pool, specific information on the timing of flooding and of drawdown in a given year, as well as in the previous year, needs to be provided as part of studies of the life histories and population biology of vernal pool wildlife.

Hydrologic Effects of Alteration by Human Activity

Many vernal pools occur in altered landscapes. A review of the literature suggests that the hydroperiod of many water bodies that function as vernal pools has been increased as a result of dredging or other artificial deepening. Landscape alteration may also decrease the duration of flooding in some pools[16]. There is some evidence from Maine that temporary pools of anthropogenic origin dry more rapidly than natural vernal pools during periods of lower-than-normal precipitation, although there are no differences in hydroperiod during average years. Similarly, vernal pools in Pennsylvania fields dry earlier than woodland pools nearby.

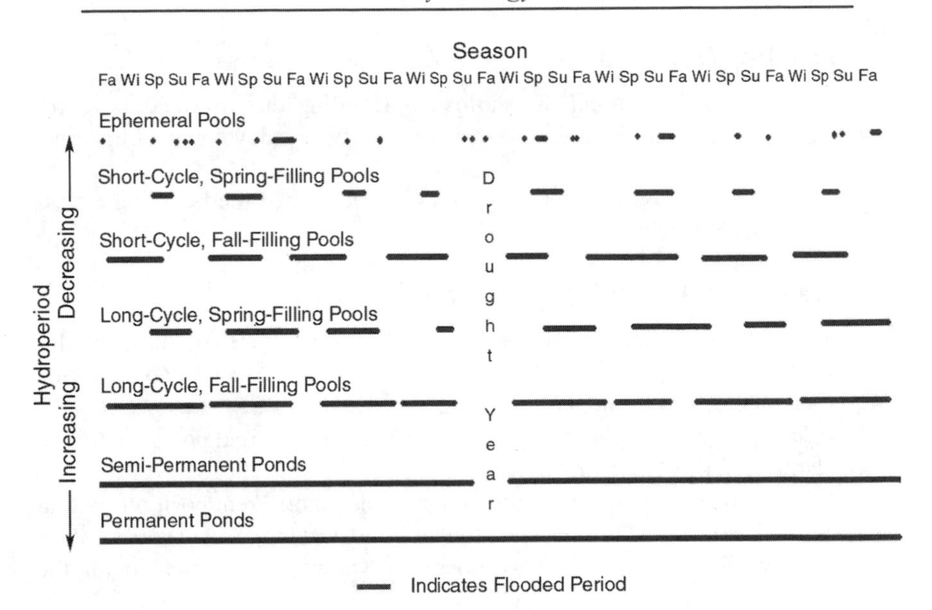

Figure 5. Vernal pools occur along a continuum of flooding, from ephemeral pools to permanently flooded, fishless ponds, as illustrated, schematically, here. Within each hydrologic class, the hydroperiod varies depending on annual weather conditions.

Some Remaining Questions about Hydrology

Water Sources

- What are the relative contributions of precipitation, runoff, and groundwater to the water supply of vernal pools in different geologic contexts?
- Are there relationships between water sources and pool biodiversity?
- Do differences in water sources contribute to differences in reproductive success in amphibians and invertebrates in vernal pools?
- Do differences in water sources result in consistent differences in water quality in vernal pools?
- Are there quantifiable relationships between differences in water sources and the hydrologic variability in vernal pools, in terms of time of flooding and flooding duration?
- What unique biological characteristics, if any, are exhibited by vernal pools that occur in floodplains and experience regular inundation by overbank flooding from rivers, streams, or large lakes?

Flooding Duration

- What are the critical thresholds for flooding duration necessary for different amphibians and invertebrate species to complete development in vernal pools?
- Can species composition be used as a predictor of the hydrologic class of a vernal pool?

Patterns of Hydrologic Variation

- Are there predictable patterns in the hydrologic cycles of vernal pools?
- Does long-term, cyclical hydrologic variation in vernal pools provide refugia for species vulnerable to predation?
- Does long-term, cyclical hydrologic variation in vernal pools contribute to population stability in amphibians?
- How do pools in different parent materials, in different positions within the watershed, and with different combinations of water sources, vary in their degree of hydrologic flashiness? How does such variation affect the biota in the pools?

Effects of Hydrologic Alteration

- What are the short- and long-term effects of groundwater withdrawals on community composition and population dynamics of amphibians and other species in vernal pools?
- When land use in the watershed changes the dynamics of surface flow, what are the effects on community composition and population dynamics of amphibians and other species?
- What changes in local and regional precipitation and temperature are predicted in association with global climate change, and how might these changes affect vernal pool hydrology?

References for Chapter 2

[1] Mitsch and Gosselink 2000. [2] Jokinen and Morrison 1998. [3] Brooks 2000, Brooks and Hayashi 2002, Cole and Fisher 1978, Sobczak et al. 2003. [4] Kiviat et al. 1994. [5] Colburn unpublished data, Jokinen and Morrison 1998. [6] Sobczak et al. 2003. [7] Brooks and Hayashi 2002, Jokinen and Morrison 1998, Pollan 1998. [8] Dexter 1956. [9] Cole and Brooks 2000, Gosselink and Turner 1978, Mitsch and Gosselink 2000. [10] Brooks 2000, Brooks and Hayashi 2002, Kenk 1949, Sobczak et al. 2003, Wiggins et al. 1980, Williams 1983. [11] Sobczak et al. 2003. [12] Brooks 2000, Brooks and Hayashi 2002, Dexter and colleagues 1943 ff., Kenk 1949, Sobczak et al. 2003, Weaver 1943, Wiggins et al. 1980, Williams 1983. [13] Masters 1968, Wiggins et al. 1980. [14] Welch 1952. [15] Wiggins et al. 1980. [16] Calhoun et al. 2003, DiMauro and Hunter 2002, Seale 1982.

Chapter 3

Vernal Pools in the Landscape: Origins, Landscape Positions, and Habitat Characteristics

Differences in the origins, landscape positions, and substrates of vernal pools contribute to a range of sizes, depths, chemical characteristics, and hydrologic regimes among pools. This between-pool habitat variability contributes to great diversity in the biological communities of plants and animals in vernal pools and is important to the conservation of these pools. This chapter considers landscape factors and major non-hydrological habitat characteristics of northeastern vernal pools.

Origins and Distributions Across the Landscape

Geologic processes created most of the depressions where vernal pools occur, and local geology influences pool hydrology (see Chapter 2). Many vernal pools owe their existence to glacial action that took place thousands of years ago. Other pools occur as the result of more recent events, including local erosion, the migration of stream channels, and human alteration of the forested landscape over the past several hundred years. Any woodland area where water can accumulate in the spring and persist for two months or more has the potential to be a vernal pool.

Origins

The origins of many vernal pools can be traced directly or indirectly to the continental glaciers that covered the glaciated northeast in the recent geologic past. The legacy of glaciation is broadly evident in the patterns of landforms, the distributions of mineral soils, and the locations of vernal pools in today's landscape. In some parts of the glaciated northeast, the landscape is pockmarked with approximately circular depressions that were created more than 10,000 years ago when the retreating Pleistocene ice sheets left behind chunks of ice buried in rock, gravel, sand, and/or mud. As the ice melted, the overlying debris collapsed to create "kettle holes." Well-sorted sands and gravels deposited as glacial outwash often contain large numbers of kettle holes. Some large, deep kettle holes whose bottoms extend tens of meters below the groundwater table filled with water over the millenia and are now lakes with diverse aquatic communities including fish. Many shallower kettle holes developed into ponds and vegetated wetlands. Other small, shallow kettle holes became vernal pools. In some areas with extensive glacial outwash deposits, high seasonal groundwater tables, and many kettle-hole depressions, there are high densities of vernal pools.

In some locations, shallow hollows were scoured by glacial action into the bedrock of mountain ridgetops or exposed ledges. Today these perched basins collect and hold rainfall and runoff and support communities of vernal pool wildlife. Depressions overlying poorly drained soils such as some glacial tills, other morainal deposits, and the fine sediments of glacial lake beds can also support perched vernal pools. Pool basins in these soils were created by melting glacial ice blocks, hollowed out by local erosion, gouged out by ice, or impounded by landslides or even by fallen trees that blocked seasonal water flows across parts of the landscape. Because perched pools must depend on precipitation and runoff for their supply of water, depressions on bedrock and very poorly drained soils generally support relatively few vernal pools (Table 3).

In a floodplain, the stream and its flow patterns over the years are the dominant factors affecting the land's surface topography. Floodplains often have many low areas that were scoured out during high water and that remain flooded after the stream has returned to its channel. Such areas may be in contact with groundwater, or deposits of fine sediments may help them hold water. Old sections of channel that were cut off as the stream changed course over the ages also retain water and some of them now function as vernal pools.

Under some circumstances, quarries, farm ponds, detention basins, and other areas excavated or impounded by human activity provide habitat for one or more vernal pool species. In working agricultural landscapes, as well as much of the eastern seaboard where forests have reclaimed abandoned farms, many natural vernal pools have been altered, and others have been created artifically[1]. Some anthropogenic pools support dynamic communities whose characteristics are comparable to those in natural pools; others may serve as "sinks" in which animals breed but fail to produce offspring that could contribute to the long-term maintenance of the species[2].

Geographic Variation

Vernal pools can be found in a wide variety of geologic settings, and some landscapes are more likely to contain large numbers of pools than others because the processes that created those landscapes favor the pooling and retention of

Table 3. Distribution of 106 Vernal Pools in Amherst, Massachusetts, on Different Types of Underlying Materials

Soil Drainage Class	**Percent of Pools**
Somewhat to excessively well drained	25
Well drained	3
Moderately well drained	41
Poorly drained	26
Very poorly drained	4
Rock outcrops	2
Gravel pits or urban areas	1
Soil Origin	
Glaciolacustrine	87
Glaciofluvial	1
Alluvial	5
Till	4
Swamp deposit	2
Artificial fill	1

Data from Stone 1992

water (Table 4). The proportion of pools in different types of settings varies markedly from one location to another (Table 5). For example, in southern and northern Maine, pools occur predominantly in isolated upland depressions within undisturbed forests. Central Maine pools, in contrast, occur most commonly in forested wetland complexes with pit and mound topography, and most of them are artifacts of human disturbances such as logging roads, ditches, and depressions in clearcuts[3]. Up to half the pools in parts of northeastern Massachusetts are associated with wetlands and surface waters. In contrast, in the glacial till and outwash deposits of outer Cape Cod, vernal pools lie in kettle depressions and swales surrounded by pitch pine and/or scrub oak forest, and fewer than 1% are associated with larger wetlands or surface water bodies. Similarly, in the 48,500-ha (120,000-ac) Quabbin Reservoir watershed in central Massachusetts, 94% of 430 pools identified through color-infrared aerial photography lie within the forest and are not associated with other water bodies[4]. In the St. Louis Moraines and Chippewa Plains areas of north-central Minnesota, pools occur over a range of glacial deposits including lake bed, outwash plain, and moraine, but pool densities are highly correlated with deposit type, occurring most densely in morainal deposits and least frequently in lake bed deposits[5].

A recent comprehensive survey of isolated wetlands in the United States by the US Fish and Wildlife Service shows that in the glaciated northeast, wetlands

Table 4. The Distribution of Eighty-Four Vernal Pools from the Glaciated Northeast according to Landscape Type

Landscape Setting	Number of Pools
Upland woodland	52 (62%)
Floodplain	6 (7%)
Vegetated wetland	5 (6%)
Field	3 (3.6%)
Artificial or altered	18 (21%)

Data from Anderson and Giacosie 1967, NJ; Broch 1965, NY; Brodman 1995, OH; Colburn, unpublished data, MA; Dexter and Kuehnle 1948, 1951, OH; Freda 1983, PA; Hinshaw and Sullivan 1990, ME; Husting 1965, MI; James 1966a, 1966b, 1967, ON; Jokinen and Morrisson 1998, CT; Kenk 1949, MI; Kleeberger and Werner 1983, MI; Nyman 1991, NJ; Pierce et al. 1984, CT; Schneider and Frost 1996, WI; Seigel 1983, NJ; Shoop and Doty 1972, RI; Stangel 1988, MA; Wacasey 1961, MI; Waldman and Ryan 1983, NY; Way et al. 1980, OH; Wiggins et al. 1980, ON; Weaver 1943, OH; Wilbur 1972, MI; Williams 1973, IN; Williams 1983, ON; Wilson 1976, NY; Windmiller 1996, MA; Woodward 1982a, CT

account for an estimated 0.5 to 44.2% of the land surface, with isolated wetlands estimated as 4.5 to 53.8% of the total. This translates into isolated wetlands occupying from less than 0.001% to nearly 20% of the land surface, depending on the local geology. In some of the study areas, such as central and northern Minnesota and northern Indiana, isolated wetlands account for more than 80% of individual wetlands. Not all isolated wetlands are vernal pools, and the data for this study generally did not include wetlands smaller than 0.4 ha (about 1 ac), so many vernal pools were omitted[6]. Still, the data from this and other studies suggest strongly that the distribution of vernal pools is closely linked to local geologic conditions.

The density of pools also varies markedly across the landscape, ranging from none to more than 12/km^2 (more than 30/mi^2) (Table 6). Within a particular area, inter-pool distances are also highly variable. A pool's nearest neighbor may be as close as 5 m (16 ft) or as distant as 2.4 km (1.5 miles)[7]. At the Quabbin Reservation in Massachusetts, 24% of pools have another pool within 100 m and 59% have another pool within 300 m (328 ft and 985 ft). As we will see later, the proximity of other pools helps determine the value of vernal pools as breeding habitats for a variety of amphibians.

Vernal Pools as Types of Wetlands

As wetlands, vernal pools defy simple classification. For instance, under the hydrogeomorphic (HGM) approach to wetland classification used by some federal agencies in the United States, most vernal pools are isolated depressional wetlands,

Table 5. The Landscape Position of Vernal Pools in Different Geographic Areas

Location/Landscape Setting	Percent of Pools
Ohio and Illinois (61 pools)[a]	
Woodland	28
Wetland	18
Floodplain	1.6
Formerly forested fields, pastures, hog wallows	41
Altered areas (roadside ditches, railway embankments, dikes)	11.5
Northeastern Massachusetts (156 pools)[b]	
Isolated in upland	ca. 50
Floodplain	< 10
Vegetated wetland associated with surface waters	4–51
Reading, Northeastern Massachusetts (58 pools)[c]	
Isolated in upland	9
Bottomland not adjacent to streams	54
Bottomland adjacent to streams	38

[a] Data from Dexter et al. 1943 ff

[b] Data from Windmiller 1995b

[c] Data from Kenney 1991

but some are riparian depressions and others are slope wetlands[8]. Under the US Fish and Wildlife Service's wetland classification method, most vernal pools are defined as seasonal palustrine wetlands due to their shallow depths and periodic drawdown[9]. They are found within a number of classes in the palustrine system, including aquatic bed, unconsolidated shore (vegetated or unvegetated), unconsolidated bottom, streambed (vegetated or unvegetated), scrub-shrub wetland, and forested wetland. Some deeper (more than 2 m or 6 ft), usually semi-permanently flooded pools fall within the littoral subsystem of the lacustrine system, with the aquatic bed, unconsolidated shore, and unconsolidated bottom classes being most commonly represented.

Soils and Substrates

Vernal pools are found on a wide variety of soils (Table 3)[10]. Not much information about soil chemistry in association with vernal pools has been published. In three pools in New York, soil pH is lower than in other wetland types that were studied. The pH of soil surrounding forty-eight vernal pools in Centre County, Pennsylvania, ranges from 3.45 to 4.33[11].

Table 6. Densities of Vernal Pools across the Landscape

Location	Number/km^2	Number/mi^2
Southern Maine forests	13.5	35
Central Maine forests	1.4	3.6
Eastham, Cape Cod, Massachusetts		
Glacial collapse valley	5.8	14
North of glacial collapse valley	0	0
Quabbin Reservation, northcentral Massachusetts	1.1	2.8
Chippewa National Forest, northcentral Minnesota		
Overall average	8	20
Lake plain land-type association	3	8
Outwash land-type association	mean 5, range 3–6	13
Ground and end moraine	mean 10, range 8–13	25
End moraine	mean 10, range 5–12	25

Data from Brooks et al. 1998, MA (Quabbin); Calhoun et al. 2003, ME; Massachusetts Natural Heritage and Endangered Species Program 2001, MA (Cape Cod); Palik et al. 2003, MN; Portnoy 1987, MA (Cape Cod)

In most vernal pools, as a result of extended flooding, a layer of organic substrate lies on top of the mineral soil. The depth of organic material appears to be a function of the hydrologic regime. In vernal pools that regularly dry for four to eight months of the year, most of the leaves and other debris that fall into the pools get oxidized and do not accumulate in the basin, so the organic deposits are shallow. New York's Natural Heritage and Endangered Species Program classifies such vernal pools as mineral soil wetlands, with a thin layer of organic material overlying mineral soil, usually on till[12]. In contrast, in vernal pools with long hydroperiods or soils that remain saturated during drawdown, the extended presence of water reduces the availability of oxygen to decomposers such as bacteria and fungi, because water contains less oxygen than air, and the decomposition of plant debris is limited by the low oxygen levels. Such pools may have organic deposits several meters deep[13].

The materials that overlie the soil in vernal pools provide food, shelter, and sites for the attachment of amphibian and invertebrate eggs. These materials vary with surrounding land use and vegetation, pool depth profiles, and hydrology. In woodland pools, leaves or needles and sticks cover the bottom. In pools with an open canopy, algal growth may be extensive[14]. Even in woodland pools, algae are often found growing on the bottom early in the spring, before trees leaf out and the canopy closes. Aquatic plants, grasses and sedges, herbaceous terrestrial plants, ferns, mosses, and/or shrubs may be present within the pool basin, depending on the extent of canopy closure and the depth and duration of flooding (see Chapter 5). Open-canopy pools and those with long hydroperiods characteristically have emergent non-woody vegetation that includes sedges, grasses, and perennial hydrophytes. Wetland shrubs may be present, more commonly in annually drying pools than in semi-permanent ones. In many pools, *Sphagnum* moss forms part of the substrate[15]. *Sphagnum* pumps hydrogen ions into water in exchange for nutrients that it needs for growth, and it can have marked effects on the water chemistry of some vernal pools.

Mud substrates are common, particularly in areas adjacent to roads or other areas disturbed by human activity or where the perimeter is only sparsely wooded, and mud may be the predominant bottom type in up to one third of the pools in some areas. Vernal pools whose watersheds are disturbed by forest clearing and development show higher algal growth, lower amounts of coarse organic debris, and higher proportions of mineral sediment than do pools in undisturbed settings[16].

Pool Size

Surface areas and depths of vernal pools span a wide range but, in general, they tend to be small and shallow. Information in the literature about the size of pools includes maximum diameter, maximum surface area, maximum and/or mean depth, and/or maximum volume. The small size of most vernal pools is an important consideration for regulatory approaches to the conservation of pools, as well as for the design of projects that involve pool creation or restoration.

Surface Area

Surface areas of vernal pools fluctuate seasonally and normally reach an annual maximum in spring (plates 2, 3). Table 7 summarizes data reported in the scientific literature for individual pools throughout the northeastern United States and southeastern Canada. Most of these pools are smaller than 0.1 ha or 0.25 ac. Unpublished size data from Massachusetts vernal pools are similar to those in the literature (Table 8). The size range is similar in semi-permanently flooded pools and annually drying pools, but semi-permanent pools tend overall to be larger than annual pools. Surveys encompassing 1,483 pools throughout the glaciated northeast report size ranges similar to those from studies of individual pools (Table 9).

Water Depth

Vernal pools are shallow. Most have a maximum depth in early spring of around 1 m, or just over 3 ft (tables 7, 8, 9). As with surface area, the maximum depths of semi-permanent pools tend to be somewhat greater than in annually flooded pools.

Table 7. Summary of Maximum Surface Areas and Depths Reported in the Literature for Individual Northeastern Vernal Pools

	Surface Area (m²)	**Depth (cm)**
	All Pools (N = 52)	All Pools (N = 40)
Range:	16–7,500	16–300
Mean:	1,168	109
Median:	660	100
	Semi-Permanent Pools (N = 11)	Semi-Permanent Pools (N = 8)
Range:	21–7,500	100–200
Mean:	2,082	118
Median:	1,500	100
	Annually Drying Pools (N = 40)	Annually Drying Pools (N = 32)
Range:	18–5,026	16–300
Mean:	814	106
Median:	599	100

Data from Anderson and Giacosie 1967, NJ; Bachman and Bachman 1994, MI; Broch 1965, NY; Brodman 1995, OH; Dexter and Kuehnle 1948, 1951, OH; Freda 1983, PA; Hinshaw and Sullivan 1990, ME; Husting 1965, MI; James 1966a, 1966b, 1967, ON; Jokinen and Morrisson 1998, CT; Kenk 1949, MI; Kleeberger and Werner 1983, MI; Nyman 1991, NJ; Pierce et al. 1984, CT; Schneider and Frost 1996, WI; Seigel 1983, NJ; Shoop and Doty 1972, RI; Stangel 1988, MA; Wacasey 1961, MI; Waldman and Ryan 1983, NY; Way et al. 1980, OH; Weaver 1943, OH; Wiggins et al. 1980, ON; Wilbur 1972, MI; Williams 1973, IN; Williams 1983, ON; Wilson 1976, NY; Windmiller 1996, MA; Woodward 1982a, CT

The maximum annual water depths of vernal pools change considerably from year to year[17] (Plate 2). The maximum depth in a Connecticut vernal pool was 48 cm (19 in) one year and 80 cm (32 in) the next. Similarly, maximum annual water depth in a semi-permanent pasture pool in Ohio varied by 80 cm over a six-year period, ranging from 55 to 135 cm (22 to 53 in); the pool dried fully during two of those years. In a cluster of fourteen vernal pools on outer Cape Cod, in Eastham, Massachusetts, annual maximum depths varied by 48 to 98 cm (19 to 38 in) in individual pools over the four-year period from 1995 through 1998. During the dry year of 1995 on Cape Cod, the maximum depth recorded in eight of the pools ranged from 0 to 20 cm, with a mean of 5 cm and a median of 3.5 cm (0 to 8 in, mean

Table 8. Maximum Size and Depth Data for Sixty-Four Vernal Pools in Massachusetts

	Surface Area (m²)	Depth (cm)
	All Pools (N = 64)	All Pools (N = 50)
Range:	10–22,500	20–200
Mean:	1,433	69
Median:	315	60
	Semi-Permanent Pools (N = 10)	Semi-Permanent Pools (N = 8)
Range:	177–22,500	70–200
Mean:	5,590	110
Median:	1,898	93
	Annually Drying Pools (N = 40)	Annually Drying Pools (N = 32)
Range:	10–3,600	20–150
Mean:	584	62
Median:	300	55

Data from Colburn, unpublished data

2 in, median 1.4 in); and in 1997, an unusually high-water year, the maximum depth in all fourteen pools ranged from 45 to 110 cm, with a mean of 70 cm and a median of 83 cm (18 to 43 in, mean 27.5 in, median 33 in).

Water Chemistry

As we have seen with other habitat characteristics, the chemistry of vernal pools encompasses a wide range of conditions. A pool's chemistry is a function of its landscape position, the surrounding geology and vegetation, the sources of water, and the land uses within the watershed[18]. Local geology is important — pools over volcanic bedrock are generally more acidic than pools over sedimentary rock, for instance, but other factors also contribute to water quality. Pools that are supplied solely by precipitation are apt to be more acidic than pools with significant inputs of surface runoff and groundwater. Pools in undisturbed watersheds, similarly, are generally more dilute and more acidic than pools affected by roads and other landscape alterations.

Table 9. Size Data from Habitat Surveys of Vernal Pools

Location and Biological Criteria, if any[1]	Number of Pools	Maximum Surface Area	Maximum Depth	Survey Reference
New Jersey Abandoned gravel pits Ao, At	12	450 m^2 mean maximum area	range 15–125 cm	Hassinger et al. 1970
New Jersey Kettles and quarries A		range 10–60,000 m^2	at least 100 cm	Nyman et al. 1988
Massachusetts Middlesex County[a] no Am	91	range 7–35,062 m^2; median 329 m^2; mean 1,709 m^2	range 15–203 cm; median 62 cm; mean 80 cm	Windmiller 1996
Massachusetts Middlesex County Am	87	range 0.3–101,502 m^2; median 730 m^2; mean 3,025 m^2	range 28–279 cm; median 81 cm; mean 97 cm	Windmiller 1996
Massachusetts Cape Cod, depressions below water table	43	range 3–20,000 m^2; mean 1,963 m^2	range 5–85 cm; median 30 cm	Portnoy 1990
Massachusetts Reading, undeveloped area A, Rs, or E	30	range 66–11,629 m^2	no depth data provided	Kenney 1991
Massachusetts Reading, developed area A, Rs, or E	28	range 14–20,000 m^2	no depth data provided	Kenney 1991
Massachusetts Central A, Rs, or E	430	28% <0.025 ha, 66.5% < 0.05 ha, 86% < 0.1 ha, 4% > 0.2 ha; median = 352 m^2	no data presented	Brooks et al. 1998
Massachusetts Connecticut Valley, isolated pools	106	range 50–3,300 m^2; median 260–300 m^2; mean 510–550 m^2	no depth data provided	Stone 1992
Massachusetts Connecticut Valley Am	13	range 1,000–24,300 m^2; m^2; median 1,000 m^2; mean 5,100 m^2		Cook 1978, 1983
Massachusetts Connecticut Valley Aj	7	range 1,000–10,000 m^2; mean <2,000 m^2	range 60–240 cm; mean 130 cm	Cook 1978, 1983

(continued)

Table 9. (continued)

Location and Biological Criteria, if any[1]	Number of Pools	Maximum Surface Area	Maximum Depth	Survey Reference
Maine Southern A, Rs, or E	195	42% > 399 m²; 20% > 1,394 m²	mean 67 cm	Calhoun et al. 2003 and unpublished data
Maine Central A, Rs, or E	55	median 36 m² (natural), 24 m² (anthropogenic); 7% > 399 m² (regulatory minimum for protection)	mean 22 cm, natural and anthropogenic pools	Calhoun et al. 2003 and unpublished data
Nova Scotia Am	41	ponds: < 5,000 m²; ditches: <25 m²	41% <50 cm, 15% 50–100 cm, 41% 100–200 cm; 1 permanent pond, no depth data	Dale et al. 1985
Quebec Rs	15	range 40-520 m²	50–150 cm, most 50–100 cm	Gascon and Planas1986
Ontario Am	16	range 3–7,563 m²; mean 114 m²	>25 cm	Clark 1986
Michigan A or Rs	55	range 4–6,354 m²	no depth data provided	Ling et al. 1986
Minnesota North-central, seasonal wetlands	206	100–2,500 m²		Palik et al. 2003
Pennsylvania	45	maximum volume 4–743 m³	no depth data provided	Rowe and Dunson 1993
Pennsylvania Rs	8	range 250–1,000 m²	60 cm	Seale 1980

[1] Some pools were identified on the basis of physical characteristics, and others were further classified by the presence of obligate vernal pool animals. For pools where biological criteria were used, the abbreviations for indicator taxa are as follows: A = *Ambystoma,* Aj = *A. jeffersonianum*, Am = *A. maculatum*, Ao = *A. opacum*, At = *A.tigrinum*, Rs = *Rana sylvatica*, E = *Eubranchipus.*

[a] Vernal pools that did not support spotted salamanders but did support other vernal pool animals.

Table 10. Water Chemistry of Vernal Pools

Location and Special Conditions	pH and Color	Alkalinity (mg $CaCO_3$/l)	Al (mg/l)	References and Number of Pools
New Jersey				Gosner and Black 1957
Pine Barrens, sphagnaceous pools	3.6–5.2, mean 4.4, tea-colored			(25 pools)
Pine Barrens, gravel pits and transient pools	5.5–6.6, mean 6.3			(10 pools)
northern areas	6.6–9.5, mean 7.8			(11 pools)
New Jersey			210–790	Freda and Dunson 1986
Pine Barrens, *Sphagnum* pools	3.8–4.2; very dark, 4.2–9.0 mg tannic acid/l			(5 pools)
darkwater pools	4.1–4.6; dark, 1.4–4.6 mg tannic acid/l		210–510	(5 pools)
dammed seepage streams	4.11–4.48; 0.5–2.4 mg tannic acid/l		110–150	(3 pools)
man-made ponds and ditches	4.38–5.85; 0.3–2.4 mg tannic acid/l		50–620	(3 pools)
Connecticut				
New London	4.9–6.2			Pierce and Harvey 1987 (4 pools)
Litchfield County	6.8–7.5			(4 pools)
Massachusetts				
Cape Cod, depressions below high water table	3.74–6.25, median 4.8; color 22–875, median 140			Portnoy 1990 (46 pools)
Massachusetts				
Connecticut Valley	4.75–6.82; color 44-356	-0.8–39	20–200	Jackson and Griffin 1991 (14 pools)

(continued)

Table 10. (continued)

Location and Special Conditions	pH and Color	Alkalinity (mg $CaCO_3$/l)	Al (mg/l)	References and Number of Pools
Nova Scotia Spotted salamander pools	3.9–7.8, mean 5.0; color 0–296, mean 77	0–46, mean 5.7		Dale et al. 1985 (59 pools)
Nova Scotia Wood frog breeding pools	4.26–7.8, mean 5.34; color 0–296, mean 80	0–99, mean 9.5		Dale et al. 1985 (58 pools)
Ontario Central, Spring snowmelt, dry in summer	4.55–6.63 nearly 1/3 of pools have pH above 6	-0.4–12.2	154 8% with toxic levels	Clark 1986a, 1986b (16 pools)
Ontario Southwest Well-buffered ponds and soils	7.2–10.2, mean 8.3	59–960 mean 195		Hecnar and M'Closkey 1996 (143 pools)
Michigan Upper Penninsula, granite bedrock	mean 5.2			Ling et al. 1986 (55 pools)
Michigan Ann Arbor, 10 m from roads	pool 1: 5.8–7.7 pool 2: 7–7.8			Kenk 1949 (2 pools)
Pennsylvania	56% < 4.5 21% >5.0, receive runoff from roads treated with crushed limestone			Sadinski and Dunson 1992 (34 pools)
Pennsylvania Pocono Ridge and Valley Province	(1) 3.8–4.5 clear-water, *Sphagnum* pools; (2) 4.5–5.5 dark-water pools with 1–10 mg tannin/l; (3) 5.5–8.55 high-pH pools influenced by human activity in watershed			Freda and Dunson 1985b (74 pools)

Table 11. Summary of Massachusetts Audubon Society Data for pH and Alkalinity in Seventy-Five Vernal Pools in Massachusetts, 1983–1988

	pH	Alkalinity ($mg\ CaCO_3/l$)
All Pools, N=75 pools		
Mean all pools	5.50	6.07
Median all pools	5.42	1.50
Minimum all pools	4.24	-10.00
Maximum all pools	8.50	153.40
All Pools Except Those in Berkshires, N=62 pools		
Mean without Berkshires	5.39	3.02
Median without Berkshires	5.38	1.30
Minimum without Berkshires	4.24	-10.00
Maximum without Berkshires	7.11	58.30
Berkshires Pools, N=13 pools		
Mean Berkshires	6.91	49.78
Median Berkshires	6.72	31.40
Minimum Berkshires	5.70	2.40
Maximum Berkshires	8.50	153.40
Altered Pools, N=14 pools		
Mean altered pools	6.39	20.70
Median altered pools	6.24	10.15
Minimum altered pools	5.31	0.90
Maximum altered pools	8.50	153.40

Acidity, pH, Alkalinity, Hardness, and Color

The water quality constituents most often studied in vernal pools are pH, alkalinity, hardness, dissolved metals, and color, and they are often related to one another. Tables 10 and 11 summarize some of the available data on these parameters.

Acids release ionized hydrogen, or H+ ions, into water. Strong acids release most of their hydrogen ions readily, and weak acids tend to hold onto most of their hydrogen ions and release only a few of them. The free hydrogen ions in the water represent active acid that is potentially able to interact with geological materials, chemicals in the water, and living organisms.

Information on the amount of acid present in water is of great interest, because high levels of active acid have biological effects. For example, in water with a lot of free hydrogen ions — highly acidic water — nutrients are generally less available to plants and animals than in neutral or basic water, and toxic metals are often released into the water from surrounding rocks and soils. Hydrogen ions in soil and water affect many biological processes and limit the distributions of some plants and animals.

There are two different measures of the amount of acid in fresh waters. One, acidity, is a chemical measure of the total amount of acid that is present and potentially available to neutralize a base added to the water. It includes both ionized hydrogen that has been released into water by acids and other hydrogen that is bound to the acids and that can be released as bases are added. Acidity is determined by the titration of a water sample with a base to a predetermined endpoint. It is not commonly measured in studies of the water chemistry of vernal pools.

In contrast to acidity, pH provides information about the amount of free hydrogen ion, or active acid, in a vernal pool. It is determined using chemicals that turn different colors depending on the amount of active acid present, or with electronic meters that can measure the amount of active hydrogen ion present. pH is one of the most commonly measured water quality variables in studies of vernal pools.

Strictly speaking, pH is the negative logarithm of the concentration of ionized hydrogen in water. A pH of 7.0 is neutral, a pH above 7 is basic, and a pH below 7 is acidic. Because it is logarithmic, each pH unit represents a ten-fold change in the amount of hydrogen ion present, and because it is a negative logarithm, pH gets smaller the more hydrogen ion is present. The lower the pH value, the greater the concentration of active acid in the water, with each pH unit representing a ten-fold change in active acid. So, a vernal pool with a pH of 4.0 has ten times more active acid than a pool with a pH of 5.0, the pool with a pH of 5.0 has ten times more active acid than a pool with a pH of 6.0, and so on.

Acids enter vernal pools from surrounding soils, groundwater, and precipitation, and through biological processes that take place in the pools. Depending on the kinds of acids and on the chemicals that are already present in the water, the pH may or may not change as acids enter a pool. Many dissolved substances can bind to or neutralize acids and prevent the pH from going down — thereby buffering the water from changes in the amount of active acid. Alkalinity, or buffering capacity, also known as acid neutralizing capacity, is a measure of how much acid can be added to water before the pH changes. Some good buffers are calcium and magnesium carbonates and bicarbonates. Waters that contain high levels of these substances are called hard waters. Hard waters tend to have higher pH and alkalinities than soft waters, which are low in dissolved materials that can neutralize acids.

Vernal pools in much of the glaciated northeast are soft, acidic, and nutrient-poor (tables 10, 11) [19]. Three classes of highly acidic pools have been identified in

New Jersey and Pennsylvania, those with clear water and with *Sphagnum* in or bordering the pool; those that have extensive mats of *Sphagnum* and water that is dark due to a high tannin content; and pools with dark water but without high densities of *Sphagnum*[20]. Tannins are complex organic acids released during the decay of plant materials. Waters with high concentrations of tannins can look like strong tea. Light cannot penetrate very far into dark water, so the growth of algae and aquatic plants is reduced in pools with colored water. Organic acids can serve as food for bacteria and some animals, they can buffer water from inputs of mineral acids, and in some cases they can bind metals and prevent them from having adverse effects on aquatic organisms. Thus, pH alone is not necessarily a good measure of chemical conditions experienced by animals in vernal pools. If water chemistry and the responses of pool animals to it are of interest, some measure of the contribution of organic acids, even if just reported in terms of color, may contribute valuable relevant information. High color is not always synonymous with very low pH. The pH of a small (1,500 m^2, 16,150 ft^2), semi-permanent, brown-water pool in Ohio is 5.8 to 6.1[21].

In areas with carbonate bedrock, such as limestone and marble, pool water is usually harder and has a higher pH than in areas with naturally acidic soils and bedrock, such as granite. Vernal pools that are subject to disturbances such as dredging, agricultural runoff, livestock use, and highway runoff are also likely to be well buffered and less acidic than pools in undisturbed settings[22]. For example, in a study of pools in Pennsylvania, undisturbed pools were acidic, and those with a pH above 5.0 received runoff from roads that were treated with crushed limestone. In Massachusetts, my colleagues and I have found that pools with pH above 7.0 are either natural woodland pools in the Berkshires section of the state, where limestone bedrock is prevalent, or they are excavated fire ponds, farm ponds, gravel pits, or ornamental ponds.

Conductivity, or specific conductance, is a measure of the ability of water to carry a current. It varies with temperature and dissolved ions and can serve as a rough measure of dissolved substances. Most woodland vernal pools have low conductivity.

A consistent pattern of seasonal variation in water chemistry has been observed in many vernal pools[23]. Annual minima in pH, alkalinity, dissolved inorganic and organic substances, color, and conductivity usually occur at the time of ice-out in early spring. As pools gradually shrink and dry, conductivity, pH, color, and dissolved substances increase. In some vernal pools, pH, alkalinity, and dissolved substances fluctuate widely in response to rainfall, while in others the chemistry is highly stable. Over the course of four years, nine pools in Pennsylvania commonly showed a pH depression of 0.35 units after rain[24]. Permanently and semi-permanently flooded pools tend to be more stable than temporary pools[25].

In the eastern half of the glaciated northeast, precipitation contains strong mineral acids derived from industrial and power-plant emissions that originate farther west. The relationship between the deposition of atmospheric acids — "acid precipitation" or "acid rain" — and vernal pools has been a concern for many years. Since the implementation of the Clean Air Act Amendments of 1990 in the United States, improvements in air quality and decreases in the deposition of atmospheric acids have been measured at meteorological stations and stream monitoring sites across the glaciated northeast[26].

Dissolved Oxygen

Like terrestrial wildlife, aquatic animals need oxygen to survive. Some species breathe air at the water's surface, but others obtain dissolved oxygen from the water by diffusion across their gills or other body surfaces. The amount of dissolved oxygen can thus be an important factor affecting the presence and survival of aquatic animals in vernal pools.

The amount of oxygen that can be dissolved in water decreases as temperatures increase. At freezing, a liter of water can hold around 14 mg of dissolved oxygen, and this decreases to about 9 mg at 21° C (70° F) and to 7.5 mg at 30° C (86° F). This is one reason why coldwater habitats tend to have more diverse communities than warmwater ones, and why many aquatic insects in all kinds of habitats emerge from the water in spring or early summer and spend the summer as adults or diapausing eggs.

Oxygen enters water from the air, and also from the surfaces of algae and aquatic plants as they photosynthesize (oxygen is one of the byproducts of photosynthesis). Because oxygen diffuses into water very slowly, currents and wind are important for mixing air and water and adding dissolved oxygen to water bodies. Respiration by plants, animals, and decomposer microorganisms uses up oxygen in the water, and without such mixing the levels of oxygen can fall very low. Low levels of dissolved oxygen often occur in still waters such as vernal pools, especially if they are in the bottom of deep hollows or are sheltered from the wind by surrounding vegetation. Oxygen also can be depleted in a pool or pond under ice, again because oxygen cannot be mixed into the water. If water is very well mixed, or if there is a high amount of photosynthesis by algae and aquatic plants, oxygen can actually occur at concentrations above saturation in the water, forming bubbles in algal mats and on plant leaves.

Dissolved oxygen levels reported from vernal pools are highly variable. In sixteen spring-filling vernal pools in Ontario, dissolved oxygen levels average 6 mg/l. Dissolved oxygen levels in shaded woodland pools may be low, particularly as the season progresses and water temperatures increase. In the absence of canopy cover, algal photosynthesis may result in significant daily fluctuations in dissolved oxygen (ranging from less than 1 mg/l very early in the morning to 4.5 mg/l in the afternoon)[27].

In a temporary woodland pool in Michigan, oxygen levels under winter ice cover ranged between 1.3 mg/l and 5.1 mg/l (mean: 3.1 mg/l) and fluctuated during the open water period between 2.3 mg/l and 9.3 mg/l (mean: 4.86 mg/l). In a second pool nearby, with an open canopy and extensive plant growth within the pool basin, dissolved oxygen under ice fluctuated from a high of 11.1 mg/l soon after flooding in early December to a low of 1.5 mg/l, accompanied by "a foul odor . . . indicating a disintegration of organic matter," at the end of January[28]. The dissolved oxygen level under the ice subsequently recovered, showing a high of 8.3 mg/l in early March and fluctuating from 3.9 to 8.0 mg/l (mean: 6.4 mg/l) during the open water period.

These results suggest that dissolved oxygen levels may limit the distributions of animals in some vernal pools under ice or when there is a high level of algal photosynthesis. Some species that can survive in low-oxygen environments, either by breathing air or through metabolic adaptations that allow cellular activity when only limited amounts of oxygen are available, may be able to live in vernal pools during periods when other species are excluded by low oxygen.

Water Temperature

Vernal pool temperatures fluctuate seasonally and diurnally[29]. At the time of spring snowmelt, water temperatures are close to freezing, but they increase rapidly, averaging 8 to 12° C (46 to 54° F) in early to mid April and typically reaching 27 to 30° C (80 to 86° F) by summer. The maximum temperatures reached and the amount of daily variation in temperature in a given pool depend on canopy cover, basin configuration, water sources, and water depths. Differences between the surface and bottom of up to 10° C (18° F) and daily fluctuations of 8 to 15° C (14 to 27° F) are common.

These seasonal and daily temperature variations have implications for the aquatic life of vernal pools. The body temperature of vernal pool animals is not maintained at a constant level but instead varies with environmental conditions. In general, the rates of biological processes increase as temperature increases. For air-breathing animals, this means that oxygen demand increases with increasing temperature. At the same time, as we have seen, the amount of oxygen that can be dissolved in water decreases with increasing temperature. Some marine invertebrates can maintain a constant metabolic rate over the range of temperatures typical of their habitat, but this has not been demonstrated for invertebrates that are found in vernal pools[30]. For many animals that depend on vernal pools, development occurs at relatively low temperatures, and the rate of growth increases as temperature increases. The life cycle is complete before temperatures reach a critical threshold. In some species there is a trade-off between size and speed of development, with a larger size being attained when maturity is delayed due to slowly increasing temperatures. The successional changes that take place in pool communities over the course of a season in part reflect the abilities of different species to tolerate high summer temperatures and low oxygen levels. This subject is addressed further in the detailed species accounts presented later in this book.

Some Remaining Questions about Habitat

- What are the relationships between landscape position, geologic substrate, and pool surface areas and depths?
- How do landscape position, geology, and land use affect water quality?
- Can the distribution of vernal pools be predicted from a knowledge of local geology?
- How do upland forests contribute to the hydrology and water quality of vernal pools?
- How do changes in watershed land use and/or vegetation affect vernal pools?
- What is the influence of atmospheric deposition of acids, metals, and other substances on the water quality of vernal pools?
- What is the contribution of organic acids to the chemistry of vernal pools? What organic acids are present? To what extent are they biologically available to microorganisms, macroinvertebrates, and amphibians? Do they buffer the pool from inputs of mineral acids? Do they bind potentially toxic metals?
- How do water temperatures, dissolved oxygen, pH, hardness, nutrients, and other water quality constituents vary vertically and horizontally in vernal pools? How do they change over time?
- How large an upland buffer is needed around vernal pools to protect hydrology, water quality, and food sources of the pools, and what kind of restrictions on activities are needed within the buffer area?

References for Chapter 3

[1] Dexter and colleagues 1943 ff. [2] DiMauro and Hunter 2002. [3] Calhoun et al. 2003. [4] Brooks et al. 1998, Burne 2001, Portnoy 1987, Stone 1992. [5] Palik et al. 2003. [6] Tiner 2003b. [7] Beatini 2003, Brooks et al. 1998. [8] Brinson 1993, 1995, 1996; Leibowitz 2003. [9] Cowardin et al.1979. [10] Palik et al. 2003, Stone 1992. [11] Horne and Dunson 1994a, Kiviat et al. 1994. [12] Kiviat et al.1994. [13] Sobczak et al. 2003. [14] Bachmann and Bachmann 1994, Cole and Fisher 1978. [15] Clymo 1964, 1983; Moore and Bellamy 1974. [16] Bachmann and Bachmann 1994, Gascon and Planas 1986, Gates and Thompson 1990, Kenney 1991, Thompson and Gates 1982. [17] Colburn unpublished data; Dexter and Kuehnle 1948, 1951; Jokinen and Morrison 1998; also see references listed in tables. [18] Freda and Dunson 1985b, Leibowitz and Nadeau 2003, Sadinski and Dunson 1992, Whigham and Jordan 2003. [19] Clark 1986a, 1986b; Dale et al. 1985; Freda and Dunson 1985b, 1986; Gosner and Black 1957; Jackson and Griffin 1991; Ling et al. 1986; Portnoy 1990; Sadinski and Dunson 1992. [20] Freda and Dunson 1985b, 1986; Gosner and Black 1957. [21] Brodman 1995. [22] Colburn unpublished data, Hecnar and M'Closkey 1996, Kenk 1949, Sadinski and Dunson 1992. [23] Colburn unpublished data, Freda and Dunson 1985b, Sadinski and Dunson 1992. [24] Sadinski and Dunson 1992. [25] Freda and Dunson 1985b. [26] Clark and LaZerte 1985; Freda and Dunson 1984, 1985a, 1985b, 1986; Horne and Dunson 1994a, 1994b; Pierce 1985, 1987; Pierce and Harvey 1987; Pierce and Shayevitz 1982; Pierce and Wooten 1992; Pierce et al. 1984, 1987; Sadinski and Dunson 1992. [27] Bachmann and Bachmann 1994,

Clark 1986a, 1986b. [28] Kenk 1949. [29] Clark 1986a, 1986b; Colburn unpublished data; Dexter and colleagues 1943ff.; Jokinen and Morrison 1998; Kenk 1949; Seale 1982; Wiggins et al. 1980; Williams 1983. [30] Prosser 1973, Warren 1971.

Chapter 4

Bacteria, Protists, Algae, and Fungi

Most of the literature on vernal pools focuses on macroscopic animals and vascular plants. Hidden in the water and the mud, however, forming films on surfaces of leaves, wood, and living plants, and numbering in the millions of individuals, are many hundreds of kinds of microscopic organisms that produce food, transform gases and dissolved substances, and break down dead plant and animal material. Remarkably complex and, in many cases, beautiful in structure (Figure 6), the bacteria, algae, protozoans, and fungi perform functions that are critical to the lives of the more visible vernal pool wildlife.

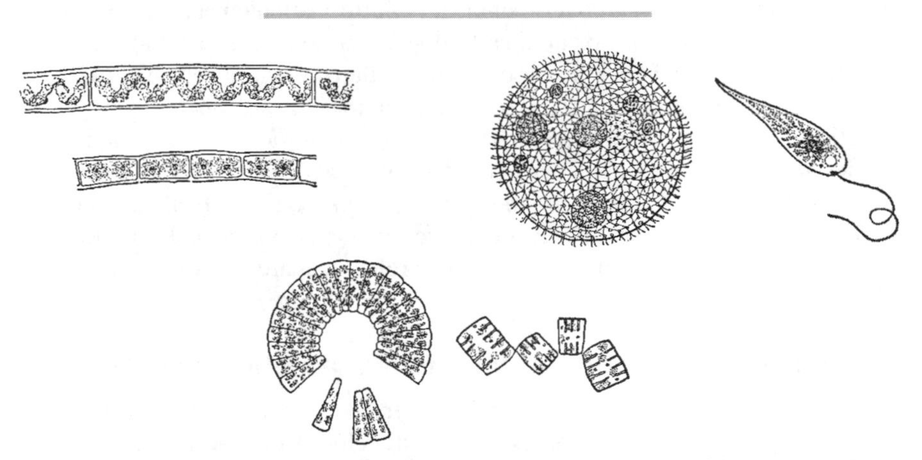

Figure 6. Algae and microorganisms of vernal pools include filamentous algae (upper left), flagellates such as *Euglena* (upper far right), colonial photosynthetic protozoans such as *Volvox* (upper center right), and diatoms such as *Meridion* (bottom left) and *Tabellaria* (bottom right). (From Morgan 1930, courtesy of the Morgan estate.)

These diminutive taxa, while unlikely to become the focus of conservation efforts or popular natural history studies, are highly diverse and are essential to energy flow in vernal pools. Bacteria and fungi play an important role in pool energetics by decomposing the tree leaves, dead herbaceous plants, and woody debris that serve as the base of the food web (see Chapter 13). Their activities release nutrients that are taken up by blue-green algae, diatoms, and green algae, whose photosynthesis also contributes to the food web in many pools. Algae are often visible as underwater films coating the bottom in vernal pools, particularly in spring when the water is clear and leaf-out has not yet occurred on the surrounding trees, and as floating clumps or even extensive mats at the surface, particularly in late spring when water temperatures start to become warm[1]. Microscopic protozoans feed on the decomposers and algae and are fed upon, in turn, by larger filter-feeders and scrapers, contributing substantially to pool productivity. Algae and decomposers are also fed on by grazing snails, insect larvae, and some tadpoles.

Microscopic taxa dominate the vernal pool biota in both numbers of species and numbers of individuals. In a small, high-conductivity, snowmelt pool in Quebec in the weeks immediately following the appearance of water in early spring, fully 76% of the biota consisted of taxa ignored in most biological surveys! Of the 276 taxa documented, 194 were protists — single-celled photosynthesizers such as diatoms, dinoflagellates, and euglenas; green algae; and protozoans such as amoebas, ciliates, and zooflagellates — and the rest were bacteria.

This chapter briefly discusses the microscopic inhabitants of temporary waters. It first considers the means by which pool microbiota deal with seasonal drying and then provides a concise summary of information for each major group. The protists represent a taxonomically diverse and complex assemblage. For convenience, I do not break the various zooflagellates, phytoflagellates, amoebas, and ciliates into their individual taxonomic groupings but instead discuss them all together under the heading "Protozoa." Most of the information presented here comes from Marshall Laird's intensive survey of a small (5 m^2, or 54 ft^2) snowmelt pool in Quebec, discussed in his 1988 study of mosquito habitats around the world; from Charles Masters' 1968 guide to temporary ponds; and from Roman Kenk's 1949 study of Michigan pools[2]. Other references are listed at the end of the chapter.

Dispersal of Microflora and Microfauna into Vernal Pools

Microscopic organisms are readily dispersed to suitable environments, and thus, newly created water bodies rapidly develop a diverse microflora and microfauna. Even the smallest ephemeral puddle supports a rich community of algae, bacteria, and protozoa soon after rain showers have ended. In vernal pools, some microorganisms probably persist from one year to the next by surviving the dry period in a desiccation-resistant state on the pool bottom, but many are carried passively into pools by insects and other mobile animals after flooding. In this way, it is likely that many vernal pools are recolonized annually by microorganisms that are transported by migrants from other habitats.

Aquatic beetles are particularly effective agents of dispersal[3]. Experimental studies with field-collected aquatic beetles have documented the transport and successful transfer of more than one hundred genera of algae, including blue-greens, greens, and diatoms; twenty genera of protozoans; unidentified genera of fungi; and water mites, rotifers, nematodes, and ticks! Other studies with aquatic and terrestrial beetles have documented transport of additional genera. Probably other mobile aquatic insects that move between water bodies, such as backswimmers and water boatmen, also serve as dispersal agents. Birds, especially ducks, have been shown to transport algae and microorganisms on their feathers and feet, as well as through their digestive tracts[4]. It is likely that amphibians and reptiles also transport microscopic organisms between water bodies.

The Major Groups of Microbiota

Bacteria

Bacteria are abundant in situations where organic materials are present, a condition that is characteristic of vernal pools, in which fallen tree leaves, other dead plant materials, and organic sediments tend to accumulate. In a tiny snowmelt pool in Quebec soon after thawing, the rapid proliferation of seventeen species of bacteria provides abundant food to nourish the rich protozoan community and contributes nutrients supporting diatoms and other photosynthetic microorganisms[5].

Protozoans — Flagellates, Amoebas, and Ciliates (Sarcomastigophora and Ciliophora)

A look through a compound microscope at a drop of water from a vernal pool will yield a great variety of protozoans, including species of well-known genera such as *Euglena, Paramecium, Spirostomum, Volvox, Vorticella,* and *Stentor.* Stalked ciliates often cover the bodies of crustaceans and aquatic insects, imparting a white, fuzzy appearance. In a small snowmelt pool in Quebec, eighty different protozoans were identified within twenty-one days of thawing[6].

Some protozoans, like algae and plants, can produce their own food through photosynthesis, using the sun's energy to convert water, carbon dioxide, and nutrients into carbohydrates, fats, proteins, and other cellular constituents. Others feed extensively on bacteria, and some also feed directly on decomposing organic matter. In turn, protozoans are fed upon by rotifers, water fleas, fairy shrimp, and other filter-feeders.

Blue-Green Algae (Cyanophyta, Cyanobacteria)

Blue-green algae are active both in well-oxygenated water and where oxygen levels are low. They become especially abundant when high levels of nutrients and/or organic materials are present. Blue-green algae grow on the surfaces of submerged plants as well as in floating mats. The species present vary depending on the relative amounts of oxygen, carbon dioxide, and nitrogen in the water. Species of *Spirogyra* do well in temporary pools with limited light penetration. Both *Spirogyra* and *Lyngbya* were found on the surfaces of submerged plants in a permanent or semi-permanent, fishless pool in southern Michigan[7].

Green Algae (Chlorophyta)

Green algae, according to Masters, "are sometimes found in sufficient numbers to color the water in temporary pools"[8]. These algae grow well in well-oxygenated waters. Among the genera likely to be found in vernal pools are *Closterium*, with more than sixty North American species, found preferentially in acidic waters of pH 5 to 6; *Euastrum; Tetmemorus; Micrasterias;* and *Cosmarium.* In an open, long-cycle pool in southern Michigan, *Cylindrocapsa geminella* and *Vaucheria sessilis* were found in low numbers both on submerged surfaces and in floating mats. Both *V. sessilis* and *V. geminata* also occurred on submerged plant surfaces in a nearby permanent or semi-permanent pool, along with *Oedogonium, Zygnema,* and *Mougeotia scalaris*[9]. All of these algal genera are found in a wide variety of freshwater habitats, including streams.

Chara is neither single-celled nor microscopic, but it is an alga. It is large and plant-like and is often classified with the green algae. It is usually found in fairly hard water, and calcareous precipitates often cover its dissected leaf-like branches, giving it the common name of stonewort. *Chara* has root-like structures called rhizoids that allow it to penetrate the pool bottom and become firmly anchored in the substrate (Figure 7). Superficially, it somewhat resembles a number of aquatic plants including water milfoils, some of the bladderworts, and coontail. The bottoms of a permanent, fishless pool in Michigan and an autumn-flooding, long-cycle pool in Amherst, Massachusetts, are densely carpeted with *Chara*, and the alga is responsible for most of the primary production within the latter pool[10].

Diatoms and Yellow-Green Algae (Chrysophyta)

Diatoms, unlike the green algae, are usually olive green, yellowish, or brownish. They incorporate silica into their cell walls and are particularly abundant in areas with siliceous bedrock and sandy soils. In contrast to other algae and most plants, they store the products of photosynthesis in the form of oil droplets, rather than starch. They can thus serve as highly nutritious sources of food for bacteria, protozoans, zooplankton, and macroinvertebrate grazers.

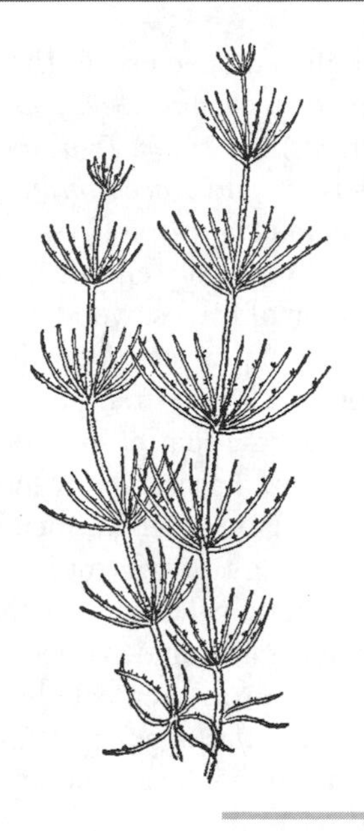

Figure 7. *Chara* is a large, upright alga found in some vernal pools, especially in areas with calcareous bedrock. (From Morgan 1930, courtesy of the Morgan estate.)

Vernal pools, especially woodland pools, have rich diatom communities in which the genera *Asterionella, Tabellaria,* and *Melosira* are particularly well represented. A 5 m^2 (54 ft^2) high-conductivity, snowmelt pool in Quebec supported 114 species of diatoms within weeks of thawing, and an adjacent 1 m^2 (11 ft^2) woodland pool shaded by tamarack and black spruce and carpeted with leaves and needles had 106 diatom species[11].

In southern Michigan, a filamentous yellow-green alga, *Tribonema vulgare* (as *T. bombycinum*), appeared in April in a short-cycle woodland pool, growing initially on submerged vegetation and later in floating mats. It was also present as the dominant species in an open, long-cycle pool in which several species of green algae were present in lower numbers[12].

Fungi

Fungi play a key role in nutrient cycling and energy flow within northeastern vernal pools (see Chapter 13). A study of decomposition of different plant materials in an annual vernal pool and two continuously flooded pools in Ontario provides some information about fungal species present, their relative abundance on different plant substrates, and seasonal changes in distribution and abundance[13]. A brief summary of these results follows.

During the dry phase, terrestrial fungi begin the decomposition process. Terrestrial fungi identified from decomposing ferns, grasses, herbaceous plants, and maple leaves in the annual vernal pool included species of *Alternaria, Cladosporium, Hyalodendron, Rhinocladiella, Epicoccum, Aereobasidium, Aureobasidium, Phoma, Gonatobotrys, Paecilomyces, Fusarium, Tripospermum, Penicillium, Hormiactis, Pestalotia, Trichoderma,* and *Acremoniella,* plus *Gonatobotrys simplex, Botrytis cinerea,* and *Acremoniella atra.*

Terrestrial fungi need abundant oxygen and do not do well under flooded conditions. Most fungi responsible for the breakdown of leaves, wood, and other detritus in flooded vernal pools fall in the broad group of aero-aquatic fungi. These fungi have life cycles with both an aquatic and a terrestrial phase. Upon flooding of the annual vernal pool in spring, aero-aquatic *Helicoon* and *Helicodendron* appeared on detritus previously colonized primarily by terrestrial fungi. *Tricellula aquatica, Alatospora acuminata,* and *Lemonniera*, members of a third group of fungi, the aquatic hyphomycetes, were present but, with the exception of *Lemonniera*, which accounted for 10% of fungal density on the grass *Agrostis* in early spring, they were never present at significant densities.

In the continually flooded pools, decomposition was dominated by aero-aquatic taxa including *Helicodendron, Helicoon, Papulaspora viridis,* and *Clathrosphaerina*. A few terrestrial fungi that depend on simple organic substrates, including *Rhizopus* and *Mucor*, were collected as well, but they were never dominant. There were some differences in the species colonizing different plant substrates, and the relative abundance of fungi varied among substrates, as well. Some shifts in composition of the decomposer community occurred over time, with the most dramatic changes seen upon flooding of the annual pool in early spring. The implications of these changes will be discussed further in the context of pool energetics in Chapter 13.

Fungi can also be pathogenic. Recently, fungi and ranaviruses have been implicated in high mortality in free-living populations of amphibians. A chytrid fungus, *Batrachochytrium dendrobatidis*, which infects the skin of amphibians, is known worldwide and has caused mass mortality in frogs and toads in Australia, Central America, and the western United States. It has been reported from toads in Maryland and from treefrogs in Illinois, but it has not caused mortality in those populations[14]. The fungus is widely dispersed by the international trade in amphibians for pets, pond stocking, and food[15]. Some biologists are concerned about the potential for this fungus, as well as viral pathogens, to be dispersed to vernal pools, and recommend that researchers and students disinfect their boots, nets, and other equipment before sampling in vernal pools[16].

Some Remaining Questions About Microbiota

- What species of bacteria, protozoans, algae, and fungi occur in vernal pools?
- What are the conditions under which different species occur?

- What are the life histories of vernal pool microbiota? Are they permanent residents, or do they recolonize vernal pools each year?
- What patterns of seasonal succession in bacteria, protozoans, algae, and fungi occur in pools of different hydrologic classes?
- What kinds of year-to-year differences are seen in the species and numbers of microorganisms?
- How do microorganisms contribute to decomposition, food quality, and overall energy flow in vernal pools?

Additional research questions focusing on algal photosynthesis, decomposition, and energy flow are considered in Chapter 13.

References for Chapter 4

[1] Bachmann and Bachmann 1994, Colburn unpublished data, Williams 1983. [2] Kenk 1949, Laird 1988, Masters 1968. [3] Darwin 1859, Fernando 1958, see also review in Milliger et al. 1971. [4] Proctor et al. 1967. [5] Laird 1988. [6] Laird 1988. [7] Kenk 1949. [8] Masters 1968, p. 21. [9] Kenk 1949. [10] Cole and Fisher 1978, Wilbur 1972. [11] Laird 1988. [12] Kenk 1949. [13] Bärlocher et al. 1978. [14] Berger et al. 1998, Collins and Storfer 2003, Speare et al. 2000. [15] Daszak et al. 1999. [16] Aram Calhoun, University of Maine, personal communication.

Chapter 5

Vegetation

Changes in the woodland plant community often provide the first evidence that a vernal pool is at hand. A gap in the tree canopy, a rosy wash of buds and branches on highbush blueberries or other shrubs along the perimeter, or the bright green of skunk cabbage signals a break in the upland woods and the presence of an aquatic system. Strictly speaking, vernal pools in the glaciated northeast are defined in terms of their physical features and their fauna. There does not appear to be a suite of plant species that can be used to define northeastern vernal pools[1]. In this, northeastern vernal pools are very different from temporary pond basins in the southeastern United States and California, which typically support characteristic associations of both plants and animals[2]. Still, flood-tolerant plants typical of wetlands in the local area often grow within the basins of vernal pools as well as along the shoreline. Plants contribute to in-pool habitat and the food web, and there is some evidence that the distributions of some animal species in vernal pools parallel those of certain plants[3].

Some of the most common plants that grow in and around vernal pools are listed in Table 12. They can be identified with the help of common field guides to trees, shrubs, grasses, ferns, and wetland plants. Some useful guides are noted in the references at the end of this chapter[4].

Vegetation Surrounding Vernal Pools

In relation to the surrounding plant community, vernal pools fall into two categories. In the first group, the pools occur as moist "islands" within an upland landscape. These pools are often distinctly separated from the surrounding uplands by narrow zones of wetland vegetation along the shore. In the second group, the pools occur as depressions within larger wetland systems, including red maple swamps, spruce-fir swamps, Atlantic and northern white cedar swamps, shrub swamps, fens, and bogs. In such pools, the plant community surrounding the pools does not necessarily differ markedly from the rest of the wetland. In both upland and wetland pools, the plants growing along the shoreline are representative of locally common wetland species that tolerate different degrees of root inundation.

Table 12. Common and Scientific Names of Plant Species Typical of Northeastern Vernal Pools

Common Name	Scientific Name
alders	*Alnus* spp.
American elm	*Ulmus americanus*
arrowwood	*Viburnum recognitum, V. dentatum*
Atlantic white cedar	*Chamaecyperis thyoides*
balsam fir	*Abies balsamea*
beggar's-ticks, stick-tights	*Bidens* spp.
birches	*Betula* spp.
black ash	*Fraxinus nigra*
black willow	*Salix*
black spruce	*Picea mariana*
bladderwort	*Utricularia gemini-scapha, Utricularia* sp.
bluejoint	*Calamagrostis canadensis*
box elder	*Acer negundo*
bulrushes	*Scirpus cyperinus*, other *Scirpus* spp.
burreed	*Sparganium* spp.
buttonbush	*Cephalanthus occidentalis*
cardinal flower	*Lobelia cardinalis*
catbriar	*Smilax rotundifolia*
cinnamon fern	*Osmunda cinnamomea*
Eastern cottonwood	*Populus deltoides*
green ash	*Fraxinus pennsylvanica*
hemlock	*Tsuga canadensis*
highbush blueberry	*Vaccinium corymbosum*
leatherleaf	*Chamaedaphne calyculata*
maleberry	*Lyonia ligustrina*
manna grasses	*Glyceria* spp.
marsh fern	*Thelypteris palustris*
northern white cedar	*Thuja occidentalis*
pin oak	*Quercus palustris*
pitch pine	*Pinus rigida*
poison ivy	*Toxicodendron radicans*
pondweeds	*Potamogeton natans*, other *Potamogeton* spp.
red maple	*Acer rubrum*

(continued)

Table 12. (continued)

Common Name	Scientific Name
red spruce	*Picea rubens*
reed canary grass	*Phalaris arundinacea*
river birch	*Betula nigra*
rushes	*Juncus militaris*, *J. canadensis*, *Juncus* spp.
sedges	*Carex* spp., *Scirpus* spp.
sensitive fern	*Onoclea sensibilis*
sheepberry	*Viburnum lentago*
sheep laurel	*Kalmia angustifolia*
silver maple	*Acer saccharinum*
smooth alder	*Alnus serrulata*
stinging nettle	*Urtica dioica*
speckled alder	*Alnus rugosa*
swamp loosestrife, water willow	*Decodon verticillatus*
swamp white oak	*Quercus bicolor*
sweet pepperbush	*Clethra alnifolia*
three-way sedge	*Dulichium arundinacea*
tussock sedge	*Carex stricta*
viburnum	*Virbunum* spp.
water lilies	*Nymphaea odorata*
white oak	*Quercus alba*, *Q. bicolor*
white pine	*Pinus strobus*
white spruce	*Picea glauca*
willows	*Salix* spp.
winterberry	*Ilex verticillata*
witherod, wild raisin	*Viburnum cassinoides*
yellow birch	*Betula allegheniensis*

Depending on local topographic conditions and the hydrologic regime of a particular pool, the ring of flood-tolerant vegetation surrounding a pool may extend for less than 1 m (3 ft) or for more than 20 m (65 ft) beyond the pool's usual high-water line.

Vegetation In Vernal Pools

Among the kinds of plants that often occur in the basin of vernal pools are ferns, mosses, herbaceous annuals and perennials, shrubs, and trees.

Ferns

Ferns known to grow in vernal pools include marsh fern (*Thelypteris palustris*), royal fern (*Osmunda regalis*), and sensitive fern (*Onoclea sensibilis*). Marsh fern is often found growing in flooded pools, while sensitive fern occurs most commonly on saturated sediments once a pool has drawn down. In some pools, these species carpet the pool basin during the dry period (Plate 2, center).

Mosses and Liverworts

Sphagnum mosses constitute a significant portion of the substrate and surrounding vegetation in some vernal pools (Figure 8). This is particularly the case in areas with acidic soils or siliceous bedrock, such as the Canadian Shield, the Upper Peninsula of Michigan, the New Jersey Pine Barrens, and much of New England and the Maritime Provinces[5].

Two liverworts, crystalwort or slender Riccia (*Riccia fluitans*) and purple-fringed Riccia (*Ricciocarpus natans*), are also reported to be associated with vernal pools (Figure 9)[6]. Both are distributed throughout the glaciated northeast. Unlike the more familiar leafy liverworts, which look a lot like mosses, the Riccias are known as thallose liverworts because their basic body form is a thin, flattened lobe ("thallus") similar to that of many green algae. Their seeds germinate on the pool bottom when water is gone. When pools are flooded, crystalwort floats just below the water surface in tangled masses of branching, 1-mm-wide ribbons. Purple-fringed Riccia has a floating thallus up to 3 cm (1.2 in) across and resembles a large

Figure 8. *Sphagnum* moss is common in many acidic pools. Shown here is a spore-capsule-bearing stalk and spore capsule at left and a stalk without spore capsules at right. (From Morgan 1930, courtesy of the Morgan estate.)

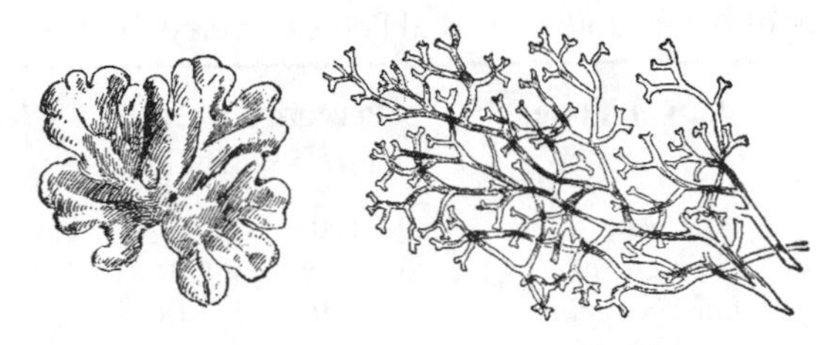

Figure 9. *Riccia* is a liverwort that often occurs in vernal pools. (From Morgan 1930, courtesy of the Morgan estate.)

duckweed. Upon pool drying, a rounded, leafy growth form of this species develops on the exposed sediment.

Higher Plants

Most higher plants found in the basin of vernal pools occur in many kinds of freshwater wetland habitats. There is some evidence that a few species may be particularly well adapted for life in vernal pools. For example, featherfoil (*Hottonia inflata*), an aquatic member of the Primulaceae that is uncommon in parts of the glaciated northeast, appears to be associated with certain types of vernal pools in Maine, Massachusetts, and Virginia[7]. Its abundance in a given year is thought to be related to the previous year's water levels. False loosestrife (*Ludwigia*), a member of the Onagraceae, occurs in habitats with fluctuating water levels[8]. Horned pondweed (*Zannichellia palustris*), a member of the Zanichellaceae, is sometimes found in vernal pools where water is hard, with high pH and high conductivity[9].

Plant Zonation and Distribution

Plant growth within the basin of vernal pools appears to reflect both the hydroperiod and the extent of canopy closure — and hence of light penetration — over the pool during the growing season[10]. In Maine, semi-permanent pools are less likely to have shrubs than pools that dry seasonally. Small, well-shaded woodland pools in New Jersey and Michigan lack algal mats or plants, while more open pools in fields or with only partial shading by trees contain much aquatic vegetation and algae[11]. Plants often occur in distinct rings or zones from a pool's edge to the center, reflecting an increased tolerance to prolonged soil saturation[12].

Both the amounts and kinds of vegetation in vernal pools are highly variable (Table 13)[13]. Most pools have relatively small percentages of their bottoms covered with vegetation, but usually at least a small amount of plant growth is present and in some pools the entire basin is covered. Shrubs are the most common plants in 106 pools in Amherst, Massachusetts, and account for half the cover in almost two-thirds of the pools; herbaceous vegetation dominates in more than a quarter

Table 13. Percent of the Bottom of Vernal Pools Covered by Plants

Location	Cover Type	Percent Cover		Reference
		Range	Median	
Quebec 15 pools	emergent submergent emergent plus submergent	0 – 100 5 – 100 10 – 130	15 30 45	Gascon and Planas 1986
Massachusetts Cape Cod, 17 pools	shrub	2 – 99		Colburn and Garretson Clapp 2004
Massachusetts Amherst, 106 pools	total: shrub, herbaceous, and mature trees	10 – 50+		Stone 1992

of the pools and mature trees dominate in more than an eighth[14]. In seventeen vernal pools on Cape Cod with a wide range of shrub cover, some of the pools without shrubs are unvegetated, and others have dense growths of perennial grasses, rushes, sedges, *Sphagnum* and other mosses, and a variety of herbaceous annual plants[15].

Vegetative Classes

Given that plant cover and composition in vernal pools vary widely, and given that detritus provided by plants growing next to or in pools contributes substantially to the food web (see Chapter 13), it can be useful to describe pools according to their dominant vegetation. A preliminary vegetative classification of vernal pools follows. It is based on my observations of vernal pools as well as on descriptions of pools in the literature. As with the hydrologic classification proposed in Chapter 2, this is suggested as a working model for use in research and conservation. Other classes may be added as the plant communities of more pools are described.

In determining the vegetative class of a pool, it is important to focus on the dominant plants immediately next to or in the vernal pool, which may be completely different from the surrounding forest. Also, it is worth noting that the plants growing in the basin of a vernal pool may change from one year to the next, depending on maximum water depths and when the pool fills and dries each year. The seeds of many herbaceous species can lie dormant in the sediment of vernal pools for many years, sprouting only when specific conditions of hydrology, temperature, and nutrients are present. Thus, the classification below is based largely on long-lived perennial plants associated with vernal pools.

Mixed-Canopy Pools

Mixed-canopy pools have a mixture of flood-tolerant plant species around the margins and a variable degree of canopy cover. Little plant growth is seen within the basin under flooded conditions. The substrate of mixed-canopy pools consists largely of leaves or needles and woody debris contributed by the surrounding trees and shrubs. Most small, natural vernal pools fall within this category (plates 4, 5).

A wide variety of plant species can be found around the borders of mixed-canopy pools. The most frequently encountered of these are included in Table 12. In deciduous forests, the plant community surrounding pools may include red maple, American elm, black ash, willows, pin oak, white oak, sweet pepperbush, poison ivy, cat brier, winterberry, highbush blueberry, maleberry, swamp azalea, alders, and various viburnums. In floodplains, silver maple, green ash, river birch, eastern cottonwood, box elder, and black willow may also be abundant. In coniferous areas the pools may be surrounded by hemlock, white pine, pitch pine, white cedar, spruce, and balsam fir. A mixture of deciduous and coniferous species occurs around pools in mixed forests. *Sphagnum,* or peat moss, is often present along the edge of pools or on tussocks within the pool. Cinnamon fern often occurs along the upper edges of pools and delineates the upper zone of soil saturation.

The tree canopy above the pool may be complete (or dense), partial, or open. Most of the basin in dense-canopy pools does not support aquatic, wetland, or terrestrial vegetation, although there may be an occasional shrub within the pool proper. In pools where a partial or open canopy allows adequate light penetration during the growing season, shrubs, emergent vegetation, or submerged aquatic plants may be present.

Red Maple Swamp Pools

Red maple swamp pools are of two types: (1) isolated depressions that are dominated by red maples growing on the shore and in the basin, but surrounded by upland beyond the shore, and (2) flooded depressions that occur within larger red maple swamp areas. The plant community immediately surrounding and within both types of pools is dominated by red maple.

The canopy in red maple swamp pools may be dense or partial, depending on the density of trees within the basin; this is influenced by depth, hydroperiod, and pool size. Red maples in wetlands have fairly shallow root systems and are readily blown over, and gaps in the canopy which let large amounts of light into the pool are sometimes produced by storms with high winds. The water level in these pools often draws down rapidly following leaf-out. The substrate is dominated by decomposing red maple leaves. *Sphagnum* is often present within the basin. When the pool is dry, the bottom may support a variety of herbaceous plants including marsh fern, and shrubs including highbush blueberry and winterberry may grow occasionally within the basin.

Coniferous Swamp Pools

Coniferous swamp pools occur as depressions within larger evergreen swamps, and as isolated depressions whose shorelines are dominated by conifers such as white cedar, hemlock, or spruce (Plate 4). The growth of mosses, ferns, herbaceous plants, and shrubs within the basin depends on the extent of canopy closure. Because most conifers are evergreens, light levels and temperatures in coniferous swamp pools may be lower than in open pools or those dominated by deciduous trees. The substrate in these pools is heavily carpeted with conifer needles, cones, and branches, the water is often highly colored by organic acids, and *Sphagnum* is often present.

Shrub Swamp Pools

Shrub swamp pools are dominated by shrubs growing within the basin. Some occur as depressions in larger shrub-dominated wetlands, and others function as small, isolated shrub swamps. Highbush blueberry, alders, leatherleaf, and winterberry are very common, and a wide diversity of other wetland shrubs, including buttonbush, which often forms monospecific stands, may also occur in these pools. During drawdown, depending on the canopy cover and the length of the dry period, a variety of mosses, ferns, and herbaceous plants may cover the area not covered by shrubs.

Decodon Pools

In southeastern New England, a high proportion of coastal vernal pools is dominated by swamp loosestrife or water willow, *Decodon verticillatus*. Such pools tend to occur in kettle depressions and are usually surrounded by a ring of shrubs that may include sweet pepperbush, highbush blueberry, winterberry, and maleberry. Water occurs around the clumps of *Decodon,* and often a moat of water occurs between the central *Decodon* zone and the pool edge, shaded by the overhanging shrub border. *Sphagnum* is often present.

Marsh Pools

Marsh pools are extensively vegetated with non-woody submerged, floating, and/or emergent perennial and annual plant species typically found in marshes and the shallow shoreline areas of lakes and ponds. Most commonly, they are long-hydroperiod vernal pools with basins that are open to the sky (plates 3, 5). In such pools the dominant plant species may include bur-reeds; grasses, particularly manna grasses, bluejoint, and reed canary grass; rushes; and sedges including three-way sedge, tussock sedge, other *Carex* spp., and bulrushes. Patches of deeper water in such pools often support water lilies, pondweeds, and bladderworts. A great variety of herbaceous annual plants can be found in these pools, with the species present varying annually with differences in hydroperiod.

Seasonal Changes In Vegetation

The plants growing in vernal pools may change dramatically in the space of only a few weeks or months. Many pools that start the season as a flooded basin covered with decaying tree leaves may later be heavily vegetated with a variety of grasses, ferns, and herbaceous annual plants. All of these plants benefit in some way from the pool's presence, whether by the concentration of water, increased depths of organic substrate, higher availability of nutrients, or slightly elevated light levels in comparison with the surrounding forest. Some pools support dense floating growths of duckweeds or water meal (species of *Lemna*, *Wolffia*, *Wolffiella*) and may appear as solid green mats in late summer. As fall advances, within-pool plants die back and nearby trees lose their leaves. The basin again is carpeted with fallen leaves and may blend in so well with the rest of the forest floor as to be easily overlooked by the casual observer.

Because plants differ in their abilities to tolerate flooding of their roots, the composition of the plant community and the amount of biomass present in different plant species changes both seasonally and from one year to the next, depending on the depth of flooding and the rate and timing of drawdown. The longer the basin remains flooded, the more the plant community is restricted to species that can withstand extended soil saturation. In contrast, when pools dry early in the growing season, seeds of plants that are less tolerant of flooding can germinate and develop in the pool. In Kendall II, a short-cycle, spring-filling vernal pool in Ontario, the vegetation underwent substantial seasonal changes, with wetland and aquatic sedges and grasses (*Scirpus atrovirens*, *Carex retrorsa*, and *Glyceria grandis*) declining and being replaced by more terrestrial species including ferns (*Onoclea sensibilis*, *Thelypteris palustris*) and ditch stonecrop (*Penthorum sedoides*) as the pool dried[16]. I have found no quantitative assessments of changes in plant community composition and cover from year to year or over a season as water levels decline in northeastern vernal pools.

Some Remaining Questions About Plants

- To what extent does the plant community reflect water quality and hydrology?
- How do plant communities in vernal pools change over time, both within and between years?
- What are the life history strategies that allow vernal pool plants to adapt to seasonal and between-year variations in pool hydrology?
- Are there direct relationships between the plant community composition and the species of aquatic invertebrates and/or vertebrates that use a pool for

reproduction or feeding? Can the plant community serve as a predictor of the fauna of a vernal pool?

- What are the relationships between plant community composition around a vernal pool and the detrital base of the food web within the pool?
- To what extent does the composition of the surrounding plant community, and of the plants growing within the pool basin, affect water chemistry and food quality?
- Are there particular classes of vernal pools that provide habitat for plant species that do not typically occur in other types of wetlands?

References for Chapter 5

[1] Cutko 1998. [2] Keeley and Zedler 1998, Kirkman et al. 1999, Rawinski 1997, Zedler 1987. [3] Colburn and Garretson Clapp 2004. [4] Cobb 1956, Magee 1981, Newcomb 1977, Niering 1985, Petrides 1958, Redington 1994. [5] Freda and Dunson 1986, Gascon and Planas 1986, Gosner and Black 1957, Portnoy 1990, Stangel 1988. [6] Conrad 1956; Kenk 1949; Masters 1968; Thomas Rawinski, Massachusetts Audubon Society, personal communication; Snure 1957. [7] Mark 1999; Mark McCollough, Maine Division of Inland Fisheries and Wildlife, personal communication; Rawinski 1997 and unpublished data. [8] Fassett 1957, Prescott 1980, Massachusetts Audubon Society unpublished data. [9] Masters 1968. [10] Calhoun et al. 2003, Freda 1983, Husting 1965, Nyman 1991, Pierce et al. 1984, Piersol 1910, Seale 1982, Williams 1983. [11] Nyman 1991, Seale 1982. [12] Bärlocher et al. 1978, Rawinski 1997, Massachusetts Audubon Society unpublished data. [13] Colburn and Garretson Clapp 2004, Gascon and Planas 1986, Stone 1992. [14] Stone 1992. [15] Colburn and Garretson Clapp 2004. [16] Bärlocher et al. 1978.

Chapter 6

Life History Strategies of Pool Animals

The mass migrations of spring-breeding amphibians, and the sudden appearance of fairy shrimp and a host of other freshwater invertebrates when a pool floods, illustrate adaptations that allow vernal pool animals to survive in habitats that contain water during only part of the year. Fundamentally, animals whose reproductive success depends on vernal pools must produce offspring that will, themselves, survive to reproduce. For vernal-pool-dependent species, this means life history strategies that provide for the successful completion of an aquatic developmental phase when water is present, for survival during the dry period, and for getting around the reality that successful reproduction may be impossible in some years when weather results in unfavorable hydrologic conditions in pools. Ultimately, the hydrologic cycle is the most important physical factor governing the biological communities of vernal pools, but interactions within and between species can also affect the ability of species to survive[1].

In vernal pools, water is available to aquatic animals only during a limited window of time which varies among pools and between years. The timing of life history stages is critical to reproductive success and is tightly linked to the timing of flooding and drawdown. Eggs must have enough time to hatch and develop under favorable conditions, and animals must have reached their drought-withstanding or drought-avoiding stages before the pool dries. Both flooding and the start of drawdown can serve as cues that direct developmental changes in animals, initiating aquatic life stages or a shift to stages adapted to drawdown, as appropriate. Timing is especially important in those pools with the shortest hydroperiods — short-cycle, spring-filling pools — where water is present for a few short months between the spring thaw and summer drawdown and drying. Species breeding in these pools must develop rapidly to avoid being stranded in a dry pool.

How long conditions in a vernal pool are favorable for aquatic life varies from one year to the next. The life history strategies of many species include a high degree of flexiblility and allow the organisms to make use of the aquatic habitat as long as it is available. This flexibility and variability is part of the fascination of vernal pools — what we find and when we find it one year may be completely different from what we found the previous year or what we will find next year. It also makes many species of animals that occur in vernal pools highly valuable study subjects for evolutionary biologists who are interested in the development and control of adaptive characteristics that allow animals to survive in variable environments.

As we have seen, although vernal pools fill to their maximum depth and volume in spring, they may also be flooded at other times of the year in response to winter thaws or summer storms. Some animals with rapid development or the ability to become dormant at any time may become active whenever water is present in vernal pools. In contrast, for many species, especially those with fairly long developmental periods, limited tolerance of temperature extremes, or specific food requirements, it is important to avoid starting development in response to flooding at times other than the spring. Many of these specialized species have drought-resistant mechanisms that require specific cues before dormancy can be broken. In this way, their development is restricted to appropriate times of the year.

Life History Patterns

The following discussion summarizes the general life history patterns seen in vernal pool wildlife. The emphasis is on hydroperiod and how animals avoid or withstand the stress of seasonal drying, because the hydrologic cycle plays such a critical role in determining whether individual species can successfully complete their development. However, water chemistry, temperature, and biological interactions also affect the distributions of animals in vernal pools, and these factors are considered in some of the detailed discussions of individual species' distributions and life history adaptations in chapters 7 to 12, and in the community discussion in Chapter 13. For further information about a wide variety of constraints imposed by temporary fresh waters including springs, intermittent streams, and ponds, and of their effects on the community composition, faunal dynamics, adaptations to drying, and life histories in aquatic insects, readers are referred to a comprehensive review published by D. Dudley Williams in 1996[2].

Biologists have looked at the life history strategies of vernal pool species from a variety of perspectives[3]. Evolutionary biologists and population ecologists are apt to consider the life span, the number of opportunities for reproduction, and the number and size of young that different species produce under different kinds of environmental conditions. A habitat-based approach defines life histories in terms of the types of pools where animals occur, based largely on hydroperiod. Other perspectives classify life history strategies according to drought-resisting mechanisms, such as resting eggs, migration away from the pools, or aestivation in the sediment, or they focus on the seasonal timing of when species are found in vernal pools.

Reproductive Strategies in Vernal Pools

In terms of population theory and evolutionary theory, two widely divergent life history strategies have been recognized, termed, respectively, r-selected and K-selected. K-selected species tend to be long-lived, to produce small numbers of large young, and provide parental care of offspring. In contrast, r-selected species have short lifespans, produce large numbers of young, and do not provide parental care. It has been theorized that animals in variable habitats will exhibit a mixture of these strategies, sometimes termed "bet-hedging"[4].

There are several basic strategies for ensuring successful reproduction in an environment characterized by periodic drying and hydrologic variability from one year to the next. Every approach has potential benefits and adverse implications. For instance, if adult energy is invested in producing very large numbers of small eggs or young, it adds to the probability that some offspring will survive to adulthood but decreases the odds of any one individual's survival. If the same amount of reproductive effort focuses on the production of few, large eggs or young, the likelihood that individual offspring will develop successfully before the pool dries is enhanced, but there is a greater risk of losing the entire brood under highly adverse conditions. Similarly, if adults are long-lived and can reproduce in more than one year, the probability that they will reproduce successfully at some point in their life cycle is enhanced. The species has greater latitude in colonizing pools where conditions are periodically unfavorable for reproduction, but energy must be invested in adult survival and there must be mechanisms by which more than one life stage can survive or avoid pool drying. In species with short life spans and only one chance to breed, on the other hand, more of the adult energy can be focused on reproduction, but year-to-year variability may periodically result in local extinction of the population. Effective mechanisms for dispersal and recolonization are, therefore, also important parts of the life history strategies of some vernal pool inhabitants.

In vernal pools, the life history strategies of many, if not most, species involve combinations of life spans, different drought-resistant stages, numbers of opportunities to breed, numbers of offspring per brood, and dispersal. Some examples of these strategies in vernal pools include the following.

- The production of large numbers of eggs or young. This strategy is seen in many amphibians including wood frogs and toads, some fingernail clams, snails, fairy shrimp and clam shrimp, and a variety of other invertebrates.
- The production of relatively few, large eggs or young, often with parental brooding. Marbled salamanders, Herrington's fingernail clam, and water fleas are among the vernal pool residents with this approach.
- Comparatively long adult life spans with opportunities for breeding in more than one year. This strategy is known as iteroparity, and animals that breed more than once are known as iteroparous. (In contrast, a strategy that involves

Key to Plate 1, at right. 1. Red Maple. 2. Pussy Willow. 3. Skunk Cabbage. 4. Wood Ducks. 5. Spring Peeper. 6. Wood Frog. 7. Wood Frog Egg Masses. 8. Spotted Salamander. 9. Blue-Spotted Salamander. 10. Salamander Spermatophores. 11. Spotted Salamander Egg Masses. 12. Marbled Salamander Larva. 13. Caddisfly Larva. 14. Dragonfly Nymph. 15. Fairy Shrimp. 16. Crawling Water Beetle. 17. Turtle Leech.

Plate 1. While snow still covers the ground in late winter or early spring, fairy shrimp and many other aquatic invertebrates are already well into their brief lives when migrating amphibians arrive to breed in vernal pools. (Painting by Barry W. Van Dusen; used with permission of Massachusetts Audubon Society.)

Plate 2. Vernal pools change dramatically through the seasons and from year to year, as may be seen in views of this short-cycle, spring-filling pool **(top)** at the beginning of drawdown after peak flooding in late spring, **(middle)** when nearly dry in early summer, and **(bottom)** with little water in the spring of a drought year. (Photographs at top and middle by Frances Garretson Clapp; at bottom by Betsy

Plate 3. In this marshy pool in coastal pitch-pine forest, **(top)** spring's high water gives way to **(bottom)** emergent sedges, grasses, and other wetland plants in mid summer. (Photographs by Betsy Colburn.)

Plate 4. (Top) Many northern and high-elevation pools are surrounded by evergreens and are densely shaded year round. **(Bottom)** In mid winter, a blanket of ice and snow covers this large, mixed-canopy, semi-permanent pool in the north woods. (Photograph at top by Aram Calhoun; at bottom by Daniel Schneider.)

Plate 5. (Top) During spring high water, this floodplain pool is connected to the rest of the floodplain forest and to the adjacent river itself. **(Bottom)** Open, semi-permanent pools often support pond weeds, water lilies, and other marsh plants. (Photographs by Betsy Colburn.)

Plates 6 (above) and 7 (opposite). Across much of the glaciated northeast, the forest was removed and the land was converted to farms during the first two centuries of European settlement. When economic and political conditions changed, farms were abandoned and the forest regenerated. Today, there is more forest in some parts of this region than there was one or two centuries ago. With forest regrowth has come a return of woodland wildlife, including those animals that breed or feed

in vernal pools. This sequence of images represents **(top, Plate 6)** the primeval forest of the glaciated northeast before European settlement, **(bottom, Plate 6)** an agricultural landscape in central Massachusetts at the peak of agricultural clearing in 1830, **(top, Plate 7)** reforestation following the abandonment of farms, and **(bottom, Plate 7)** the modern second-growth forest. (Art on Plate 6 and top, Plate 7: used with permission of Fisher Museum, Harvard Forest, Harvard University; photograph at bottom of Plate 7 by David R. Foster.)

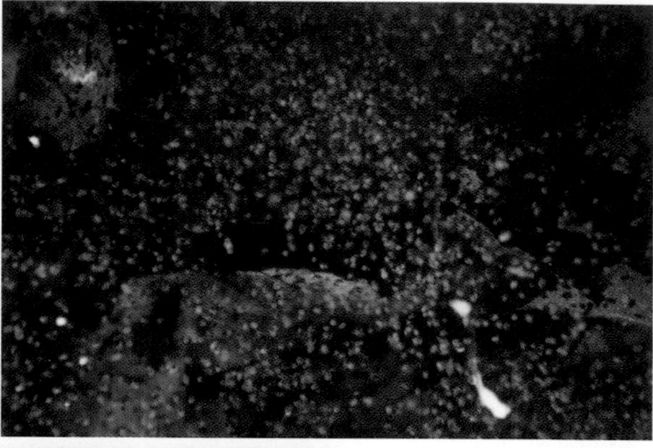

Plate 8. Many types of invertebrates can be found in vernal pools, including **(top)** snails and fingernail clams, which feed, respectively, on algae and fine detritus in the water; **(middle)** water fleas, or cladocerans, which occur in numbers large enough to be considered "swarms" in early spring, and which provide food to newly hatched salamander larvae as well as to invertebrate predators; and **(bottom)** isopods, or aquatic sow bugs, which shred leaves into small pieces. (Photographs at top and bottom by Leo Kenney; in middle by Scott Jackson.)

Plate 9. Fairy shrimp are brightly colored, with **(top)** distinct physical differences between the sexes, including modified antennae in males and egg sacks in females, but **(bottom)** they blend in well against the background of flooded leaves in early spring. (Photographs by Scott Jackson.)

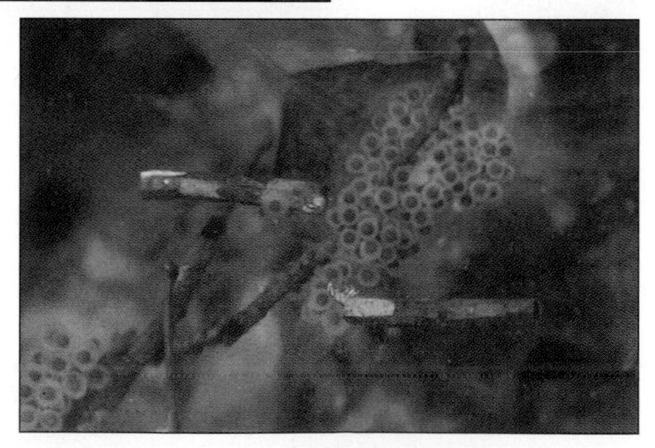

Plate 10. Caddisfly larvae, in their cases constructed of leaves and other debris, are among the most visible aquatic insects in vernal pools in early spring. Common examples include **(top)** northern case makers (Limnephilidae), often called "log-cabin caddisflies," and **(middle, bottom)** giant case makers (Phryganeidae). (Photographs at top by Scott Cooper; at middle and bottom by Scott Jackson.)

Plate 11. The wood frog has the broadest geographic distribution of any amphibian that depends on vernal pools for breeding habitat, and it is one of the first species to breed once the ice starts to melt in the late winter or spring. **(Top)** The dark mask is a distinguishing feature of the wood frog. **(Bottom)** Females choose oviposition sites and lay their eggs in communal clusters. (Photograph at top by Scott Jackson; at bottom by Leo Kenney.)

Plate 12. Several species of frogs, treefrogs, and toads, including **(top)** green frogs and **(bottom)** Fowler's toads, breed in vernal pools as well as in other waters. (Photographs by Leo Kenney.)

Plate 13. The spotted salamander is something of a "poster child" for vernal pools, due to its large size, bright yellow or orange spots, and wide distribution. Its large, baseball-sized egg masses are visible in pools long after the adults have returned to their forest burrows. (Photograph by Gary Meszaros.)

Plate 14. Spotted turtles **(top)** forage in vernal pools in spring, and painted turtles **(bottom)** can be found basking along the edges of flooded pools in spring and summer. (From Morgan 1930, courtesy of the Morgan estate.)

Plate 15. Isolated woodlots, wetlands, and the remnants of vernal pools are visible in an agricultural landscape that occupies what was formerly a forested area in the glaciated northeast. (Photograph by Betsy Colburn.)

Plate 16. This vernal pool, seen **(top)** surrounded by forest and **(bottom)** after the forest was clearcut to the pool's edge, could have been protected by advance planning and the implementation of best management practices for forestry. (Photographs by Daniel Schneider.)

breeding only once is known as semelparity, and single-breeding animals are known as semelparous.) Multiple breeding is seen in vernal pool amphibians (wood frogs live three to five years, and mole salamanders live two decades or more), the vernal pool fingernail clam *Sphaerium occidentale* (adults live two years, as compared to one year for most other species of fingernail clam), and some parasitic leeches.

- Complex life cycles that depend on adult mobility and dispersal ability, precise timing of breeding, and specialized drought-resisting mechanisms in the eggs. This strategy is seen in a number of vernal pool beetles.
- Short life spans, coupled with rapid development and the production of large numbers of drought-resistant eggs or cysts that hatch at different intervals after flooding, thus spreading each generation over several seasons. This strategy is characteristic of many crustaceans and aquatic insects.
- Continuous breeding as long as water is available and other conditions are favorable, often with the production of multiple generations each year, and the production of drought-resistant eggs, cysts, or aestivating stages as the pool dries. Among the species using this strategy are flatworms, some snails, some crustaceans, and some aquatic insects.

Life History in Relation to Pool Type

Some of the species found in vernal pools are well adapted to breed successfully in short-cycle pools. Animals such as wood frogs and fairy shrimp hatch from eggs early in the season, grow rapidly, and are in their drought-resistant stage by late spring or early summer, before the pools dry. In contrast, spotted and blue-spotted salamanders reproduce more successfully in pools with a longer hydroperiod and are more likely to breed in long-cycle pools that retain water well into the summer, or in semi-permanent pools. Similarly, many aquatic insects and other invertebrates vary widely in the amount of time they need to spend in flooded pools in order to complete their development. For example, the ruby meadowhawk, a common red dragonfly of mid to late summer, develops successfully in short-cycle pools, while the closely related band-winged meadowhawk is found most typically in semi-permanent pools.

Life History in Relation to Adaptations to Pool Drying

The comprehensive review of life history strategies in temporary pond animals, "Evolutionary and ecological strategies of animals in annual temporary pools," by Glenn Wiggins and his colleagues, published in 1980, is the definitive work on this subject[5]. These authors look at vernal pool wildlife in terms of the species' life history adaptations to drying. Summarizing the literature and their own studies on a series of pools ranging in hydrologic class from short-cycle, spring-filling to semi-permanent, they distinguish the following four general ecological strategies in vernal pool animals.

Group 1 — Overwintering Residents. Overwintering residents are incapable of active dispersal. They survive during drought by means of resistant eggs, cysts, or burrowing into the sediment. Taxa in this group include molluscs; crustaceans, including branchiopods, crayfish, ostracods, and copepods; hydroids; flatworms; and annelids, both oligochaetes and leeches.

Group 2 — Overwintering Spring Recruits. Overwintering spring recruits require the presence of water for reproduction. They become active and grow when the pool floods, and adults emerge before the pool dries and oviposit in or on the water. Adults may disperse to pools other than the one they emerged from. These animals survive drought by aestivating in the dry basin as eggs, larvae, or adults, and include some beetles, polycentropodid caddisflies, water mites, siphlonurid and leptophlebiid mayflies, and flies including chironomids, tabanids, stratiomyiids, and ceratopogonids.

Group 3 — Overwintering Summer Recruits. Overwintering summer recruits oviposit in the dry basin and aestivate as eggs or larvae. Among these taxa are limnephilid and phryganeid caddisflies, lestid damselflies, libellulid dragonflies, mosquitoes, phantom midges, marsh flies, and some chironomid midges.

Group 4 — Non-Wintering Spring Migrants. Non-wintering spring migrants actively leave vernal pools at or before drying and spend the dry period elsewhere, then return to the pools in spring. These migrants include amphibians, some of the beetles, large aeshnid dragonflies, baetid mayflies, and water bugs.

Life History in Relation to When Species Occur in Vernal Pools

In his 1983 study of a short-cycle, spring-filling vernal pool in southern Ontario, D. D. Williams divides the fauna into five groups, characterized by different general life history patterns and the timing of species' occurrence in the pool[6]. This classification is based on the time during which animals are active in the pool and does not consider the period in which they may be present in a dormant state as eggs, cysts, or aestivating individuals.

Group 1 — Persistent Species. Persistent species are always present in the pools, either in the water or within the dry sediment during drawdown. Members of this group include beetles in the genus *Hydroporus*, some chironomid midges, molluscs, oligochaete worms, and some cyclopoid copepods.

Group 2 — Emerge-On-Flooding Species. Emerge-on-flooding species become active within a few days of flooding. Many of these animals are present in the dry pool basin as eggs or cysts that hatch when the pool fills. They typically complete their life cycles within four to six weeks. This group includes crustaceans, including fairy shrimp, cladocerans, and a harpacticoid copepod; some water mites; and insects, including mosquitoes, some chironomids, some water beetles, a species of dragonfly, and caddisflies.

Group 3 — Late-Spring Species. Late-spring species do not appear until two to five weeks after spring flooding and most complete their life cycles in about five weeks. Included in this group are wood frogs and mole salamanders — based on the appearance of their larvae, not their egg masses — clam shrimp, damselflies, some chironomids, some breeding beetles, and some non-breeding migrant beetles.

Group 4 — Drying-Phase Species. Drying-phase species appear after the pool has started to dry, approximately two to three weeks before drying is complete. Among these late-season taxa are mayflies, some beetles, and some chironomids, all of which complete their development rapidly.

Group 5 — Dry-Phase Species. Dry-phase species are found in the pool basin only during the dry period and include myriapods, terrestrial beetles, and arachnids.

A New Synthesis of Life History Strategies in Vernal Pools

Site-specific hydrologic conditions and other habitat characteristics are reflected in the species present in a vernal pool and their life history strategies. Fundamental biological factors — for instance, inherent differences between fairy shrimp and frogs — also constrain the dispersal and drought-withstanding mechanisms that allow animals to survive in vernal pools. Still, there are some distinct patterns of life history strategies that can be used broadly to describe and classify vernal pool wildlife throughout the glaciated northeast.

Migrants *versus* Permanent Residents

Vernal pool wildlife may be classified as either migrants or permanent residents. Migrants include breeders, which use the pools for reproduction, and non-breeders, which use the pools as feeding areas but reproduce elsewhere. Migrants spend a significant part of their life cycle away from the vernal pools, and their survival is affected markedly by the conditions that exist in their non-pool habitats. In contrast, permanent residents remain in or near the pools, and their survival is thus determined by local conditions, such as the amount of moisture that remains in the substrate during drawdown. A few insect species that fall in between these two extremes forage away from the pools as adults but remain near the pools and lack a clearly defined non-pool-dependent life stage.

Life History Classification of Vernal Pool Animals

Combining information on whether species are migrants or permanent residents, their drought-resisting mechanisms, the timing of breeding, and the cues that govern when eggs hatch, the following broad categories of life cycles in vernal pool animals may be recognized.

Migratory Breeders

"Migratory breeders" take advantage of the flooded phase of vernal pools for breeding, but they leave the pools during the drawdown phase. The best-known users of this strategy are the mole salamanders, wood frogs, and other woodland amphibians that are specialized for breeding in temporary ponds. When it is time to breed, these species tend to return to the pools they came from as juveniles, and they show high fidelity to their breeding pools[7]. As is discussed in more detail in Chapter 11, individuals that have been displaced from breeding pools ignore intervening water bodies and return, often over great distances, to the original location.

The evolutionary strategy of spending an early life stage in one habitat, migrating elsewhere to feed and grow, and returning to the natal habitat to breed is seen in many unrelated species of animals, including Atlantic and Pacific salmon as well as vernal pool amphibians[8]. In Pacific salmon, 95% of returning individuals home back to their natal stream; in wood frogs, 80 to 90% return to their natal pool to breed. The rest disperse into other waters. In species that breed more than once, individuals consistently return in successive years to the site of first breeding.

The homing strategy increases the probability of successful reproduction, since the habitat has already produced at least one generation. At the same time, if all individuals produced by a pool consistently homed back to their natal habitat, there could be a tendency over time for the population to become inbred, with reduced genetic variability, and the population would be at risk in the event that something happened to make the pool unsuitable as habitat. The dispersal strategy ensures genetic exchange among populations and provides for continuation of part of a local population if the natal habitat should be destroyed or otherwise become unsuitable. A proportion of the young wood frogs or salamanders that were originally produced by a given vernal pool can be expected to migrate elsewhere when it becomes time for them to breed for the first time, and they will contribute their genes to a different breeding population at a different pool. If the original pool is destroyed or altered in some way, a portion of the genetic material of its amphibian populations will be retained in other pools.

Other migratory breeders include limnephilid caddisflies (subfamily Limnephilinae), whose adults aestivate in tree holes and caves during the summer months; some predaceous diving beetles, backswimmers, and water boatmen, which spend part of their life cycles in permanent water bodies; a number of water mites which leave the pools as parasitic larvae attached to aquatic insect hosts that are themselves migratory breeders, spend the winter in permanent waters attached to the hosts, and are carried along in the spring when the hosts disperse into vernal pools; and some mosquito species, which overwinter in caves, buildings, and other sheltered areas. Homing has not been demonstrated in vernal pool invertebrates.

Among migratory breeders, there are three basic life history patterns.

"Early spring migrants" migrate to the pools and deposit their eggs in late winter or early spring, right about the time that ice is melting and the pools have filled to their maximum. The wood frog and most of the mole salamanders fall into this category, as do several kinds of mosquitoes, including species of *Culex*. In such species, the timing of the breeding migration is critical. The eggs of early spring migrants begin to develop upon deposition. Conditions in the pool need to be appropriate for egg deposition, and there must be enough time for the eggs to mature and hatch, and for the young animals to complete their development, before the pool dries. Animals using this strategy include species with rapid development that use spring-filling, short-cycle pools, as well as species with relatively long developmental periods that use pools with a longer hydroperiod. This group is comparable to Group 4, non-wintering spring migrants, of Wiggins and his colleagues and includes some taxa listed in Group 2 and Group 3 of Williams.

"Spring-summer migrants" move to flooded vernal pools somewhat later in the season, after water temperatures have warmed and food has become abundant. Backswimmers, water boatmen, several species of dragonflies, and some beetles fall in this category, as do toads and gray treefrogs. Among species that use this strategy, several different developmental patterns are possible: embryonic development may begin immediately, as in the early-spring migrants; or, the eggs may hatch subject to appropriate cues of water chemistry and/or temperature (thermally-cued hatching); or, the eggs may overwinter in the basin and hatch the following year (delayed hatching). Some spring-summer migrants with rapid development rates may breed continuously during the flooded period and produce several generations before the pool dries and the adults move on to other habitats. In water mites that are carried to vernal pools as larvae that have overwintered on migratory insect hosts, the remaining stages of development into adults, mating, and egg deposition and hatching must be completed before new larvae are ready to attach to hosts and be carried out of the pool. Spring-summer migrants contribute markedly to the biodiversity of vernal pools and to the successional patterns that have been observed in vernal pools through the season (see Chapter 13). These species are included in Group 4 of Wiggins and colleagues, and Group 3 and Group 4 of Williams.

"Fall migrants," also known as dry-phase migrants, move into vernal pools in fall and oviposit within the dry basin. Marbled salamanders and limnephilid and phryganeid caddisflies use this strategy. The eggs of many fall migrants can resist desiccation for an extended period of time. Hatching may occur upon or shortly after flooding (hatch-on-flooding development) or be delayed until stimulated by appropriate cues of photoperiod and temperature. The young animals obtain a head start over those species in which oviposition is delayed until spring. This group includes some species found in Group 3 of Wiggins et al. and Group 2 of Williams.

Non-Breeding Migrants

"Non-breeding migrants" include a variety of predaceous diving beetles and several species of turtles that use vernal pools as feeding areas during much of the flooded period. In such species, feeding in vernal pools may represent an important phase of the life cycle and contribute to egg development and growth.

Snakes, birds, and mammals that make incidental use of vernal pools also fall into this category. Williams includes most of these species in his Group 3; some, including the turtles, fall within his Group 2.

Permanent Residents

Like migratory breeders, "permanent residents" vary in their life history patterns. Unlike migrants, permanent residents do not leave the vicinity of the vernal pool for extended periods of time. This group includes some aquatic insects that are aerial as adults but do not travel long distances, species that are present in the pool basin as aestivating adults or juveniles during unfavorable conditions ("aestivators"), and species that spend the dry phase as eggs or desiccation-resistant cysts.

"Aestivators" are species that spend hot or dry periods in a state of dormancy. They are typically inactive during drawdown and have reduced metabolic rates and a higher tolerance to desiccation during aestivation than when they are active, but they emerge and resume feeding and other activities immediately upon flooding in spring or fall. Some become active if pools flood in response to summer storms, but others break dormancy only if water temperatures and other conditions are favorable. Examples of aestivators include beetles in the genus *Hydroporus*, fingernail clams, pulmonate snails, rotifers, tardigrades, and nematodes. Breeding in these species may be continuous during the flooded phase, or it may be precisely cued to thermal and other conditions within the pool. This group corresponds to Williams' Group 1 and some members of Wiggins et al.'s Group 1.

Desiccation-resistant eggs and cysts of permanent residents may hatch immediately upon flooding, or they may exhibit delayed development. In "hatch-on-flooding" species, the eggs hatch and the cysts become activated upon flooding in late fall, winter, or early spring. This group includes flatworms, fairy shrimp, some mosquitoes, and some beetles. The eggs of many members of this group require a period of cold-conditioning or other stimuli before they will be ready to hatch, thereby preventing emergence in response to short-term flooding as a result of summer rainfall. This group is equivalent to Williams' Group 2 and includes some members of each of Wiggins et al.'s three overwintering groups. "Thermally cued hatchers," like hatch-on-flooding species, withstand drought as eggs or cysts, but in these species hatching does not occur immediately upon seasonal flooding. Instead, water temperatures, and possibly other cues such as photoperiod, stimulate egg development and hatching. Most insects and crustaceans in vernal pools fall in this category. Typical examples include spreadwing damselflies (*Lestes*) and clam shrimp. Successive hatches of different species through the season contribute both to gradual increases in the number of species and to shifts in species composition in vernal pools from flooding to drawdown. All four of the groups identified by Wiggins and his colleagues, as well as Williams' Groups 3 and 4, include species using this strategy.

Colonization and Dispersal

As recently as 9,000 to 15,000 years ago, the glaciated northeast was buried under a continental ice sheet. The deciduous, mixed, and coniferous forest ecosystems we know today developed gradually following glacial retreat. These forests have not remained static. Hurricanes, occasional fires, and climatic variation have worked with succession over time to create the forested landscape. In the past 200 to 400 years, the most profound changes to northeastern forests have been associated first with forest clearing and farming, second with land abandonment and reversion to secondary forest, and third with new clearing, road construction, and forest fragmentation (plates 6, 7). Losses of core species such as American chestnut and invasions of insects such as gypsy moths have also taken their toll.

In geologic terms, the landscape of the glaciated northeast is new. How did animals make their way to vernal pools following glaciation, how are populations replenished following local extinctions, and how do populations expand to newly created pools? Further, if pools are isolated, and if many species either are permanent residents or show a high degree of homing during breeding, how is genetic exchange maintained among pools? These questions are closely related.

As to how wildlife got to vernal pools originally, we can assume that frogs, toads, treefrogs, and salamanders walked and hopped there, gradually spreading across the landscape as forests moved in along the front abandoned by the retreating glaciers. How does this jibe with the high site fidelity and homing back to the natal pool reported for vernal pool animals? Homing ensures that a pool that has been proven suitable for reproduction will receive a heavy reproductive investment on the part of its progeny. But, remember, a proportion of individuals in each generation "strays" to other pools. Straying ensures perpetuation of the population if the original pool is lost, and it also maintains gene flow between pools. Further, straying allows colonization of new habitats as they become available.

The populations of vernal pool amphibians were probably greatly reduced during the peak of agricultural clearing in the eighteenth and nineteenth centuries. It is likely that they were largely restricted to the 20% of the landscape that remained forested, at least in the southern and eastern portions of the glaciated northeast, and later rebounded as the forests returned.

Similar dispersal across the landscape presumably occurred for turtles, crayfish, and mobile insects such as water beetles, water bugs, caddisflies, and dragonflies. Some insects, in fact, appear to specialize in colonizing newly flooded water bodies.

For small pool animals such as fairy shrimp, clam shrimp, water fleas, ostracods, copepods, snails, fingernail clams, worms, and water mites, as well as water-loving plants, dispersal to isolated pools in a matrix of upland forest is far more chancy. Mobile insects, crayfish, birds — especially shorebirds and ducks — and to a lesser extent mammals and turtles, appear to play an important role in transporting invertebrate eggs, cysts, and other resting stages, as well as living individuals, between aquatic sites[9]. Viable eggs or embryos of fairy shrimp, water fleas, and crayfish, and seeds of aquatic and wetland plants, are transported via the digestive systems of ducks and shorebirds that feed on them. Amphipods, snails, and fingernail clams travel as hitch-hikers on the feathers and feet of birds. Fairy shrimp eggs are unaffected by passing through the digestive tract of crayfish. Turtles carry leeches on their algae-covered shells as they move between wetlands. For some water mites, both dispersal and, in some species, avoidance of desiccation depend on the insect hosts of the parasitic larvae. By traveling seasonally to perennial waters with their hosts, these mites avoid summer drawdown and disperse to new pools the following spring (see Chapter 10). Smaller invertebrates, including rotifers and nematodes, are transported by water beetles that move between pools.

It seems quite likely that, in some cases, the species present in vernal pools represent the outcome of random dispersal events, and not the full biological community that could potentially occur there. This is particularly likely in landscapes that have been disturbed historically, where species that were present prior to European settlement and land clearing may have been eliminated and have not yet recolonized even though the forest has returned.

Some Remaining Questions About Life History Strategies

- What are the similarities and differences in the life history strategies of closely related species in vernal pools?
- How do the life history strategies of different species address the environmental variability and stresses associated with seasonal drying that are characteristic of these habitats?
- Are certain life history strategies more characteristic of species found in pools with a high degree of between-year variability, and other strategies more common in pools with somewhat predictable water regimes?
- How rapidly are newly flooded pools colonized?
- How important are regular dispersal and recolonization for populations of species with limited mobility and life spans of one year or less?
- How far do migratory species travel between overwintering sites and vernal pools?
- How will increasing forest fragmentation affect the distribution and dispersal of vernal pool animals?

References for Chapter 6

[1] Eyre et al. 1992, Schneider and Frost 1996. [2] Williams 1996. [3] Stearns 1992; Wiggins et al. 1980; Williams 1983, 1996. [4] Hornbach et al. 1980; Schwartz and Jenkins 2000; Stearns 1976, 1992, 2000; Way et al. 1980; Williams 1996. [5] Wiggins et al. 1980. [6] Williams 1983. [7] Berven and Grudzien 1990, Husting 1965, Shoop 1968, Whitford and Vinegar 1966. [8] Berven and Grudzien 1990, Hasler and Scholz 1983. [9] Green et al. 2002, Maguire 1963, Proctor 1964, Proctor and Malone 1965, Proctor et al. 1967.

Chapter 7

Non-Arthropod Invertebrates

Invertebrates account for the overwhelming majority of the animal kingdom, and within their small basins vernal pools support examples of most of the animal phyla that occur in fresh water, including hydras, sponges, rotifers, gastrotrichs, horsehair worms, flatworms, annelids, molluscs, and arthropods (Table 1, Appendix). All of these taxa have limited mobility and are permanent residents of vernal pools, although many are readily dispersed by birds, turtles, and other migratory species. These animals play important ecological roles in a pool's economy, whether as consumers of detritus and algae, as predators on smaller animals, as parasites on larger animals, or as food for other invertebrates and amphibians. This chapter considers non-arthropod invertebrates; the arthropods, which are particularly well represented in vernal pools, are discussed at length in chapters 8 to 10.

Relatively few non-arthropod invertebrates are able to tolerate the periodic disappearance of water that is characteristic of vernal pools. Several species of flatworms, molluscs, and leeches are able to thrive when pools contain water and to survive the dry period by aestivating, encysting, or producing desiccation-resistant eggs. Suprisingly, the biodiversity, life history adaptations, and ecology of most other non-arthropods in temporary waters have not been well studied. The following summary reviews the distributions and life histories of some of the more common species of freeliving flatworms, fingernail clams and snails, and leeches, and presents less detailed discussions of the biology of some of the other non-arthropod invertebrate inhabitants of vernal pools. Some references that describe the habitat, life history, and ecology of species that may be encountered in vernal pools, as well as some keys for the identification of specimens, are listed at the end of this chapter[1].

Flatworms (Platyhelminthes: Turbellaria)

As a group, freeliving flatworms are abundant in freshwater habitats, where they can reach densities of several thousand individuals per square meter[2]. These are small, flattened, unsegmented worms with simple eyes, an extendable pharynx or proboscis that they use in feeding, and a ciliated epithelium that provides the primary means of locomotion. There are several orders. The Tricladida includes the large planarians, which generally attain lengths greater than 10 mm (0.4 in). The so-called freshwater microturbellarians, which are usually only a few millimeters (less than an eighth of an inch) in length, are members of the Catenulida, Lecithoepitheliata, Proseriata, and Neorhabdocoela. Triclads are omnivores, and microturbellarians are predators that feed on a wide range of small aquatic animals including cladocerans and mosquito larvae[3]. Despite their small size and apparently simple structure, flatworms are very effective predators and can be important in controlling mosquito populations in some pools, particularly in early spring[4].

Several relatively large species, including two large triclads, *Hymanella retenuova*, and *Phagocata velata,* regularly occur in vernal pools in the glaciated northeast. *Hymanella* is restricted to vernal pools, while *P. velata* occurs in a wider range of aquatic habitats. These two species often co-occur in vernal pools and are difficult to distinguish from one another. This has led to some confusion in the literature, especially in some older papers. For some time it was thought that there were two closely related species of northeastern *Phagocata*, *P. vernalis* in vernal pools and *P. velata* in streams and springs[5]. *P. vernalis* is now recognized as a synonym of *H. retenuova, P. velata* is known to occur in vernal pools along with *H. retenuova*, and the flatworms identified as *P. vernalis* in Kenk's studies of flatworms in vernal pools in Michigan in the 1940s were actually *P. velata*[6]!

The microturbellarian fauna of northeastern vernal pools is incompletely known. *Dalyellia viridis* is a common neorhabdocoel. Readily recognized by its bright green color — due to the presence of symbiotic algae in its integument — this species occurs in both permanent and temporary waters. Several other rhabdocoels and some allelocoels are known from annual vernal pools, and others are likely to be found in semi-permanent pools (see Appendix)[7]. More than twenty-five species of microturbellarians are known from European vernal pools[8].

Distribution

Hymanella retenuova is common throughout the midwestern United States, and *Phagocata velata* is widely distributed throughout glaciated northeastern North America. *Dalyellia viridis* is a cosmopolitan species that is found in Europe as well as North America.

Life History

The flatworms that inhabit vernal pools have developed several different mechanisms for surviving through the summer dry period. *Hymanella retenuova* produces a desiccation-resistant cocoon in which its eggs are encased and held during the dry months. *Phagocata velata* fragments into pieces, each of which secretes a mucus film that hardens to form a round cyst; upon flooding the mucus dissolves and small worms that have developed from the fragments emerge. *Dalyellia viridis* produces large eggs that resist desiccation.

Aestivating eggs of *H. retenuova* appear to hatch soon after flooding inundates the cocoon. No information is available about the specific factors that stimulate hatching. Young animals are found in fall-filling pools in Michigan in December, soon after flooding, and grow in the cold water and the adults disappear by late May[9]. In spring-filling vernal pools in southern Ontario, individuals appear by mid April, and the populations peak in May to early June. This species is believed to feed on fairy shrimp, and its population cycle appears to follow closely behind that of the fairy shrimp in the pools[10]. As the water temperatures become warm and pool drying approaches, specimens can often be found bearing a distinct, reddish cocoon, approximately 1.8 x 1 mm (0.07 x 0.04 in). This cocoon, which contains the eggs, will later be deposited onto the pool bottom where it will remain until the pool refills. By late May most individuals are found with egg cocoons, and no adults are found by mid June. While many planarians have egg cocoons, this desiccation-resistant cocoon is an adaptation to the temporary pond habitat unique to this species.

The timing of the life cycle of *P. velata* is similar to that of *H. retenuova*, but the details of reproduction and aestivation differ. In *P. velata,* the mucus-encysted fragments of the adult worm undergo differentiation into small flatworms, which emerge once the pool is flooded. Prior freezing does not seem to be required for emergence. Sexual reproduction is rare and appears to be restricted to cold-water conditions[11]. In pools that contain water in late fall and winter, sexually reproductive adults can sometimes be found. This flatworm is one of the most characteristic winter inhabitants of a small fall-filling vernal pool in Michigan[12]. As the water temperatures start to warm in spring, the animals cease reproductive activity and prepare for encystment. They usually lose some of their color, and it is possible to see lines that appear to divide the body into segments. The body divides along these lines by fission, and fragments of the body are pinched off. The fragments secrete a layer of slime that hardens to form a desiccation-resistant cyst.

Unlike *Phagocata* and *Hymanella*, *Dalyellia* does not emerge immediately upon flooding. It appears that a period of diapause is required before the eggs will hatch. The specific factors that stimulate hatching are unknown.

Most other microturbellarians in vernal pools probably aestivate as eggs, but *Stenostomum* sp. and *Catenula lemnae* may aestivate as adults in mucus-covered cysts, and the allelocoele *Prorhynchus stagnalis* may either encyst or burrow into the sediment following the retreating groundwater table[13].

Co-Occurrences of Freeliving Flatworms in Vernal Pools

For ecologists, the co-occurrence of closely related species, or of species that have similar behaviors or feeding strategies, is of interest because it is usually presumed that competition will ultimately favor one species over another. Co-occurrence, then, suggests that the animals use different food, occur in different parts of the habitat, or are temporally offset in terms of their peaks of abundance and their life cycles, thereby minimizing potential competitive interactions. Another possibility is that food and space are not limiting. In habitats such as vernal pools, where environmental stresses place severe constraints on the survival of animals, the presence of closely related species provides especially useful opportunities for research into the mechanisms that allow the species to coexist.

Phagocata velata and *Hymanella retenuova* have commonly been reported from the same vernal pools[14]. These two species appear to use different parts of the habitat. In collections of leaves and other materials in Ontario vernal pools, only *Phagocata* was visible initially, but after the sample was left in water overnight both species were found to be present. As many as five species of flatworms have been found coexisting in Michigan vernal pools.

Flukes (Platyhelminthes: Trematoda)

Flukes are parasitic flatworms in the platyhelminth class Trematoda. Most are only a few centimeters (an inch or less) in length as adults. Some species, including blood flukes, liver flukes, and the schistosomes that cause bilharziasis, or river blindness, are medically important because they cause disease in humans or domestic animals. Flukes are also important internal and external parasites on fish, amphibians, reptiles, and aquatic invertebrates. They are not discussed in other studies of vernal pools, nor are they covered in most books on freshwater invertebrates, but they are ecologically important in vernal pools because of their effects on their molluscan hosts and on amphibians, and probably on other invertebrates, as well. They figure prominently in current research on the causes of amphibian declines and of deformities in anuran amphibians, and they are important subjects for research in evolutionary biology and community ecology because of their complex relationships with their hosts. For these reasons, they are discussed, briefly, here. Unlike the discussions of other taxa, this summary focuses broadly on the life history strategies and ecological effects of trematodes, and it does not review the species found in vernal pools and their distributions. An excellent review of trematodes and other parasites of amphibians and reptiles, with a comprehensive list of pertinent references, is included in a protocol for monitoring parasite populations, published electronically by Environment Canada in 2001[15].

Both orders of trematodes, Monogenea and Digenea, occur in vernal pools. The monogenetic trematodes have a single host during their life cycle, and the digenetic trematodes have from two to four hosts. Flukes can have significant effects on their hosts, including increased mortality, decreased growth rates, altered behavior, and decreased size[16]. The life cycles of flukes are closely tied to the life cycles of their hosts, and the complicated mechanisms by which the parasites use host behavior and physiology to maximize their own growth and dispersal, as well as the strategies by which hosts minimize the impacts of the parasites, have long fascinated population biologists, evolutionary ecologists, and parasitologists. There are many variations in the patterns of life cycles in both orders, only a few of which are described below.

As a group, most monogenetic trematodes are ectoparasites on fish. Most of the species found in vernal pools are external parasites on amphibians or invertebrates, although a few are internal parasites. As adults, these flukes somewhat resemble a cross between free-living flatworms and leeches, as they have oval, flattened bodies with one or more suckers with which they attach to the bodies of their hosts. An infestation by large numbers of these parasites can make a host look as though it has a fuzzy film over the surface of its body.

A few monogeneans are internal parasites. Adults of the monogenetic trematodes *Polystoma nearcticum* and *Pseudodiplorchis americanus* live within the urinary bladders of treefrogs, such as the gray treefrog, and desert spadefoot toads, respectively; these or related species may also parasitize frogs and toads in vernal pools. Reproduction in the adult parasites is stimulated by the host's sex hormones, and the worms release their eggs into the water when the adult amphibians move into breeding pools. The parasites' eggs hatch into ciliated larvae known as onchomiracidia, which seek out host tadpoles and attach to their gills. When the tadpoles are ready to metamorphose, the larval trematodes either migrate down the gut from the gills into the bladder (*Pseudodiplorchis,* some species of *Polystoma*), or enter the water and swim into the urinary opening and into the bladder (*P. nearcticum*). They spend the next several years feeding on the host and becoming reproductively mature. When the host is ready to breed, so is the parasite, and the cycle begins anew[17]. These trematodes are migratory breeders that, like some species of water mites, escape pool drying by traveling away on their hosts before drawdown and return to breed when their hosts migrate to the pools in spring.

The digenetic trematodes are all internal parasites of vertebrates, and they include most of the species of medical and economic importance. The life cycles of these flukes are very complex, with several life stages and at least two different hosts (Figure 10). The following is a simplified overview of the reproductive strategies found in these parasites. It is based on a variety of excellent references, and much more detail can be found in them and in literature cited therein[18].

The primary, or definitive, host in which the adults of digenetic trematodes live is a vertebrate, usually a carnivore. It may be a frog, toad, salamander, snake, lizard, turtle, mammal or bird. The first intermediate host is a snail. Lymnaeids, physids, and planorbids all serve as intermediate hosts for trematodes in vernal pools.

Adult digenetic trematodes mate and produce eggs that pass into the water with the primary host's excrement. The eggs hatch into ciliated larvae known as miracidia, which seek out snails and infect them. Within the snails, the miracidia lose their cilia and develop into mother sporocysts, which are non-mobile but capable of producing large numbers of the next life stage of the worm. From this

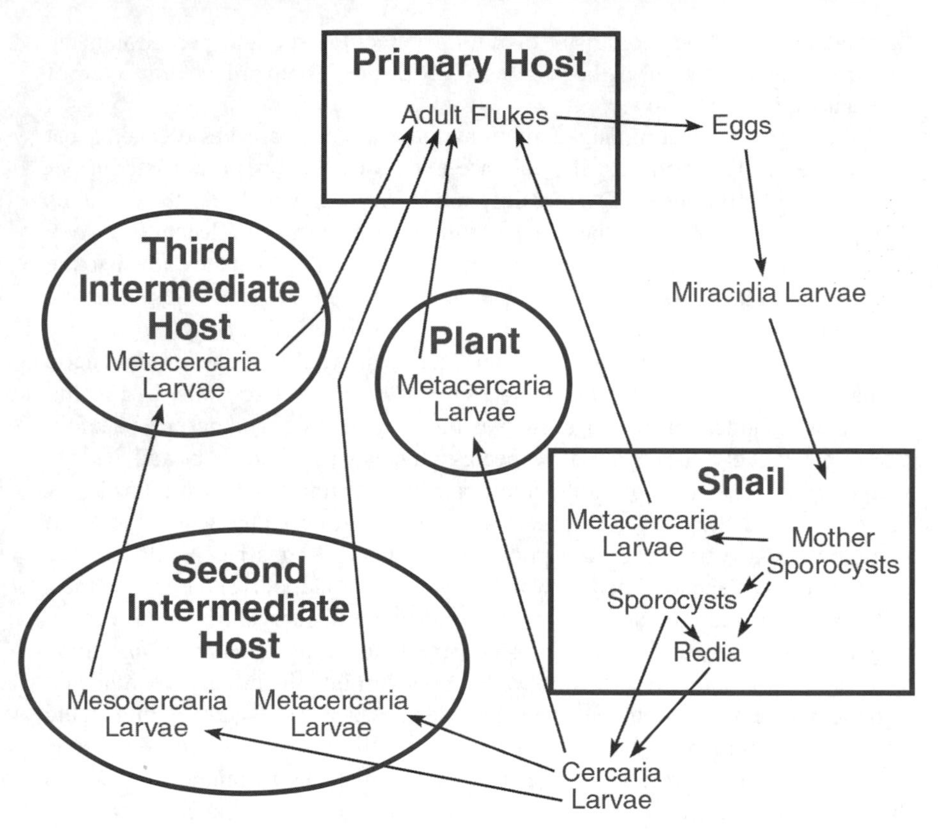

Figure 10. Trematode worms have complex life cycles that include many different stages of development and require from one to four hosts. (Modified from Gogorin 2001.)

point on, the life cycles vary depending on the number of intermediate hosts and the species of worm.

In some flukes, the snail is the only intermediate host. In some of these species, the sporocysts produce large numbers of dormant cysts known as metacercaria. Metacercaria are part of the life cycle of most flukes, and they allow the worm to survive for a long period of time until it is eaten by its definitive host. If the snail is eaten by the definitive host bird, mammal, or other vertebrate, the metacercaria break dormancy and develop into adult flukes.

In most digenetic trematodes, there is a second intermediate host which may be a bivalve mollusc, another species of snail, an aquatic insect such as a dragonfly or damselfly nymph, a crustacean such as an amphipod or crayfish, or an amphibian. In these worms, depending on the species, the mother sporocysts in the snails produce large numbers of either daughter sporocysts or redia larvae, which differ morphologically from each other but have the same basic function. The redia are mobile and usually leave the mother sporocysts and migrate to the host snail's hepatopancreas, a structure that combines the physiological functions of a liver and a pancreas. Here, they produce large numbers of cercaria, which are free-swimming larvae that leave the infected snails synchronously in response to environmental cues such as temperature or light and actively seek out and attack the next host. Because each of the developmental stages in the snail produces many individuals of the next stage, the original number of eggs is multiplied many-fold — sometimes by many orders of magnitude — by the time the cercaria leave the host snails. Once the cercaria locate the second host, they usually attach to it with a sucker, burrow into the host's skin, lose their swimming tails, migrate into the host's tissues over a period of several hours, and encyst as metacercaria. When the second intermediate host is eaten by a host bird or mammal, the metacercaria break dormancy and develop into adult trematodes, and the life cycle is completed and can begin again.

Some cercaria show a high degree of host specificity, and the timing of their release and their behavior in the water contribute to their ability to find and attach to the secondary host. In other species, a variety of second intermediate hosts may be parasitized. Swimmer's itch is caused when cercaria larvae whose hosts are non-human vertebrates try to burrow into human skin — the cercaria cannot penetrate into the body, but they cause an immune reaction that results in itching and intense discomfort.

The cercaria of some trematodes do not attack a second host animal but instead swim to plants and form metacercaria on the vegetation. When the plants are eaten by an herbivore such as a deer or a sheep, the cysts break dormancy and the flukes migrate to the liver, kidney, or other tissues within the body.

A few species of flukes have three intermediate hosts, the snail and two others. In these worms, the cercaria enter the second intermediate host and form mesocercaria, which in turn develop into metacercaria when the second intermediate host is eaten by the third intermediate host. As a hypothetical example, a young wood frog infected by mesocercaria might be eaten by a garter snake in which the fluke develops into metacercaria, and which in turn is preyed on by a red-shouldered hawk that is the definitive host for the fluke.

Amphibians are the second hosts for many species of trematodes in vernal pools. Most of the species of amphibians found in vernal pools, including wood frogs, green frogs, bullfrogs, leopard frogs, spring peepers, American toads, blue-spotted salamanders, tiger salamanders, and red-spotted newts, are parasitized by one or more species of fluke[19]. The cercaria attack the aquatic tadpoles or salamander larvae, and the metacercaria encyst in the muscles and other parts of the body. Different species of trematodes encyst in different locations within the bodies of their intermediate hosts.

The definitive host of some flukes is an adult salamander or bullfrog, a garter or water snake, a heron or other water bird, a raccoon, or another vertebrate predator that feeds on the developing amphibians in vernal pools. In other cases, the metacercaria are transmitted to their final hosts after the amphibians have transformed into terrestrial juveniles, traveled into the forest, and been eaten by animals that hunt on land instead of in water. The strategy of using an aquatic animal with a terrestrial dispersal stage in its life cycle as an intermediate host allows the flukes to disperse away from the water and to take advantage of a variety of definitive hosts that feed on terrestrial animals.

Parasitism by trematodes affects amphibian populations in vernal pools in a variety of ways. Experimental studies have found that the encysted metacercaria, when present in large numbers, can affect the development of limbs and contribute to deformities such as extra legs[20].

This effect appears to be associated only with certain trematode parasites in the genus *Riberoia*, whose metacercaria encyst near the limb buds of the tadpoles, and it seems to be influenced by the presence of pollutants in the water. Trematodes also affect the growth, mortality, and size at metamorphosis of amphibians in temporary waters[21].

Molluscs

Two groups of freshwater molluscs are regularly found in vernal pools: fingernail clams (Bivalvia: Veneroidea: family Sphaeriidae, also known as Pisidiidae) and air-breathing snails (Gastropoda: Basommatophora: families Lymnaeidae, Physidae, and Planorbidae) (Plate 8, top; figures 11, 12). Both groups have soft bodies that are protected by shells made mostly of calcium carbonate, a muscular foot, and a specialized layer of tissue known as the mantle which secretes the shell. The blood contains oxygen-holding proteins, hemoglobin in the planorbids, and hemocyanin in the fingernail clams, lymnaeids, and physids, and this contributes to the ability of fingernail clams and snails to survive in water bodies with low dissolved oxygen levels. Because their shells contain calcium carbonate, which dissolves in acidic water, molluscs are uncommon in soft water with low pH and low alkalinity, such as is typical of pools over the granitic bedrock of the Canadian Shield and much of northern New England, areas of sandy glacial outwash including the Coastal Plain, and waters that are dominated by *Sphagnum*, but they are often very abundant in well buffered water bodies such as those found over limestone or other calcareous bedrock.

Molluscs in vernal pools include herbivores, detritivores, scavengers, and predators. They are parasitized by scyomyzid fly larvae, serve as intermediate hosts for trematodes including liver and blood flukes and a number of other species that parasitize amphibians, and they are fed on by waterfowl, aquatic beetles, predaceous water bugs, and salamander larvae. In some habitats, molluscs can be present at densities exceeding 1,500/m^2 (140/ft^2), and they play an important role in

Figure 11. Fingernail clams bear live young that are miniature replicas of their parents. (Photograph by Holly Jensen-Herrin.)

the energy flow of the pool[22]. The species that inhabit vernal pools are all permanent residents. Although they are not highly mobile themselves, they are readily dispersed by birds and insects. Individuals aestivate as juveniles or adults during the dry period and emerge when the pool floods.

Fingernail Clams

Fingernail clams are tiny bivalves that typically reach an adult size and shape similar to that of a human fingernail (Plate 8, top; Figure 11). They are also known as pea clams or pill clams. The family is diverse and has representatives in every type of freshwater habitat. There are three genera, *Sphaerium*, *Musculium*, and *Pisidium*, all of which are represented in vernal pools. Fingernail clams have gills that serve as both respiratory organs for oxygen exchange and filters for straining food, including microscopic animals, algae, fine particulate detritus, and dissolved and suspended organic material, from the water. In ponds, they can consume more than 8% of the primary production, and some species, such as *Pisidium casertanum*, are highly efficient at filtering bacteria from organic sediments[23].

Distribution[24]

Five species of fingernail clams are common in northeastern vernal pools. Only one of these, Herrington's fingernail clam, *Sphaerium occidentale*, is restricted to pools that dry up annually. The others occur in a variety of permanent and temporary waters.

Herrington's fingernail clam reaches an adult length of 8 mm (0.32 in). This species occurs from Newfoundland south and west throughout the glaciated northeast. It is generally restricted to habitats that dry periodically and is especially common in woodland vernal pools, ditches, and swamps, especially in areas with calcareous deposits. The swamp fingernail clam, *Musculium partumeium*, reaches an adult shell length of up to 13 mm (0.5 in) and can be found from New Brunswick to Saskatchewan in Canada and throughout the glaciated northeastern United States. It is usually found in muddy-bottomed water bodies, particularly in floodplain pools near large streams and in spring- and fall-filling vernal pools. The pond fingernail clam, *M. securis*, is smaller, reaching an adult shell length of 8 mm (0.32 in). This species can be found from Newfoundland and Nova Scotia throughout the glaciated northeast in perennial and temporary waters, usually with muddy substrate and emergent or submerged vegetation. The ubiquitous pea clam, *Pisidium casertanum*, is a small animal whose shell rarely exceeds 5 mm (0.2 in) in length. It is the most widely distributed freshwater mollusc in the world and occurs in temporary and permanent freshwater habitats including lakes, ponds, vernal pools, streams, swamps, and other types of vegetated wetlands throughout the glaciated northeast.

Life History

Fingernail clams are all hermaphroditic, and they commonly self-fertilize. The eggs develop internally, on the surfaces of the gills, and the young clams are fully formed miniatures of the adults when they are released into the water (Figure 11). In vernal pools, the young-of-the-year are released from the parental brood sacs in late spring to early summer (May to July). Depending on the species and specific habitat conditions, adult fingernail clams release from two to twenty-five young per brood, and they may produce several broods during a breeding season[25]. The numbers of young produced exceed the carrying capacity of the habitat, and mortality in juveniles is as high as 95%[26]. Fingernail clams survive pool drying by burrowing into the damp sediment and aestivating. Aestivating clams are more tolerant of drying than animals that are not aestivating, and the physiological changes that prepare individuals for aestivation take place before the pools dry[27]. Different species aestivate at different stages of growth.

Sphaeriids have highly flexible life histories and readily adapt their growth and reproduction to environmental changes[28]. For example, when raised in the laboratory under an artificial regime of twelve hours of light per day, *S. occidentale* and *M. securis* grow and reproduce continuously, but in nature they undergo a summer diapause in vernal pools. *Musculium* in vernal pools produce only one generation per year in spring-filling pools, but in fall-filling and semi-permanent pools they opportunistically grow and reproduce in fall, as well. The variety of life history strategies and the degree to which they can change adaptively in response to environmental variation have made fingernail clams popular research organisms for evolutionary biologists.

Herrington's fingernail clam is the only species of *Sphaerium* found in vernal pools, and it is particularly well adapted for life in temporary waters[29]. It produces more young and resists drying better than other fingernail clams found in vernal pools. An average of six to twelve young are released per brood, and adults can produce several broods per breeding season. This species can survive pool drying at all stages from newly released juvenile to adult, although mortality from drying is high in the largest individuals, and all size classes can be found in pools at any time. It has a relatively long life span of three years, in comparison with an average of one year for other species, and this compensates for its slow growth rate, the short period of time that is available for its growth in seasonally drying waters, and the high mortality of juveniles.

Newborn Herrington's fingernail clams start to grow once they are released from the parental brood sac and are ready to aestivate before the pool starts to dry. In spring-filling pools, they remain dormant in the sediment through the winter, resume growth in the spring, and aestivate again when the pool dries. Once they have reached a size of around 5.5 mm (0.22 in) they are ready to release young, and they can release several broods before the pool dries. The length of time required to reach reproductive size can vary with the duration of flooding, water temperatures, and food. *S. occidentale* may become large enough to reproduce after two years. In fall-filling pools, they may grow again briefly when the pool floods, and reproductive adults may produce one or more fall broods, but they become dormant again once the water becomes cold and do not resume growth until spring. Some individuals may be able to experience three seasons of reproduction: in spring after two years of growth, in fall, and again the following spring[30]. This life history strategy allows the clams to grow whenever water is present and to produce enough young that there is a high probability that individuals will reproduce themselves successfully.

The swamp fingernail clam and the pond fingernail clam have maximum life spans of twelve to fifteen months and are less tolerant of pool drying than Herrington's fingernail clam. Under natural light regimes, the spring-released juveniles of both species of *Musculium* do not grow but instead undergo a period of diapause until late summer. The adults continue to grow and reproduce until pool drawdown, and then die. In contrast to *S. occidentale*, only newly released juveniles of *Musculium* can survive pool drying, and even they experience very high mortality[31]. The diapause helps ensure that the clams are in the drought-resistant newborn stage when the pool dries, maximizing their chances of survival until the pool refloods. This diapause occurs in permanent waters as well as vernal pools, suggesting that these species may have evolved in temporary waters and subsequently moved into perennial water bodies[32].

Because the only *Musculium* that can survive pool drying are newborn individuals that do not grow in late spring and summer, all growth and reproduction in vernal pools must occur after the pool refloods. In spring-filling pools, these clams do not start to grow until water temperatures have warmed in mid to late spring. They need to reach a minimum size before they are able to release young, and the period available for reproduction may be as short as a few weeks or as long as several months, depending on when the pool dries. The metabolic rates of the dormant juveniles increase from fall through winter, preparing the animals for development during the short period when water is present[33].

In fall-filling pools, the young *M. securis* and *M. partumeium* stop aestivating and start to grow when water levels rise in late summer or early fall. If the pool floods early enough, individuals can grow large enough to produce a litter of young in the fall[34]. Similarly, in permanent ponds and lakes, these species break diapause and begin to grow in the fall, and they usually produce a fall brood of young[35]. The clams stop growing when the water gets cold and resume their growth in the spring. In semi-permanent and permanent waters, members of the previous year's spring-released generation of the swamp fingernail clam have reached reproductive size by early spring and die in early summer, while fall-released individuals reach reproductive size later and die in mid to late summer[36].

Although the two species of *Musculium* have similar life histories overall, they differ in their strategies for maximizing survival during pool drawdown[37]. *M. securis* produces large numbers of relatively small newborns. This helps ensure that some young clams will be alive when the pool floods again, even if mortality rates are very high. *M. partumeium,* in contrast, produces fewer newborns containing higher energy reserves in temporary pools than in permanent ponds. In this way, the probability that an individual juvenile will survive drying is enhanced.

The ubiquitous pill clam, like other sphaeriids, shows high flexibility in its life history. It aestivates during drawdown in vernal pools[38].

Snails

Snails in vernal pools include members of the families Lymnaeidae, Physidae, and Planorbidae. In snails, the muscular foot secretes mucus on which the animal glides, and the mantle forms a chamber that can hold a bubble of air and through which oxygen can diffuse into the blood. Snails are grazers. They are equipped with a specialized toothed structure called a radula that they use to scrape algae and films of bacteria and fungi from the substrate, and they also feed on aquatic plants, detritus, and decomposing animals[39]. Some hang upside-down from the water's surface and feed on the organic compounds, pollen, microorganisms, and other living materials there. The presence of snails in streamside pools may

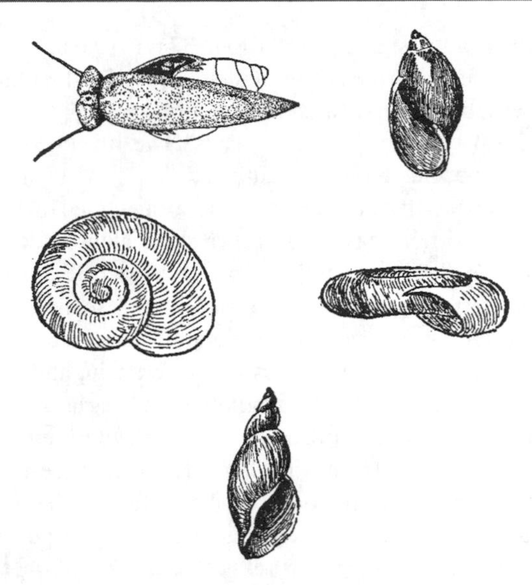

Figure 12. Pond snails — including physids such as *Physa gyrina* (top), planorbids including *Gyraulus parvus* (center), and lymnaeids such as *Stagnicola elodes* (bottom) — feed on algae and decaying detritus and serve as intermediate hosts for trematode worms. (From Morgan 1930, courtesy of the Morgan estate.)

influence the growth rate of herbivorous tadpoles, presumably through competitive interactions[40]. The common pond snail *Stagnicola elodes* (as *S. palustris*) grazes on the surface of the water and consumes everything it encounters — it has been demonstrated to prey on mosquito larvae in short-cycle woodland pools in Ontario[41].

The three common families of pond snails can be told apart by differences in their shells (Figure 12). Lymnaeid snails have tall, pointed shells with what is known as dextral, or right-handed, symmetry because they twist in a clockwise direction and have the opening on the right side. Lymnaeids vary in size from less than 1 cm (0.4 in) to 5 cm (2 in). The physids have shells that twist in a counter-clockwise direction and have the opening on the left, or sinistral symmetry, and the spire in most species is not as sharp as in the lymnaeids. The planorbids have a flat whorl without a spire.

Distributions[42]

Most lymnaeids are widely distributed throughout the glaciated northeast, but a few have more restricted distributions. The common stagnicola, *Stagnicola elodes* (formerly known as *Lymnaea palustris* and *L. elodes*) (Figure 12, bottom), is a large snail with an adult shell length of 15 to 32 mm (0.6 to 1.2 in). It is common throughout the glaciated northeast in shallow eutrophic habitats from marshes and sloughs to permanent ponds and vernal pools, especially in shallow waters with dense stands of emergent aquatic plants. The blade-ridged stagnicola, *S. caperata*, is a small species with an adult shell length of 17 mm (0.7 in). It is a typical vernal pool species that also occurs in the spring-filled margins of permanent waters. In the glaciated northeast it is known from New York and Indiana. The striped stagnicola, *S. reflexa*, has a tall, narrow shell up to 37 mm (1.5 in) in length. Some authors consider it to be a variant of *S. elodes*. This snail is found in permanent and temporary habitats with vegetation and a mud substrate in southern Ontario and Quebec, throughout the Great Lakes drainage basin, and in the upper Ohio and Mississippi drainage basins. The amphibious fossaria, *Fossaria parva*, and the modest fossaria, *F. modicella* (also known as *F. humilis* and *L. humilis*), both attain adult shell lengths of 10 mm (0.4 in). Both of these species are highly amphibious, and they can often be found on the surface of the mud next to vernal pools and other temporary waters as well as in a variety of permanently flooded habitats throughout the glaciated northeast. *Fossaria parva* is unusual among snails in that it can be found in acidic, calcium-poor vernal pools. *Pseudosuccinea columella* is another large species, with an adult shell up to 25 mm (1 in) long. It occurs from Nova Scotia to Wisconsin and south through the rest of the glaciated northeast, most commonly in permanent waters but also in temporary ponds and wetlands. This snail tolerates a wide range of conditions including acidic waters that are low in calcium, and it is an important host for flukes that parasitize amphibians[43].

Most of the physids occur in a wide variety of aquatic habitats over a broad geographic range. The polished tadpole snail, *Aplexa elongata* (commonly identified as *A. hypnorum*, a European species), is the only physid that is largely restricted to temporary waters. It is typical of woodland vernal pools throughout eastern Canada and from western New England to the upper Midwest. It has a narrow, elongate shell up to 20 mm (0.8 in) long with a sharp spire. This species occurs most commonly in pools with hard water and abundant detritus, upon which it feeds. It often hangs upside-down on the water's surface, grazing on materials floating on the water. The tadpole snail, *Physa gyrina*, attains a shell length of 24 mm (1 in) and is highly tolerant of low dissolved oxygen and pollution (Figure 12, top). It can be found on silt and detritus in temporary and permanent waters including floodplain pools, vegetated wetlands, streams, ponds, and lakes. Its range extends from Quebec west in Canada and from New York west throughout the central and western United States. The vernal physa, *P. vernalis*, is a small physid with an adult shell length of 11 mm (0.4 in). This species occurs in temporary and permanent habitats including both soft, acidic waters and hard waters in southern New England and New York, and it probably also occurs in Ohio, Michigan, and southern Canada. The eastern physa, *P. heterostropha*, is a medium-sized snail up to 16 mm (0.65 in) long. It is found in virtually every kind of freshwater habitat including temporary pools, usually on muddy substrates with detritus and vegetation, from the Canadian Atlantic Provinces to New Jersey, New York, and Pennsylvania. It tolerates a range of chemical conditions including both hard and soft, acidic waters. It is the most common physid found in New England. The pumpkin physa, *P. ancillaria*, reaches an adult shell length of 20 mm (0.8 in). It occurs in a wide variety of aquatic habitats in eastern Canada, New England, Pennsylvania, and New York. It is most common on decaying vegetation and organic debris in permanent waters, but it sometimes occurs in vernal pools.

All of the planorbids found in vernal pools are widespread and occur in a wide range of habitats. Say's toothed planorbid, *Planorbula armigera* (also known as *P. jenksii*), is one of the most common snails in vernal pools from New Brunswick to the upper Midwest (Plate 8). The shell of this snail is up to 8 mm (0.3 in) in diameter and has a pair of down-jutting teeth set back from the opening. The teeth are visible in empty shells, and in living animals they appear as a pair of lines on the upper surface of the outer whorl of the shell, not far from the opening. The modest gyraulus and the flatly coiled gyraulus, *Gyraulus parvus* (Figure 12, center) and *G. circumstriatus*, both have an adult shell diameter of 5 mm (0.2 in) and look superficially similar to *Planorbula*, but they lack the pair of teeth within the shell opening, and their shells are less heavy and do not become as large. These two species differ in how level the tops of the whorls are, with *G. parvus* having the innermost whorls depressed below the outer whorls, and *G. circumstriatus* having all whorls at the same level. Both of these snails occur in permanent and temporary waters throughout Canada and the rest of the glaciated northeast. The tiny nautilus snail, *G. crista* (also known as *Armiger crista*) has a maximum shell diameter of 2 mm (0.08 in). The shell is readily identified by a series of ridges or fins located at regular intervals across the whorls of the shell. This species occurs from Ontario and New York south and westward. The keeled promenetus, *Promenetus exacuous*, is another species that is easy to identify, because the outer whorl of the shell is not rounded but instead comes to a sharp, keeled angle. This snail, whose shell reaches a diameter of 8 mm (0.3 in), is very common in vernal pools, and also in permanent waters, throughout the glaciated northeast.

Life History

Life history information is unknown for several of the species of snails found in vernal pools. Like most of the sphaeriids, most snails live approximately one year and have a single generation per year in vernal pools, although they may produce several broods during a reproductive period, and in permanent waters some species may reproduce more than once[44]. All are hermaphroditic, and they both mate with other individuals and self-fertilize. In some species the eggs may develop parthenogenetically, i.e., without being fertilized. Snails lay their eggs within a gelatinous mass on the surfaces of the pool bottom or on plants, and the young snails hatch out looking like small editions of the adults.

In most vernal pool snails, it is the young-of-the-year that aestivate during the dry period. Typically, the juvenile animals burrow into the sediment during the summer, then re-emerge to feed, grow, and breed when the pools flood. Both young-of-the-year and young adult *Fossaria modicella* overwinter in floodplain sediments in Wisconsin[45].

A variation on the general strategy is seen in the common stagnicola. Adults follow the retreating water toward the center of the pool, but the juveniles of this species migrate away from the pool, climbing onto nearby woody vegetation to aestivate during the summer. They resume activity with fall rains, descending back to the ground and burrowing into the leaf litter along the pool margin. This behavior may reduce mortality from predation and parasitism by scizomyid flies[46].

Vernal pool snails grow rapidly and mature quickly[47]. In the highly amphibious lymnaeid *Fossaria modicella*, oviposition is stimulated by increased water temperatures; this occurs soon after flooding in spring-filling pools. The newly hatched young attain sexual maturity at a small size (4 mm, or 0.16 in) becoming reproductive in as few as three weeks. If conditions are favorable, growth and reproduction continue until a shell length of 9 to 10 mm (0.35 to 0.4 in) is reached. Similarly, the planorbid *Gyraulus parvus* matures four to five weeks after hatching and starts to breed two to three weeks later, reproducing continuously until the pools dry or the water becomes cold.

The available literature suggests that many gastropods in short-cycle vernal pools emerge from the sediments upon flooding, grow, and reproduce, and that members of the next generation burrow into the sediments to aestivate when the pool dries without having reached reproductive size. In long-cycle pools, there may be time for a second generation to be produced. Whether, as in some sphaeriids, the ability to survive the dry period is inversely related to shell size has been insufficiently addressed.

Co-Occurrences of Vernal Pool Molluscs

Commonly several species of snails and fingernail clams co-occur within a vernal pool. In Massachusetts, *S. occidentale* rarely occurs with other sphaeriids, although in an Ontario vernal pool it is found in association with *Musculium securis*, *M. partumeium*, and *Pisidium rotundatum*, as well as the gastropods *Promenetus exacuous*, *Planorbula armigera*, *Physa gyrina*, *Aplexa elongata*, and *Stagnicola elodes*[48]. It is not unusual for a vernal pool to support several species of sphaeriids, one or two physids, two or three planorbids, and a lymnaeid. There is some evidence of interspecific competition in freshwater molluscs, and usually congeneric species do not co-occur or, if they do co-occur, they are found at different depths or in different microhabitats[49].

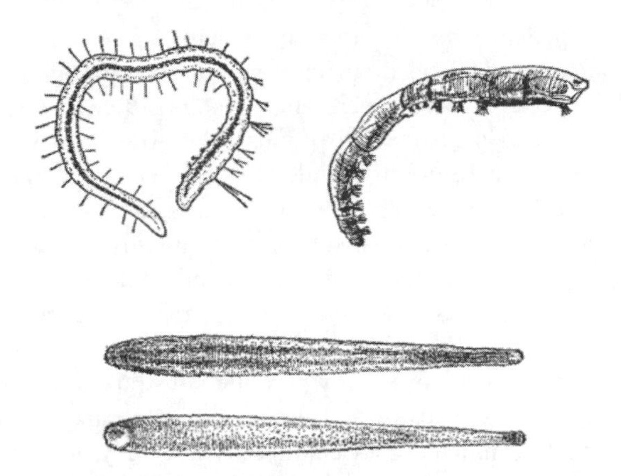

Figure 13. Annelids in vernal pools include oligochaetes such as *Nais* (upper left) and *Chaetogaster* (upper right), which feed on sediment in the bottoms of vernal pools, and leeches (center and bottom) such as *Helobdella*, which prey on other animals. (From Morgan 1930, courtesy of the Morgan estate.)

Segmented Worms (Annelida)

Representatives of both major orders of freshwater annelids, the leeches and the oligochaete worms, occur in northeastern vernal pools (Figure 13).

Leeches (Hirudinea)

Leeches are flattened, segmented worms with a shape that varies from long and slender to broadly oval (Plate 1, Figure 13). A sucker surrounds the mouth and there is a second, larger one at the posterior end of the body. The suckers allow leeches to crawl, inch-worm fashion, anchoring first one end of the body and then the other, with the strong muscles of the body alternately extending one end ahead, and then contracting to bring the rest of the body up to it. Some leeches use thc samc kinds of alternating contractions of the body musculature to swim through the water. Although people tend to think of leeches as blood-sucking parasites or predators, only some species feed on surface mucus or on blood and other body fluids. Some species are scavengers, and others prey on insects, snails, and crustaceans. Leeches serve as prey for dytiscid beetles, water bugs, and odonate nymphs[50].

Most of the leeches found in vernal pools have broad geographic distributions and occur in a wide variety of habitats. Such species include *Batrachobdella picta, Helobdella stagnalis, H. fusca*, *Erpobdella punctata*, *Macrobdella decora, Placobdella parasitica*, and *P. ornata*. When they occur in vernal pools, these leeches feed on the variety of vertebrate and invertebrate life there[51]. *Placobdella parasitica* is parasitic on snapping turtles and other turtles. *Erpobdella punctata* has wide feeding plasticity, preying on frogs and invertebrates in vernal pools (it feeds on fish in permanent waters). *Batracobdella picta* preys on amphibians, particularly in small ponds. The American medicinal leech, *Macrobdella decora*, has been documented to feed heavily on the eggs of wood frogs; apparently it is not able to penetrate the egg masses of mole salamanders. Invertebrates, including arthropods, snails, and annelids, are the usual food of *Helobdella stagnalis*, but this species has been found attached to tiger salamanders (*Ambystoma tigrinum*) on parts of the body where lymphatic sacs are near the surface.

In vernal pools, these leeches appear to be restricted to long-cycle or semi-permanent pools, or at least to pools where the substrate remains saturated for most of the year. Some species of leeches burrow into damp mud when vernal pools dry and create a mucus-lined chamber where they aestivate[52]. Although they lose a considerable amount of weight, they are able to aestivate for a month or more in such chambers. There are no records in the literature of leeches successfully aestivating in spring-filling, short-cycle vernal pools. It has been suggested that leeches may be transported into vernal pools by birds, and transport by turtles is also a possibility[53].

Leeches are hermaphrodites and require cross-fertilization from another individual. The eggs are deposited on the bottom in a cocoon. Some species live for only one year, but others have life spans of two or more years. Both single-breeding and multiple-breeding (semelparous and iteroparous) reproductive strategies are found, with some species showing a high degree of flexibility depending on site-specific conditions of food, temperature, and predation[54].

Two species of leeches appear to be particularly well adapted for life in vernal pools[55]. The erpobdellid *Mooreobdella bucera* (as *Dina*) is abundant in spring in autumnal and semi-permanent vernal pools in Michigan, the type localities for this species. Its feeding habitats and life history are unknown. A small glossiphoniid leech, *Marvinmeyeria lucida*, parasitizes molluscs in temporary and permanent waters from the western Prairie Provinces of Canada to Ontario and Michigan. It has been reported as infesting *Stagnicola elodes*, *Physa heterostropha*, *Aplexa elongata*, *Planorbula armigera*, and *Promenetus exacuous*. In a short-cycle autumnal pool in Ontario, this leech switches in early summer from its preferred winter host, *A. elongata*, and preferentially infests the large fingernail clam *Sphaerium occidentale*. Apparently lacking the ability to withstand desiccation itself, the leech remains within the pericardial space of the aestivating clam through the dry period.

Oligochaeta

Oligochaetes in vernal pools can be thought of as similar to aquatic earthworms. Species in vernal pools include members of the families

Aeolosomatidae, Enchytraeidae, Lumbricidae, Lumbriculidae, Naididae, and Tubificidae[56]. They are abundant in decaying leaf litter in woodland pools. Oligochaetes feed by ingesting sediment and digesting from it algae, organic detritus, and bacteria. They can be important in mixing and controlling the physical and chemical characteristics of sediments[57].

Although relatively little information is available on desiccation-resistance in aquatic oligochaetes, and experimental physiological studies are particularly wanting[58], it appears that several strategies allow members of this group to survive pool drawdown. *Eiseniella tetraedra* (Lumbricidae) is a facultatively aquatic oligochaete found in short-cycle woodland pools in Ontario[59]. It aestivates in a mucus-lined chamber in the sediment. Two species of *Aeolosoma* (Aeolosomatidae) emerged from flooded soil samples taken from two Michigan vernal pools[60]. Several species in this family encyst at low temperatures, coiling their bodies tightly and becoming covered with a hard mucus coat; presumably the same response occurs in response to pool drying[61]. *Lumbriculus variegatus* (Lumbriculidae) also encysts, but it uses a different encystment strategy, fragmenting into small pieces that develop a mucous coating. All of these species emerge when the pool floods.

Members of the Naididae normally reproduce asexually[62], but in temporary waters they aestivate as eggs in desiccation-resistant cocoons. The eggs do not hatch immediately when the pool floods, but warming temperatures in spring appear to stimulate hatching[63].

The life cycles of enchytraeids and tubificids in vernal pools are unknown.

Other Taxa

Most of the animals discussed in this chapter thus far represent taxa whose occurrence in and adaptations to vernal pools have been relatively well documented. Other invertebrate groups sometimes found in vernal pools, including hydras, sponges, rotifers, gastrotrichs, nematodes, tapeworms, and horsehair worms, may be present in significant numbers[64]. Most of these animals are small, inadequately collected by the techniques used by most researchers investigating the biota of vernal pools, and difficult to identify to the level of species. The literature provides little or no information on the distributions and life history strategies of members of these groups within northeastern vernal pools.

Rotifers (Rotifera)

Rotifers are microscopic multi-celled animals that were discovered by Antony von Leeuwenhoek 300 years ago, in the course of his pioneering work with the microscope. Sometimes known as "wheel animalcules," their name is based on the presence of rings of cilia on the top of their heads that, when beating, look like wheels spinning. The beating of the cilia creates water currents that allow the rotifers to filter microorganisms and small particles of detritus from the water.

With few exceptions, rotifers occur in fresh waters, and they are considered by some biologists to be the animals most specialized as a group for freshwater life. They are found wherever there is water, including damp mosses, birdbaths, and rain gutters, as well as in lakes, ponds, streams, and freshwater wetlands. Rotifers are important members of the freshwater zooplankton of lakes and ponds and can reach densities of many thousands per liter of water. Some feed on filamentous algae or colonial phytoflagellates such as *Volvox*, and others are predaceous on small aquatic animals. Rotifers, in turn, are an important food source for small crustaceans such as copepods and for insect predators such as phantom midges.

Philodina roseola, Rotaria citrina, and *Epiphanes senta* are among the species reported to be common in northeastern pools[65]. To date, major studies of northeastern vernal pool communities have not examined rotifers. The group is a significant component of the zooplankton in temporary waters in Europe, where they are abundant in temporary waters such as floodplain pools. In a study of eighteen temporary ponds in Spain, thirty-seven species of rotifers were identified, twenty-one of which were from short-hydroperiod pools[66] . This study involved only two sampling dates and did not identify all collected animals, so the diversity of rotifers in the pools is sure to be greater than that reported. It seems likely that species diversity is equally high in vernal pools in North America.

Rotifers as a group are well adapted for life in temporary waters. Most species produce resting or "winter" eggs that are quite resistant to desiccation. Resting eggs can accumulate in large numbers in sediments, and they are capable of extended dormancy[67]. In many species, including all members of the Class Bdelloidea, only females are known, and the eggs develop parthenogenetically, without fertilization[68]. In *E. senta*, a member of the Class Monogononta, the population consists only of females while water is present and conditions are favorable for growth, and eggs develop parthenogenetically into more females. When the pool starts to dry, the unfertilized eggs develop into males, and they fertilize the next generation of eggs. These fertilized eggs develop into resting eggs that lie in the sediment until conditions are again favorable for growth[69].

Rotifers in the Class Bdelloidea, many of which inhabit the aquatic film on the surface of mosses, are known to respond to environmental stresses including high temperatures, freezing, and desiccation by entering an "anhydrobiotic state" — a dormant state that involves the loss of nearly all of their body water[70]. In anhydrobiosis, most or all metabolic activity stops completely, in contrast to aestivation, in which metabolism is merely slowed down. The anhydrobiotic state can be entered at any stage in the life cycle, providing an advantage to animals that live in a highly unpredictable environment, and the animals can remain desiccated for decades without harm. Anhydrobiosis is restricted to members of the Bdelloidea, including *Philodina* and *Rotaria*. It would be interesting to know the extent to which rotifers from vernal pools experience anhydrobiosis, or whether resistant eggs are the primary mechanisms for withstanding seasonal drought.

Hydras (Cnidaria)

Freshwater coelenterates are few and poorly studied in comparison with their large and diverse marine relatives. For the most part, freshwater species do not have a free-swimming medusa stage, but only the stalked, sessile hydroid stage. Members of this group appear to lack the ability to survive out of water, so they are not typically found in vernal pools. Two hydras, *Hydra oligactis* (as *Pelmatohydra*) and *H. viridissima* (as *Chlorohydra*), were collected in southern Michigan from a fishless permanent pool that supported a variety of vernal pool indicator organisms including fairy shrimp[71]. I have collected specimens of an unidentified hydroid in a semi-permanent vernal pool on Cape Cod. Hydras feed on protists, tiny crustaceans including cladocerans, ostracods, and copepods, and other small aquatic organisms. In large numbers they can have a dramatic effect on the numbers of prey organisms such as *Daphnia*.

Sponges (Porifera)

Sponges also occur in vernal pools, although they do not appear to occur in short-cycle pools and may be restricted to semi-permanent or permanent waters. One species, *Eunapius fragilis*, is widespread throughout the glaciated northeast and occurs in a wide variety of habitats including semi-permanent pools. This sponge tolerates a wide range of water chemistry and is especially common in hard waters with relatively high pH. Sponges withstand unsuitable conditions, including cold and drought, by producing resting structures known as gemmules. These are hard, round structures that are highly resistant to drying and thermal extremes. In *E. fragilis*, the gemmules form a basal layer that underlies the thin, circular mass of the living sponge and survives when the animal is killed by the onset of winter or periodic drawdown. Living cells within the gemmules differentiate into a new sponge when conditions again become favorable for growth[72].

Some Remaining Questions About Non-Arthropod Invertebrates

- Which species occur in vernal pools? What is the distribution of these species in the glaciated northeast?
- How are these species distributed along gradients of hydrology and water chemistry?
- What are the adaptations of molluscs, annelids, and other non-arthropods to seasonal drawdown in vernal pools?
- To what extent do trematodes affect the populations of invertebrates and amphibians in vernal pools?
- How do non-arthropod invertebrates interact with each other and with other components of the fauna?
- What are their ecological roles within the pools?

References for Chapter 7

[1] Clarke 1981; Jokinen 1992; Smith 1992, 2001; Thorp and Covich 1991. [2] Kolasa 1991. [3] Barnes 1969, Kolasa 1991, Pennak 1989, Smith 1992. [4] Edman and Clark 1988, Kolasa 1991. [5] Kenk 1944. [6] Ball et al. 1981, Kenk 1944. [7] Kenk 1949, Kolasa 1991, Pennak 1989, Smith 1992, see Appendix. [8] Spandl 1925, cited in Kenk 1949. [9] Kenk 1944, as *P. vernalis*. [10] Ball et al. 1981. [11] Kenk 1949, Ball et al. 1981. [12] Kenk 1949. [13] Kenk 1949. [14] Ball et al. 1981, Kenk 1949. [15] Goater and Goater 2001. [16] Barnes 1969; Bernot 2003; Goater and Goater 2001; Kiesecker and Skelly 2000, 2001. [17] Barnes 1969. [18] Department of Pathology, University of Cambridge 2004; Digiani 2002; Fuchs and Mannesmann 2003; Gorbushin 2002; Lockyer et al. 2004; Sessions et al. 1999a, 1999b. [19] Gillilland and Muzzall 2002, Goater and Goater 2001. [20] Gillilland and Muzzall 2002; Johnson et al. 1999; Kiesecker 2002; Sessions 1998; Sessions et al. 1999a, 1999b; Sessions and Ruth 1990. [21] Kiesecker and Skelly 2000, 2001. [22] Brown 1991, Mackie et al. 1978. [23] see review in McMahon 1991. [24] Clarke 1981, Smith 1992. [25] Clarke 1981, Hornbach et al. 1980, Mackie et al. 1978, McMahon 1991. [26] Hornbach et al. 1980, Mackie et al. 1978, Way et al. 1980. [27] McKee and Mackie 1980. [28] Hornbach et al. 1980, Mackie 1979, McKee and Mackie 1981, McMahon 1991, Way et al. 1980. [29] Heard 1977; Mackie 1979; McKee and Mackie 1980, 1981. [30] McKee and Mackie 1981. [31] McKee and Mackie 1980, 1981; Way et al. 1980. [32] Mackie and Qadri 1974, Mackie et al. 1976, McKee and Mackie 1981, Way et al. 1980. [33] McKee and Mackie 1980, 1983. [34] Hornbach et al. 1980, McKee and Mackie 1980, Way et al. 1980. [35] Hornbach et al. 1980, Mackie 1979, Mackie and Qadri 1974, Mackie et al. 1976, McKee and Mackie 1981, Way et al. 1980. [36] Hornbach et al. 1980, Way et al. 1980. [37] Hornbach et al. 1980, McKee and Mackie 1981, Way et al. 1980. [38] Boycott 1936, McMahon 1991. [39] Brown 1991. [40] Holomuzki and Hemphill 1996. [41] James 1961. [42] Clarke 1981; Jokinen 1985, 1992; Mozley 1928, 1932; NatureServe 2003. [43] Jokinen 1991, 1992; Kiesecker and Skelly 2001. [44] Brown 1991; Clarke 1981; Hunter 1975; Jokinen 1985, 1992. [45] Colburn 1975. [46] Jokinen 1978, 1992. [47] DeWitt 1953, 1954; Eckblad 1973; McCraw 1960. [48] Smith 1992, McKee and Mackie 1979. [49] Brown 1991; Matthew Burne, Massachusetts Natural Heritage and Endangered Species Program, personal communication; Colburn 1975; Jokinen 1992; Mackie et al. 1978; Massachusetts Audubon Society data; McKee and Mackie 1979; McMahon 1991; Smith 1992; Wiggins et al. 1980. [50] Davies 1991, Pennak 1978. [51] Cory and Manion 1953, Pennak 1978, Platt et al. 1993. [52] Pennak 1978. [53] Wiggins et al. 1980. [54] Davies 1991, Pennak 1978. [55] Kenk 1949, Klemm 1976, McKee and Mackie 1979. [56] Kenk 1949, Wiggins et al. 1980, Williams 1983, see Appendix. [57] Brinkhurst and Gelder 1991, Massachusetts Audubon Society unpublished data, Strayer 1990. [58] Brinkhurst and Gelder 1991, Kenk 1949, Wiggins et al. 1980. [59] Wiggins et al. 1980. [60] Kenk 1949. [61] Brinkhurst and Gelder 1991. [62] Brinkhurst and Gelder 1991. [63] Kenk 1949. [64] Kenk 1949, Masters 1968. [65] Masters 1968. [66] Fahd et al. 2000. [67] Wallace and Snell 1991. [68] Wallace and Snell 1991. [69] Ricci 2001, Schroeder 2002. [70] Caprioli and Ricci 2001, Crowe 1971, Ricci 2001, Ricci and Caprioli 1998, Wallace and Snell 1991. [71] Kenk 1949. [72] Smith 1992.

Chapter 8

Crustaceans

Ask any biologist who studies vernal pools to name the most typical and common vernal pool invertebrates, and you can be sure that the answer will include at least one crustacean. Fairy shrimp and clam shrimp are restricted to fish-free, temporary waters. Ostracods or seed shrimp, cladocerans or water fleas, and copepods are well represented in vernal pools, and in some pools isopods or aquatic sowbugs, amphipods or scuds, and crayfish are important components of the fauna. Plates 8 and 9 and figures 14 to 19 illustrate some of the most common vernal pool crustaceans. Identification of crustaceans to genus or species generally requires the use of a microscope and a good key[1].

Because of their limited dispersal ability, most crustaceans in vernal pools can be considered to be permanent residents, with life history strategies designed to withstand the range of conditions experienced within the pool basins. In all groups, there are one or more common and widespread species and a number of uncommon ones with more limited distribution. There is little information on current patterns of occurrence of these animals in much of the glaciated northeast.

Fairy Shrimp (Branchiopoda, Order Anostraca)

Anostracan crustaceans, commonly known as fairy shrimp, are among the most distinctive and beautiful invertebrate indicators of vernal pools (Plate 9; figures 14, 15). They get their name from the way that they appear seemingly by magic in tiny woodland pools, sometimes showing up suddenly in sites where they have not been seen for years. They are present for periods of weeks or months and then disappear rapidly once the water begins to get warm.

It is easy to recognize fairy shrimp, as no other animals have the same combination of stalked eyes, upside-down swimming behavior, and orange, green, blue, red, and/or bronze coloration. As they swim on their backs, their many pairs of feathery legs beating in synchrony are conspicuous, and thanks to these appendages, fairy shrimp were formerly known as phyllopods, or leaf-footed animals. Adult body lengths range from 1.5 to 4 cm (0.6 to 1.5 in). Females have small heads and two egg sacs ("ovisacs") at the junction of the thorax (the body section that

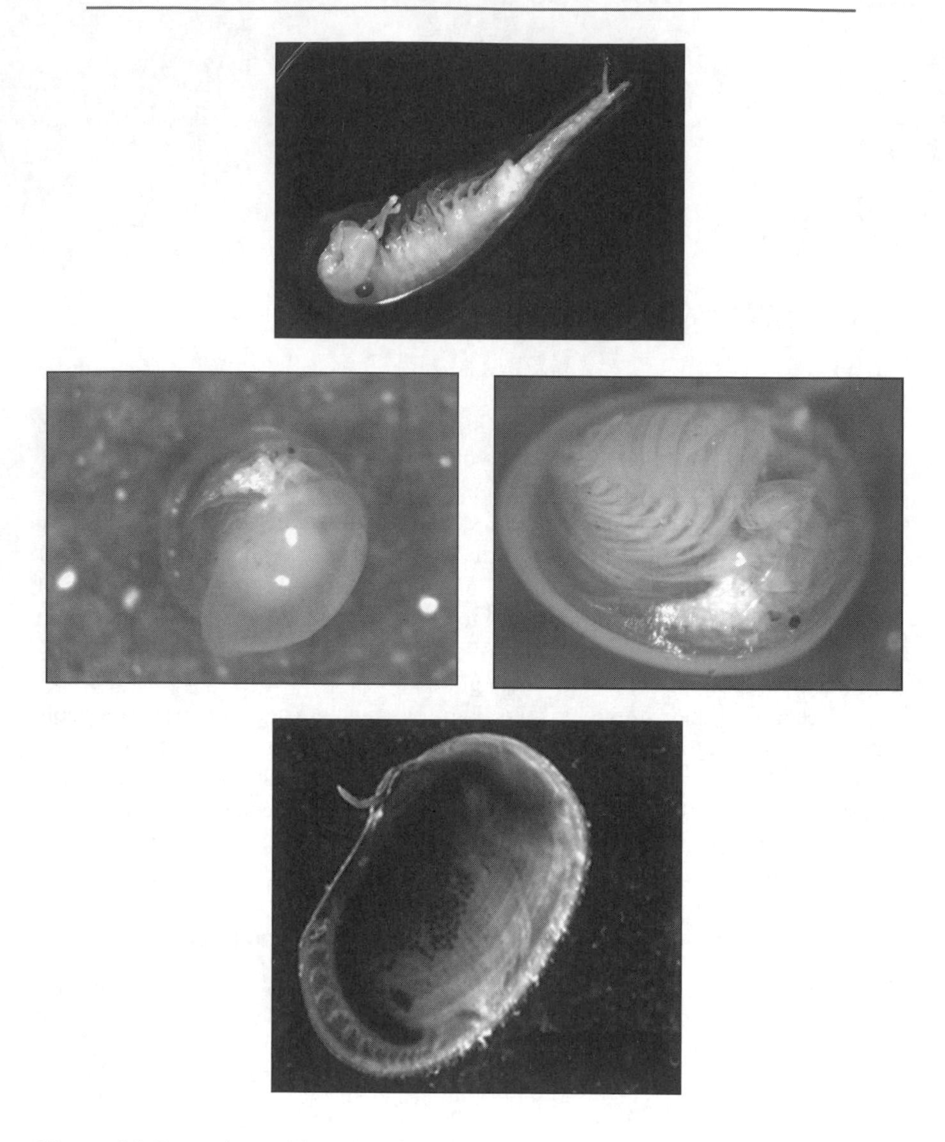

Figure 14. Large branchiopod crustaceans are typical of vernal pools and depend on ephemeral waters for their survival. Examples include the fairy shrimp *Eubranchipus neglectus* (top), common west of the Appalachians; the common clam shrimp *Lynceus brachyurus* (center), shown on the left with its valves partly opened and on the right with one valve removed; and the rare clam shrimp *Caenestheriella gynecia* (bottom). (Photographs by Mike Gray [top], Holly Jensen-Herrin [center], and Steve Weeks [bottom]. Top and bottom photographs reprinted from Weeks 1997, courtesy of Steve Weeks.)

bears the ten pairs of swimming legs) and abdomen. The heads of mature males appear greatly swollen due to the presence of large, branched second antennae, which function as claspers to grasp females during mating (Plate 9, Figure 15). Juveniles lack the ovisacs and large antennae that distinguish the sexes as adults, but they are readily recognized as fairy shrimp by their coloration, feathery legs, and behavior.

Distribution

Fairy shrimp are found in temporary waters worldwide, from deserts to high mountains to the subarctic. Seven families of anostracans occur in North America. The species most commonly found in vernal pools of the glaciated northeast belong to the family Chirocephalidae, genus *Eubranchipus*. Along the northern and southern borders of the glaciated northeast, respectively, representatives of two other families, the Branchinectidae and Streptocephalidae, are occasionally found, as well.

Fairy shrimp found in vernal pools in the glaciated northeast include the knob-lipped fairy shrimp, *Eubranchipus bundyi* (also known in some of the older literature as *Chirocephalus bundyi*), the vernal fairy shrimp, *E. vernalis,* the intricate or smooth-lipped fairy shrimp, *E. intricatus*, and the neglected fairy shrimp, *E. neglectus*. Other species of *Eubranchipus* occur along the western and southern borders of the glaciated northeast, and in the past these were occasionally found in vernal pools with the more common species.

In the northern part of the glaciated northeast, including northern New England, the upper midwestern states, and southern Canada, *Eubranchipus bundyi* is the most common fairy shrimp. Farther south, in New England it is replaced largely by *E. vernalis*, and west of the Appalachians by *E. neglectus*. There is considerable confusion in the literature about these two species, both of which have short antennal appendages. Much of the classic work on fairy shrimp in the lower Midwest was believed to be dealing with *E. vernalis*, but recent research has shown that this species is restricted to east of the Appalachians, and that the common species west of the mountains is *E. neglectus*[2].

The common species from Illinois west is the ethologist's fairy shrimp, *E. serratus*, which is rare in Indiana and absent from Ohio. Also historically reported from the upper Midwest and, like *E. serratus*, a species of the prairies and Great Plains, is the ornate fairy shrimp, *E. ornatus*[3]. Farther south, with a range extending well beyond the extent of glaciation, the dominant species is *E. holmani*. It is rarely collected in the glaciated northeast. This fairy shrimp was originally decribed from collections made in New Jersey in the 1800s, and it has been reported from a variety of other locations in the glaciated northeast including Connecticut and Long Island, New York. It was found rarely in Ohio in the 1940s and 1950s, but it has not been collected there since and may well be extirpated[4]. Less common is *E. intricatus*, known in the glaciated northeast only from Massachusetts and Vermont, and also found in the western Prairie Provinces of Canada.

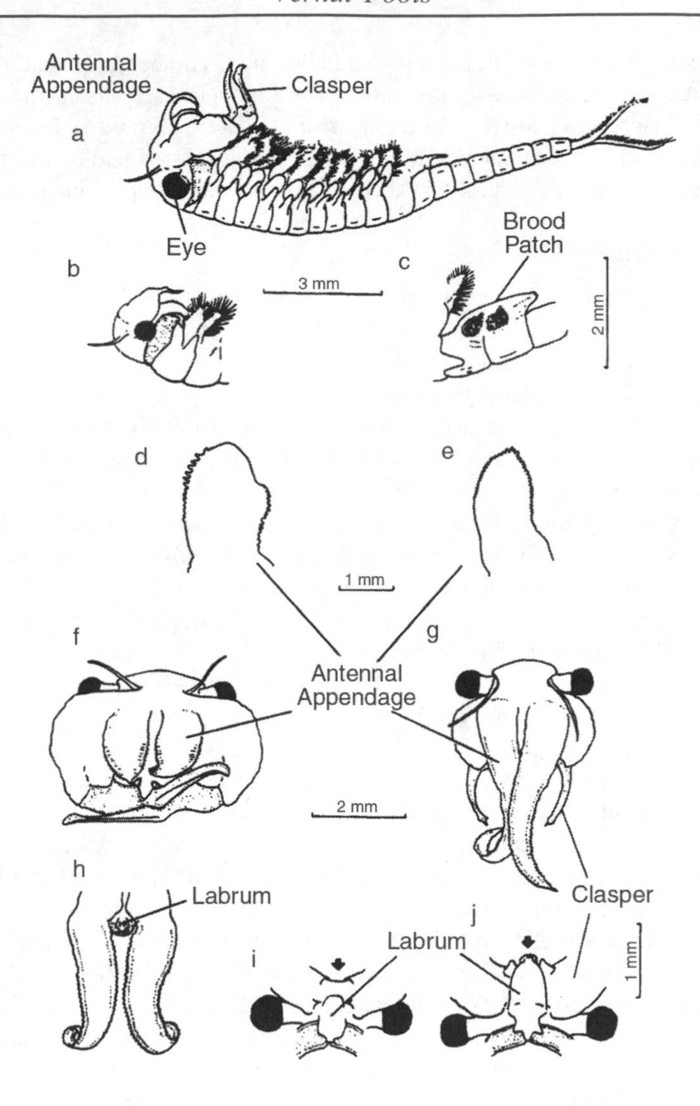

Figure 15. The antennal appendages of male fairy shrimp are used to identify the species. Here are pictured (a) a male fairy shrimp, *Eubranchipus vernalis;* (b) the anterior of a female *E. vernalis;* (c) the first two abdominal segments of a female *E. intricatus;* (d) the antennal appendage of a male *E. neglectus;* (e) the antennal appendage of a male *E. vernalis;* (f) the head of a male *E. vernalis;* (g) the head of a male *E. intricatus;* (h) the antennal appendages of a male *E. bundyi;* (i) a ventral view of the head of a male *E. intricatus;* and (j) a ventral view of the head of a male *E. bundyi.* (All images except (d) and (e) are from Smith 1992, courtesy of Douglas Smith. Images (d) and (e) have been drawn from photographs in Belk et al. 1991.)

The family Branchinectidae contains a single genus, *Branchinecta*, with nine species. One of these, *B. paludosa*, the circumpolar fairy shrimp, is known from subarctic Nova Scotia, Quebec, and Labrador in North America, as well as from Asia and Europe. In the United States this species has been collected in Alaska and the northern Rocky Mountain states. It may occur in vernal pools along the northern border of the glaciated northeast.

The family Streptocephalidae contains a single genus, *Streptocephalus*, with seven largely southern and Caribbean species. *Streptocephalus seali* is the most widely distributed North American anostracan, with a range that encompasses the Canadian Prairie Provinces and most of the United States, except for New England, and includes Illinois, Minnesota, New Jersey, and New York[5]. This fairy shrimp occurs in a variety of kinds of pools and ponds within its range[6]. In the late 1800s this species occurred with *E. holmani* in a seasonally flooded swale in New Jersey[7]. No other published information is available about *Streptocephalus* from northeastern vernal pools. In Kentucky, *S. seali* occurs later into the summer than does *Eubranchipus*, and in South Carolina it hatches in response to summer flooding[8].

Identifying Fairy Shrimp

Due to an error in an early paper, most of the available keys to freshwater invertebrates misidentify two of the common fairy shrimp species of the glaciated northeast[9]. Species are identified on the basis of the male antennae. Geographic location is also a good clue to which species are present in a pool. The following descriptions and illustrations should help in telling the species apart.

The claspers of all male *Eubranchipus* have an enlarged base, off the end of which extends a simple unbranched or L-shaped structure. There is also a fleshy pair of flaps coming off the claspers near the base, between the eyes. These are known as the antennal appendages, and most species of *Eubranchipus* can be identified based on the shape of the antennal appendages (Figure 15).

A short antennal appendage (Figure 15a,d,e,f) is characteristic of *E. neglectus* and *E. vernalis*, which occur west and east of the Appalachian Mountains, respectively. There does not appear to be any geographic overlap of these two species. The appendage in *E. vernalis* is more pointed, longer relative to its width, and with small tubercles along its entire perimeter. In contrast, the antennal appendage in *E. neglectus* is more thumb-like in outline, rounded at the apex, thicker relative to its length, and with a smooth margin near the tip.

A long, extendable antennal appendage, looking much like an elephant's trunk or the blow-ticklers popular at children's parties or on New Year's eve, is found in the northern *E. bundyi*, the southern *E. holmani*, and the very uncommon *E. intricatus* (Figure 15g,h). In the Bundy's and intricate fairy shrimps, the antennal appendage is symmetrical, decreasing gradually in size from base to tip, and it has small serrations or tubercles extending along the outer margin from near the base to the tip. In Holman's fairy shrimp, the appendage lacks these tubercles or tooth-like extensions on its basal half, and it is less symmetrical in shape. The Bundy's and intricate fairy shrimps can be distinguished from one another by microscopic examination of the head. Bundy's fairy shrimp has a bump on the end of its "nose" (labrum is the technical term); the intricate fairy shrimp lacks such a bump on its labrum (see Figure 15h,i,j).

Streptocephalus males have more complex claspers, with a couple of branches and a long, folded, claw-like extension off the basal section. *Branchinecta*, in contrast, has a simple clasper and no antennal appendages.

Life History

Fairy shrimp hatch from diapausing eggs, or cysts, that lie on the bottom of vernal pools until hatching is stimulated by flooding. Following a resting period, the eggs hatch in late fall, winter, or early spring as the pools flood. Summer flooding does not stimulate hatching in *Eubranchipus*. There is some disagreement in the literature as to the specific conditions required for eggs to be ready to hatch[10]. Apparently a period of egg maturation, which may require drying and, in some species, exposure to freezing temperatures, is needed.

Once eggs have passed through the proper conditions, they hatch when flooded. Hatching in *E. bundyi* occurs in response to the decrease in dissolved oxygen that occurs at the soil-water interface when water floods the soil. There is considerable evidence that not all eggs of a given clutch or generation hatch in response to a given flooding event. Thus, if water levels fluctuate in spring, with pool shrinkage and reflooding, several hatches may occur[11].

It is likely that some eggs remain in the sediment for several years without hatching, even under favorable conditions. In fairy shrimp from desert pools, eggs remain viable for decades and possibly for hundreds of years, with eggs from any one female hatching over a series of flooding events. It has been estimated that there are more than 9,000 dormant eggs per square meter of pool bottom (more than 960 per square foot), and that about 3% of the eggs hatch during any given flooding event. Thus, for fairy shrimp as a group, the principal short- and long-term survival mechanism involves extended dormancy and what has been termed "bet-hedging." This maximizes the chance that some individuals of a given generation will be able to develop to maturity and contribute to the next generation of eggs, some of which will, in turn, remain in the egg bank for many years[12].

Crustaceans go through a number of different larval stages as they develop. Depending on the species, some of these stages are completed within the egg, and others are free-living. In *Eubranchipus*, hatchlings emerge from the resting eggs at a well-developed stage, as advanced metanauplii with up to ten swimming legs (*Branchinecta* hatches at an earlier stage, as a nauplius) [13]. The juveniles thus look much like the adults. They pass through several molts, adding appendages and gradually maturing. Adults also pass through several molts. Young individuals are usually orange to salmon colored, with mature males taking on a greenish cast and females becoming bluish. The tips of the cerci ("tails") of *E. vernalis* are a bright fluorescent white.

The life cycle is highly variable. Complete generation times for *E. neglectus* range from as few as three weeks to as many as twenty-four weeks, depending on the onset of flooding and spring water temperatures[14]. There is a single generation per year.

Fertilization of eggs occurs right after females molt. Mature males use their clasping second antennae to grasp females just in front of the ovisacs. If the females do not resist, the males then fertilize the eggs within the ovisacs, subsequently releasing the females[15]. In *Eubranchipus bundyi*, males appear to patrol territories and to be approached by receptive females who otherwise remain concealed in vegetation and litter[16].

The fertilized eggs undergo preliminary embryonic development and then enter a state of diapause. They are released into the water and fall to the pool bottom, where they resist desiccation and lie in the sediment throughout the summer. Females may produce several clutches, releasing the fertilized eggs into the water between broods.

Fairy shrimp are most readily found in flooded vernal pools in early spring. Sometimes young animals and mature adults can be seen under ice in fall-filling pools. *Eubranchipus* do not generally tolerate warm water temperatures, and as a rule fairy shrimp are not found in northeastern vernal pools much beyond the middle or end of May, or whenever water temperatures reach 20 to 22° C (68 to 72° F). This life history pattern removes fairy shrimp from the pools before most salamander larvae have hatched and invertebrate predators have become abundant.

Feeding

Fairy shrimp are filter-feeders. Most filter-feed while swimming on their backs. Some species have structural modifications that suggest that they can scrape algae and microorganisms from the surfaces of the substrate and emergent vegetation. Some species of *Artemiopsis* (a genus of Arctic fairy shrimp closely related to *Eubranchipus*) and *Branchinecta* roll over and scrape surfaces with their thoracic appendages[17], and *Eubranchipus holmani* also shows some scraping behavior[18]. Anostracans were previously assumed to be largely herbivorous, feeding on algae; more recent work indicates that they are omnivorous, eating anything from algae to small crustaceans[19]. They are vulnerable to predation and are fed on heavily by invertebrate predators such as dytiscid beetle adults and larvae, aquatic Hemiptera, and waterfowl. They cannot survive in waters that contain predatory fish.

Habitats

Fairy shrimp occur only in waters that do not contain fish, and they are characteristically found in temporary or semi-permanent ponds. When they occur in permanent water bodies there is always a summer drawdown that exposes the shoreline. In a temporary pond that was deepened by bulldozing and subsequently stocked with fish, fairy shrimp persisted after excavation but were rapidly eliminated after fish were added[20].

The characteristics of a vernal pool that provide good habitat for fairy shrimp are not well understood. Pools that dry too rapidly after filling, lack appropriate food, or have unsuitable water quality will not support anostracan populations. In autumn-filling pools in Ontario, the depth of unfrozen water during the winter, and the presence of algae in spring, may be important factors governing the presence of *Eubranchipus bundyi*[21]. In a long-term study in Ohio, when vernal pools froze completely the fairy shrimp that were present disappeared, but a subsequent hatch occurred when the ice melted[22].

It seems that the environmental conditions required by the eggs for maturation and hatching are the primary factors determining habitat suitability. The eggs of *Streptocephalus seali* apparently do not withstand excessively dry conditions and the distribution of this species may be limited by local hydrology and soil moisture[23].

Fairy shrimp occur sporadically and unpredictably, appearing consistently in some locations from year to year, but remaining absent from other pools except at irregular intervals. Of 127 temporary ponds examined over a four-year period in Ohio, 28% contained fairy shrimp (largely *Eubranchipus neglectus*, reported as *E. vernalis*) during at least one year, and some contained them every year[24]. The results of continued study over a ten-year period suggest that the flooding history of a pool during the previous season and the precipitation during the winter and spring affect hatching of species of *Eubranchipus* in vernal pools[25]. Similarly, the presence of different species in vernal pools in Alberta varies from year to year in response to differences in precipitation patterns[26].

Most fairy shrimp of the glaciated northeast are apparently intolerant of pollution, especially oxygen depletion and siltation, although some species can be found in fairly turbid waters. Unlike some fairy shrimp of more arid areas, *Eubranchipus* do not tolerate salt, high alkalinities, or warm temperatures. The two common species in the southern half of the glaciated northeast are widespread and tolerant of a range of disturbance, but they appear to differ in their ability to tolerate siltation and turbidity. In Ohio in the mid 1900s, *E. neglectus* was the most common species, but its populations were absent or greatly reduced in waters with high amounts of suspended clay or low in dissolved oxygen. Recent surveys found it to be widespread in a range of habitats including highly disturbed areas[27]. Provided that turbidity is low, this species does well in a wide variety of pond types including depressions in cultivated fields. In an Alabama study, *E. neglectus* was abundant in a vernal pool with clear water but present in only low numbers in an adjacent, turbid pool fed by runoff from an agricultural field, in which *E. holmani* was abundant[28]. In Massachusetts, my colleagues and I have found *E. vernalis* in vernal pools ranging from roadside excavations and turbid farm ponds to undisturbed woodland pools. Both *E. intricatus* and *E. bundyi* appear to be restricted to clear waters[29].

Clam Shrimp

Clam shrimp (Branchiopoda, orders Laevicaudata and Spinicaudata), or conchostracans, like fairy shrimp, are specialized for life in temporary waters that lack fish predation. Representatives of the orders Laevicaudata and Spinicaudata occur in vernal pools in the glaciated northeast. Several species are known only from the type locality, and little is known about the habitat requirements and life histories of many members of this group.

These small crustaceans are characterized by a bivalved carapace that encloses the body and appendages. They look somewhat like white, tan, or transparent peas, lentils, or lima beans swimming rapidly through the water (Figure 14, center, bottom). The species in the glaciated northeast swim along the bottom, in vegetation, in soft mud, or along algal mats, processing detritus from the surface or collecting plankton. The swimming motion of the appendages draws water into the shell and food particles are then strained from the water by the fringed legs. Like the fairy shrimp, clam shrimp are feeding continuously as they swim. *Eulimnadia agassizii* develops a bright green color as it feeds on algae[30]. Some species can remove large particles including rotifers.

In the Order Laevicaudata, the shells are more or less spherical and lack growth lines. *Lynceus brachyurus*, the species most likely to be encountered in vernal pools in the glaciated northeast, is small, 3 to 5 cm (less than 0.25 in) in diameter, and more or less spherical (Figure 14, center). It appears to cruise through the water without a visible source of locomotion, but on close examination, the rapidly moving legs within the enclosing shell are seen to be the motive force. The shells of clam shrimp in the Order Spinicaudata resemble clam shells and have growth lines reflecting molts as the animal grows (Figure 14, bottom). These animals tend to be larger than *Lynceus*, with some species attaining shell lengths of 12.5 mm (0.5 in).

Distribution

As a group, clam shrimp are rare north of southern Canada[31]. They tend to appear in vernal pools later in the season than do fairy shrimp, when water temperatures are warmer.

Lynceus brachyurus occurs widely throughout the northern United States and southern Ontario and Quebec. It is also known from Europe and Asia.

In contrast, most members of the Spinicaudata have a highly restricted distribution and appear irregularly in both space and time. Some occur in vernal pools, but they can also reproduce successfully in ephemeral pools that retain water for no more than several days to a week. In general, members of this group are warm-water crustaceans, often appearing in pools in summer or early autumn, as opposed to cold-water taxa found commonly in early to late spring in vernal pools. Species found in the glaciated northeast include: *Limnadia lenticularis*, known in North America from Florida and five locations in Massachusetts; *Eulimnadia diversa* and *Eulimnadia thomsoni*, both known only from Illinois; *Eulimnadia agassizii*, known from Penikese Island, Massachusetts, three other Massachusetts locations, North Carolina, Georgia, and Ontario; *Eulimnadia inflecta*, known from Illinois and Ohio; *Eulimnadia stoningtonensis*, known from a single location in Stonington, Connecticut, and possibly also occurring in Woods Hole, Massachusetts; and *Caenestheriella gynecia*, originally described from Ohio and recently found in Pennsylvania and Massachusetts (Figure 14, bottom)[32].

Life History

Clam shrimp hatch as nauplii from aestivating eggs that lie dormant in the pool sediment until appropriate conditions stimulate hatching. The carapace appears after the third molt and the young animals gradually come to look like small adults through successive molts. Development is rapid. *Eulimnadia* and *Limnadia* reach maturity in only a few days at high water temperatures. Mature adults of *Eulimnadia agassizii* (or possibly *E. stoningtonensis*) were found in July, 1956, in an ephemeral pool at Woods Hole, Massachusetts, only four days after heavy rains, and again in July, 1961, seven days after flooding[33]. In both cases the pool was dry the following day. Because they develop so rapidly, *Eulimnadia* and *Limnadia* can inhabit ephemeral waters that do not meet our definition of vernal pools, but they also occur in vernal pools.

In contrast, *Lynceus brachyurus*, the most common and widely distributed conchostracan in vernal pools, completes its development in twenty-six days in Illinois and is present in vernal pools for several months, provided that the pool remains flooded[34]. Adult males use specialized abdominal hooks to clasp females, and they fertilize the eggs within the brood pouch. The fertilized eggs are released into the water or are carried to the pool bottom as the females die. Females are not known to produce more than a single brood.

As with fairy shrimp, it appears that the diapausing eggs of conchostracans hatch in response to environmental stimuli, with each species having its own specific cues that promote hatching. As a result, different communities of these crustaceans may occur in a vernal pool from one year to the next. Drying is not required for the hatching of eggs, at least under laboratory conditions. However, the sporadic appearance of clam shrimp in nature suggests that environmental factors are important in determining whether eggs will hatch in a given year. During a ten-year period in northeastern Ohio, *Lynceus brachyurus* appeared only once in a vernal pool that regularly supported populations of *Eubranchipus neglectus*[35]. In Massachusetts, I have often found *Lynceus* in vernal pools that support populations of *Eubranchipus vernalis*, but several years may pass in which fairy shrimp are present but the clam shrimp does not appear.

Habitats

Habitat characteristics that are important for conchostracans of northeastern vernal pools are unknown. Members of the Spinicaudata often occur in highly ephemeral pools, and the absence of fish predation appears to be a requirement for these clam shrimp, but *L. brachyurus* reportedly has been collected in permanent water bodies containing fish, as well as in vernal pools. However, declines in this species appear to be associated with losses of temporary waters, and in Indiana *Lynceus* has been placed on the state's "watch list" because of concerns that the species may be at risk there [36].

Cladocerans

Cladocerans (Branchiopoda, Order Anomola), sometimes referred to as water fleas, are a large and diverse group of small crustaceans (figures 16, 17). They are important components of the fauna of vernal pools all over the world. In Italy, nineteen species of cladocerans were identified in twenty-nine short-cycle, autumn-filling pools; fourteen species occur in an Australian temporary pond[37]. The richness of the cladoceran fauna of vernal pools in the glaciated northeast is comparable (see Appendix). Most cladocerans in vernal pools occur in other types of aquatic habitats, but one, *Daphnia ephemeralis*, appears to be restricted to temporary forest pools with abundant leaf litter and high concentrations of humic acids[38].

Species common in vernal pools range in length from 0.2 to 3 mm (0.008 to 0.12 in) long. They have a bivalved carapace, a propeller-like antenna that is used for swimming, and leaf-like legs with which they filter-feed continuously on small particles and microscopic organisms in the water. Superficially, some of them look like very small clam shrimp.

Like other branchiopods, cladocerans are filter-feeders. The modified antennae propel the animal through the water, and the legs create water currents that draw water between the valves of the jointed carapace. Fine particles of detritus, algae, and zooplankton are strained out of the water and ingested. Cladocerans can reach high densities in vernal pools, often occurring in such high numbers that the water appears clouded by their presence (Plate 8, center). They are important food sources for a wide variety of aquatic predators, including water beetles and salamander larvae.

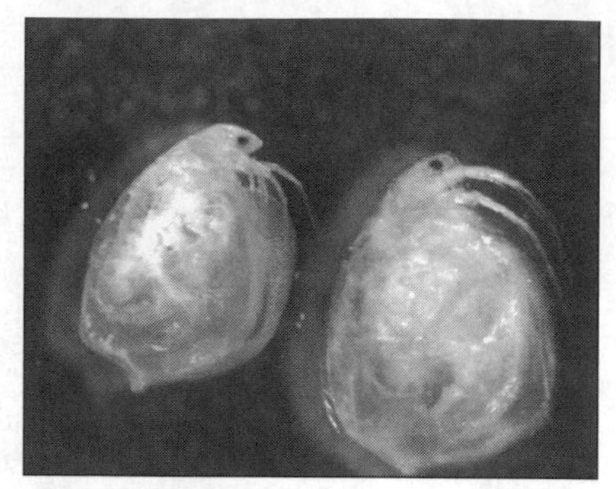

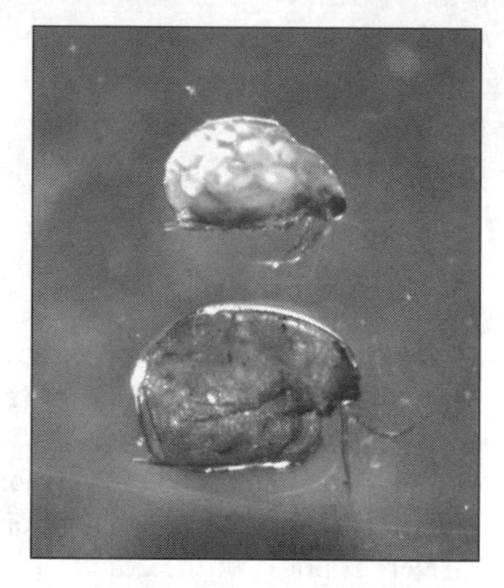

Figure 16. The cladocerans, or water fleas, in vernal pools include *Daphnia* (upper left), *Simocephalus* and *Ceriodaphnia* (upper right), and *Scapholeberis* (bottom). (Photographs by Holly Jensen-Herrin.)

Distribution

The cladocerans that occur in vernal pools generally are widely distributed and have broad habitat plasticity, and most occur in both permanent and temporary waters. The most common species in vernal pools are *Daphnia ephemeralis* and *D. pulex, Ceriodaphnia reticulata, Simocephalus vetulus* and *S. exspinosus, Scapholeberis kingi, Moina* spp., *Alona guttata,* and *Chydorus sphaericus*[39]. All of these occur throughout the glaciated northeast and, with the exception of *D. ephemeralis*, they all occur in a wide variety of habitats.

Life History

Cladocerans have interesting life cycles. Under favorable conditions, all of the adults are females, and they reproduce parthenogenetically, producing diploid

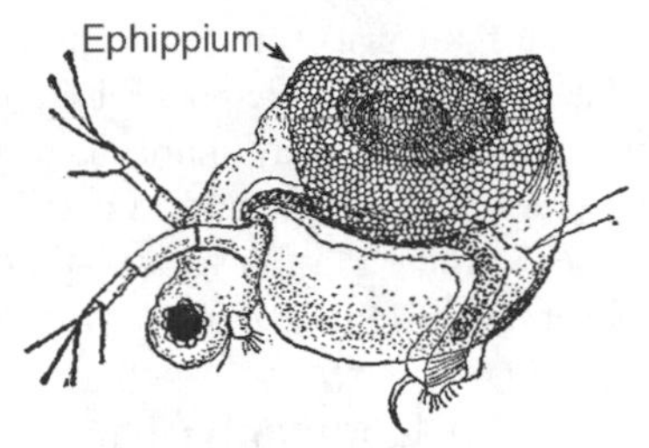

Figure 17. *Ceriodaphnia* is one of the most common genera of cladocerans in vernal pools. The valves of this individual are modified into an ephippium containing a resting egg. This ephippium will rest in the sediment of the dry pool, and the egg will hatch when the pool refloods. (From Morgan 1930, courtesy of the Morgan estate.)

eggs that develop into more females without being fertilized. The eggs develop in a brood chamber protected by the carapace, and under the magnification of a good hand lens or a microscope the developing young are readily visible through the valves (Figure 16, top left). The young are released looking like miniature adults, and they feed and grow and eventually start to produce eggs themselves.

As environmental conditions change, some of the eggs develop into males, and the females start to produce eggs that need to be fertilized and that will rest in the bottom sediments until conditions again become favorable for development. These diapausing eggs provide the population with a way to avoid adverse environmental conditions, such as high temperatures, predation, or pool drying. The eggs are protected by modifications of part or all of the carapace. In *Daphnia* and related genera, part of the enclosing valves darkens and hardens into a distinctive structure known as an ephippium (Figure 17). Cladocerans carrying ephippia look a bit as though they have a dark saddle blanket spread over their backs. The ephippia and their enclosed eggs fall to the pool bottom and diapause until environmental cues stimulate hatching. All vernal pool species appear to use this strategy to avoid seasonal drawdown[40].

Photoperiod, temperature, and carbon dioxide tension interact to stimulate hatching of the diapausing ephippial eggs, and local conditions determine whether hatching occurs in autumn or spring in fall-filling pools. In fall-filling vernal pools studied near Ann Arbor, Michigan, in the 1940s, *D. pulex* was identified as being present from November until the pools dried. However, based on more recent studies of pools in Michigan and Ontario, it seems likely the species in the pools during the fall and winter was *D. ephemeralis*, and that it was replaced by *D. pulex* later in the spring. *Daphnia ephemeralis* appears to be a cold-water species specialized for life in dark-water woodland pools. Its eggs hatch in cold water after pools fill in late autumn or early spring, and adults complete reproduction well before water temperatures start to warm in April. In pools where both species occur, *D. pulex* hatches later, and adults are present later into the season[41].

Many cladocerans change their body shape in response to environmental variables including water chemistry and predation pressure. Changes such as the development of sharp spines at the end of the body and of helmet-like structures on the head have been observed in *Daphnia* in temporary and permanent ponds.

Copepods

Copepods (Copepoda, Orders Calanoida, Cyclopoida, and Harpacticoida) represent another group of small crustaceans that are a visible component of the fauna of many vernal pools. Most are tiny, with the largest species reaching a length of 2 to 3 mm (0.075 to 0.1 in). Generally, copepods are shaped like inverted tear-drops, with a large pair of appendages hanging off the point of the drop and a pair of antennae curving off the top (Figure 18). In addition, females carry a pair of large, rounded egg sacs. Identification requires the use of a compound microscope and a good key[42].

Species found in vernal pools occur in a wide variety of aquatic habitats, but a few appear to be specialized for life in vernal pools. *Cyclops haueri* is a relatively rare species that is apparently restricted to vernal pools, and *Megacyclops latipes* is also typical of this habitat[43]. Copepods are well represented in temporary waters worldwide, with nine species in seven genera common in pools of central and southern Italy, eleven species in six genera in an Australian pool, and two species in southern California vernal pools[44]. At least fourteen species are known from vernal pools in the glaciated northeast (see Appendix).

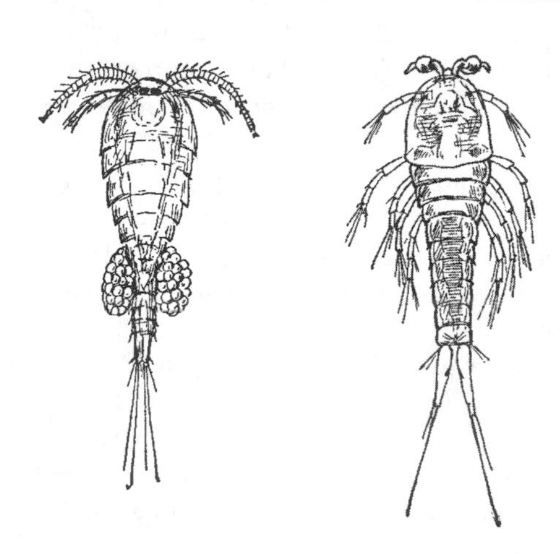

Figure 18. Copepods such as *Cyclops* (left) and *Canthocamptus* (right) prey on a host of small animals in vernal pools. (From Morgan 1930, courtesy of the Morgan estate.)

Calanoid copepods are abundant in many vernal pools from spring through drying. Large blue diaptomids, especially *Diaptomus stagnalis*, are commonly found in vernal pools in late spring. Some cyclopoids are evident through the winter in fall-filling and semi-permanent pools, increasing in numbers as water temperatures warm in spring. Others, such as *Mesocyclops leukarti* and *Platycyclops fimbriatus*, appear to remain dormant until water temperatures reach a threshold in late spring or early summer[45]. Similar patterns are seen in harpacticoids. In Michigan pools, *Canthocamptus staphylinoides* was abundant in winter and, in spring-filling pools, in early spring, and disappeared after May. In contrast, *Attheyella americana* reached its peak abundance in May and June[46].

Most calanoids are filter-feeders. Cyclopoids are predators on other small crustaceans, protozoans, and immature insects, including mosquito larvae, and they also feed on algae and organic detritus. They can be very important predators in some water bodies. Some large cyclopoids have been used effectively in mosquito control. Harpacticoids are scrapers and feed on algae, detritus, and associated microorganisms on the pool bottom[47].

Distribution

Most vernal pool copepods occur across the glaciated northeast and some, including *Macrocyclops albidus* and *Cyclops vernalis*, are among the most abundant North American species (see Appendix). The large diaptomid *Diaptomus stagnalis* is widely distributed in the north-central states south to the Gulf of Mexico, and in the western Prairie Provinces of Canada. This species is common in ponds and has been reported from temporary ponds in Kentucky, South Carolina, and Illinois[48].

Life History and Habitats

Vernal pool copepods use either desiccation-resistant eggs or encystment to withstand drying. As for all resident vernal pool animals, one of the great challenges facing copepods is that of completing the life cycle and reaching the diapause stage before the pool dries. One study of different life history strategies in copepods of a South Carolina temporary pond hypothesized that emergence strategies would be related to hydroperiod so as to allow time for the production of new diapausing individuals. Instead, the timing of emergence from aestivating eggs or cysts did not consistently allow enough time for development and production of the next generation of diapausing forms. Apparently, reserve diapausing individuals in the sediment provide an important safeguard that protects populations against early pool drying[49]. This strategy of maintaining an "egg bank" and/or a pool of diapausing individuals that do not all hatch or emerge in response to the same stimulus is seen for many vernal pool crustaceans. It is similar to the seed bank that exists for many plant species. In some species, eggs or diapausing individuals can remain viable in the sediment for decades.

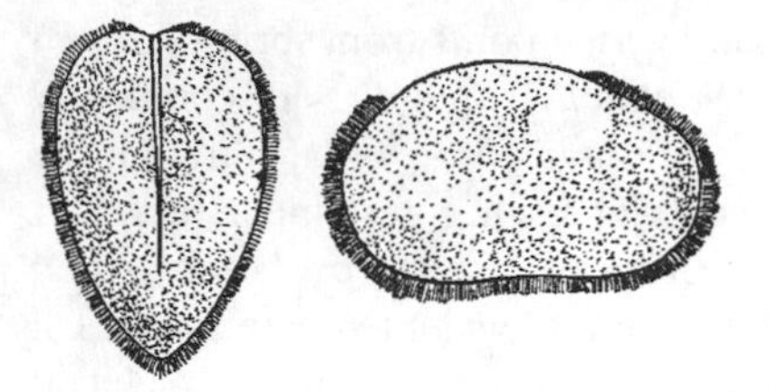

Figure 19. Ostracods, or seed shrimp, including *Cypris,* feed on organic materials in the sediment. (From Morgan 1930, courtesy of the Morgan estate.)

Copepods occur in a wide variety of vernal pool types, including semi-permanent, grassy ponds, heavily vegetated basins, and dense-canopy woodland pools. Based on surveys in Connecticut, copepods appear to be particularly characteristic of perched pools with long hydroperiods[50].

Ostracods

Ostracods (Ostracoda, Order Podocopida), or seed shrimp, are tiny, bivalved crustaceans that look much like sesame seeds and are widely distributed in lakes and ponds worldwide. The enveloping valves are often preserved in lake and bog sediments after the animals die and are used by paleoecologists to infer past temperatures and water quality. Some species are common in organic detritus in the bottom of vernal pools (Figure 19). Most species feed on detritus, bacteria, and algae, but several, including the common and widely distributed *Cypridopsis vidua*, feed on the soft tissues of snails[51].

Ostracods make up a notable component of the fauna of many vernal pools, and it has been suggested that perched pools with long hydroperiods provide particularly good habitat for them[52]. The list of ostracods found in vernal pools is quite diverse (see Appendix). Ostracods can be challenging for non-specialists to identify, and even experts disagree on the taxonomy of some species. This may explain why most of the literature on northeastern vernal pools does not consider ostracods. It is likely that the distribution of ostracods, particularly in long-cycle and semi-permanent pools, is greater than indicated here. This group is also well represented in temporary waters in California[53].

Distribution

As with the cladocerans and copepods, most ostracods found in vernal pools do not appear to be restricted to this habitat. Among the ostracods that appear to be typical — and largely indicative — of vernal pools are *Candona inopinata*, found from Tennessee to Michigan, Illinois, and Ohio; *C. decora*, known from Illinois, Michigan, and Ontario; *C. candida*, distributed broadly from the Pacific Northwest to New England; *C. suburbana* and *C. fossulensis*, known from temporary pools in Illinois; *C. albicans*, an inhabitant of muddy pools and marshes from Nova Scotia through Ohio and Illinois to California; and *Bradleystrandesia fuscata* (formerly *Cypricercus fuscatus* and synonymous with *C. reticulatus*), widespread in temporary waters throughout North America and found in small, grassy temporary pools in Michigan, Illinois, Ohio, New York, New Jersey, Massachusetts, and Ontario[54]. *Cypridopsis vidua* is the most common and most widely distributed freshwater ostracod. It occurs in Europe and northern Asia as well as in North and South America, has broad habitat plasticity, and has been recorded from permanent waters and a variety of types of vernal pools. When it occurs in temporary waters, it seems to be restricted to long-cycle pools. The related *Sarcypridopsis aculeata* is known from Ontario pools. *Cypris crassa* occurs in grassy swamp pools from Ontario to Virginia. *Cypris subglobosa* is another widely distributed species common in temporary waters in Europe, Asia, and California as well as the glaciated northeast. *Cypria ophthalmica* is distributed widely in mixed woods and boreal forests in southern Canada and the northeastern United States, inhabiting waters with much decaying material. This species, unlike many ostracods, is tolerant of acidic water. *Cypria palustera* is another species that occurs in temporary pools, as well as in marshes and ponds from the southeast interior plains of Canada to Massachusetts[55].

Life History

Ostracods tend to be inactive at cold temperatures and, therefore, most species do not appear in vernal pools until water temperatures have warmed. Many species reproduce continuously until the pool dries, but some have only one generation per year. Ostracods generally make use of either desiccation-resistant eggs or encystment to withstand drying, and the eggs can persist in the sediment as an egg bank for decades. A few species have been reported to survive by burrowing into moist sediment. They survive the winter in a state of torpor, ready to resume development and prepare to reproduce when the pool refloods.

Isopods, Amphipods, and Crayfish (Malacostraca, Orders Isopoda, Amphipoda, and Decapoda)

Isopods and amphipods are small, shrimp-like crustaceans about 1 to 2 cm (0.5 to 1 in) long (Plate 8, bottom; Figure 20). Isopods are grayish, with flat bodies, distinct segments, and long antennae. They are usually found crawling along the bottom in decaying leaves and other detritus. Amphipods are pink, gray, or pale green, with laterally compressed bodies. They swim rapidly in quick bursts, and are commonly associated with aquatic vegetation.

Both isopods and amphipods are less well able to withstand pool drying than other vernal pool crustaceans[56]. No members of these orders are typical of vernal pools. They are often found in vernal pools that receive inflow from permanent waters or vegetated wetlands, or that retain saturated sediments or an area of

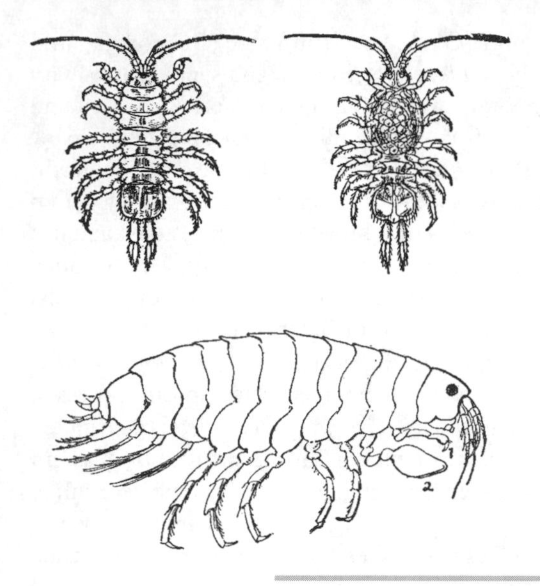

Figure 20. Malacostracan crustaceans including isopods (top) and amphipods (bottom) can be abundant in semi-permanent pools, but they do not appear to be able to survive in pools that dry completely. (From Morgan 1930, courtesy of the Morgan estate.)

continuous flooding. The frequency of occurrence of isopods and amphipods increases with hydroperiod. Studies of isopods and amphipods in second-growth and old-growth forests in New York found individuals in short-cycle, fall-filling pools in the old-growth forests, but not in areas that had been cut historically[57]. In Massachusetts, amphipods and isopods occur more frequently than expected by chance in vernal pools in areas that had not been cleared[58]. These results suggest that historic land use, specifically forest cutting, may help explain the absence of these crustaceans from most annual vernal pools.

Similarly, crayfishes (*Cambarus fodiens, Orconectes immunis*) occasionally occur in vernal pools but they, too, require continuously saturated sediments or burrows that extend below the water table during a pool's dry phase[59].

Co-Occurrences of Crustaceans in Vernal Pools

The fauna of most vernal pools that have been studied comprehensively includes a number of crustaceans in most or all of the groups discussed above. As noted in Chapter 7, ecologists are greatly interested in how animals that are closely related to one another, exhibit similar feeding behaviors, or have comparable life history strategies, manage to coexist. Because many crustaceans in vernal pools are filter-feeders or depend on decomposing detritus for their food, the occurrence of several species in the same vernal pool raises interesting questions about possible competition. How is it that vernal pools support such enormous numbers of crustaceans, and such a variety of species with similiar life styles?

The occurrence of several species in the same genus in a single vernal pool is common in the Copepoda and Ostracoda[60]. Their peaks of abundance and life cycles are often offset temporally, minimizing potential competitive interactions.

It is not unusual for several genera of branchiopods to co-occur in vernal pools, and sometimes more than one species in a genus is present. Co-occurring populations reported from vernal pools include: *Eubranchipus vernalis*, *Limnadia lenticularis*, and *Eulimnadia agassizi* in Westfield, Massachusetts; *Streptocephalus seali* and *Eubranchipus holmani* in New Jersey; *Lynceus brachyurus* and *Eubranchipus neglectus* in Ohio; *L. brachyurus, Daphnia pulex, Eubranchipus intricatus* and *E. vernalis* in Massachusetts; and *L. brachyurus, D. pulex, Eubranchipus bundyi*, an unidentified chydorid, and *Simocephalus vetulus* in Michigan. In Ontario, *D. pulex, E. bundyi, S. vetulus* (?), *Simocephalus exspinosus* (?), and *Scapholeberis kingi* occurred together in a short-cycle, spring-filling, woodland vernal pool over three years of sampling. In Michigan pools, *Daphnia ephemeralis* appeared early in the season and was replaced later by *D. pulex*, with some overlap[61].

Although fairy shrimp and clam shrimp occur in the same vernal pools, the fairy shrimp tend to hatch earlier and to complete their development before the peak in clam shrimp populations. Mature clam shrimp remain in vernal pools after fairy shrimp have disappeared and appear to be less sensitive to predation[62]. Cladocerans such as *D. pulex* also remain in pools well into the summer, declining only as the pool dries.

In most vernal pools where fairy shrimp have been studied, only a single species has been reported, but there are records of pools containing two, three, or four species. Co-occurring species of *Eubranchipus* include *E. neglectus* and *E. bundyi* in Ohio and Michigan; *E. neglectus* and *E. holmani* in Ohio; *E. vernalis* and *E. holmani* on Long Island, New York; *E. serratus* and *E. neglectus* in Illinois; *E. serratus* and *E. bundyi* in Illinois; and *E. vernalis* and *E. intricatus* in Massachusetts[63]. Usually one species is dominant and the other is present in relatively low numbers, commonly reflecting the broader distribution patterns. Thus, in Ohio, the northern species *E. bundyi* rarely occurs by itself but is often found in low numbers in pools with large populations of *E. neglectus*. The opposite is seen in Michigan, where *E. bundyi* is common and *E. neglectus* is comparatively rare[64].

Factors regulating co-occurring species of fairy shrimp have not been assessed in northeastern vernal pools. A fourteen-year study of a temporary pool in aspen parkland in Alberta shows that *E. intricatus, E. serratus*, and *E. bundyi* occur together[65]. In some years, one of these is absent but is replaced by one or two species of *Branchinecta*. Which species are present, and which ones are most abundant in a given year, seem to depend on the precipitation the previous summer. I have preliminary data on *Eubranchipus vernalis* and *E. intricatus* in Massachusetts suggesting that the peaks in abundance and reproductive activity in these species are offset temporally. *E. intricatus* is not found every year and appears to hatch in high-water years following dry summers; when it does occur, mature adults tend to be found later than those of *E. vernalis* in the same pools.

In two temporary ponds in Alabama, *E. neglectus* and *E. holmani* coexist in a short-cycle, winter-flooded pool fed by agricultural runoff, but *E. neglectus* occurs alone in a nearby, short-cycle, winter-flooded woodland pool fed by groundwater. Where the two co-occur, densities of *E. holmani* are thirty times greater than those of *E. neglectus*, and overall anostracan densities are nearly three times those in the pool where *E. neglectus* occurs alone. The life cycles of these species are not offset, but differences in feeding behavior appear to allow them to coexist[66]. In contrast, in Louisiana pools where *Streptocephalus seali* and *Eubranchipus moorei* both occur, the cues that stimulate hatching differ in the two species, with the result that they are not present in the pools at the same time[67]. Potentially, similar mechanisms regulate co-occuring fairy shrimp in northeastern vernal pools.

Some Remaining Questions About Crustaceans

- What species of large branchiopods, copepods, cladocerans, and ostracods occur in pools, and where?
- What are the habitat characteristics associated with the presence of different species of crustaceans in vernal pools?
- How do physical habitat disturbance, hydrologic variation, thermal changes, and variations in water quality affect habitat quality and reproductive success in crustaceans of vernal pools?
- Do pools that support rare species have any unique characteristics?
- What are the stimuli for hatching in resting eggs?
- How long do the eggs of crustaceans in temperate regions remain viable in the sediment?
- How much genetic segregation is there among sub-populations of crustaceans in vernal pools within clusters, within a geographic area, and overall?
- How do closely related species coexist in vernal pools?
- What actions are needed to ensure the conservation of vernal pool crustaceans?

References for Chapter 8

[1] A user-friendly key to macrozooplankton of the glaciated northeast, with good photographs of cladocerans and other small crustaceans, was recently produced by the University of New Hampshire: Aliberti et al. 2003. Excellent photographs of cladocerans can also be found in Rowe and Hebert 1999. Other photographs, illustrations, and keys to genera and species of freshwater crustaceans can be found in Eddy et al. 1982; Edmonson 1959; Pennak 1978, 1989; Smith 1992, 2001; and Thorp and Covich 1991. [2] Belk et al. 1998. [3] Hartland-Rowe 1965. [4] Belk et al. 1998, Moore 1959. [5] Biernbaum 1989, Dexter 1953. [6] Eddy et al. 1982, Pennak 1978. [7] Packard 1883. [8] Biernbaum 1989, Cole 1959. [9] Belk

et al. 1998. [10] Weaver 1943. [11] Broch 1965, Zinn and Dexter 1962. [12] Belk 1998, Belk and Cole 1975, Hildrew 1985. [13] Daborn 1976b, 1977. [14] Dexter 1959. [15] Belk 1984. [16] Dodson, unpublished, cited in Dodson and Frey 1991. [17] Daborn 1977. [18] Modlin 1982. [19] Dodson and Frey 1991. [20] Dexter 1956. [21] Ferguson 1939. [22] Dexter 1946. [23] Belk and Cole 1975. [24] Dexter 1946. [25] Dexter and Kuehnle 1951. [26] Donald 1983. [27] Dexter and Kuehnle 1948. [28] Modlin 1982. [29] Pennak 1978. [30] Zinn and Dexter 1962. [31] Dodson and Frey 1991. [32] Belk 1972, Chengaloth 1987, Smith 1995, Smith and Gola 2001, Weeks 2002b, Weeks and Marcus 1997. [33] Zinn and Dexter 1962. [34] Kenk 1949, Mattox 1939, Williams 1983. [35] Dexter and Kuhnle 1951. [36] Pennak 1978, Weeks and Marcus 1997. [37] Crosetti and Margaritora 1987, Lake et al. 1989. [38] Rowe and Hebert 1999, Schwartz and Hebert 1987. [39] Kenk 1949, Wiggins et al. 1980, Williams 1983. [40] Wiggins et al. 1980. [41] Kenk 1949, Higgins and Merritt 2001, Innes 1997, Rowe and Hebert 1999, Schwartz and Hebert 1987. [42] See, for example, Aliberti et al. 2003; Pennak 1978, 1989; Smith 2001; Williamson and Reid 2001; Wilson 1959. [43] Cole 1959, Medland and Taylor 2001, Pennak 1978, Yeatman 1959. [44] Ebert and Balko 1987, Lake et al. 1989, Mastrantuono 1994. [45] Marten 1999; Pennak 1978, 1989. [46] Kenk 1949. [47] Marten 1999; Pennak 1978, 1989. [48] Brewer 1964, Cole 1959, Taylor et al. 1990, Wilson 1959. [49] Medland and Taylor 2001. [50] Jokinen and Morrison 1998. [51] Delorme 1970b, Hoff 1943, Pennak 1978. [52] Jokinen and Morrison 1998. [53] Ebert and Balko 1987, Keeler-Wolf et al. 2004. [54] Tressler 1959. [55] Delorme 1970a, 1970b; Ferguson 1944; Keeler-Wolf et al. 2004; Tressler 1959. [56] Wiggins et al. 1980. [57] Batzer and Sion 1999. [58] Musgrove and Colburn 2003. [59] Kenk 1949, Wiggins et al. 1980. [60] Hoff 1942; Kenk 1949; Maier 1989, 1992; Wiggins et al. 1980. [61] Colburn unpublished data, Dexter and Kuhnle 1951, Dodson and Frey 1991, Higgins and Merritt 2001, Kenk 1949, Packard 1883, Smith 1995, Wiggins et al. 1980. [62] Wiggins et al. 1980. [63] Colburn unpublished data; Dexter 1953, 1956; Dexter and Kuehnle 1948, 1951; Packard 1883. [64] Dexter and Kuehnle 1948. [65] Donald 1983. [66] Modlin 1982. [67] Brtek 1967, Moore 1963.

Chapter 9

Insects

Of the millions of species of insects in the world, only a small percentage lives in water, and relatively few of these occur in vernal pools. Even so, aquatic insects represent one of the largest groups of animals in vernal pools in terms of both the numbers of individuals and the numbers of species (Table 1, Appendix). Caddisflies, water beetles, water bugs, damselflies, dragonflies, mosquitoes, phantom midges, crane flies, midges, horseflies, fishflies, mayflies, and other kinds of aquatic insects can be among the most abundant animals living in vernal pools, and they have important roles to play in the pool economy. Some insects transform detritus into food for other pool animals. Others serve as prey for salamanders and other carnivores. Still others are predators themselves and can affect the reproductive success of other invertebrates and of amphibians in vernal pools. Some useful guides to the identification of aquatic insects are listed at the end of the chapter[1].

Because caddisflies and water beetles tend to be the most visible and most commonly observed insects in vernal pools, they are discussed first, followed by accounts of dragonflies and damselflies, water bugs, true flies, and other insects.

Caddisflies (Trichoptera)

Caddisflies are closely related to butterflies and moths. Their caterpillar-like larvae, most of which have distinctive, portable, tubular cases, are a visible and important component of the fauna of northeastern vernal pools (Plate 10, figures 21 to 23). Several species are characteristic of short-cycle, spring-filling pools (see Appendix). Caddisflies evolved in coldwater streams, and most species do not tolerate high temperatures or low levels of dissolved oxygen. The development of the tubular case allowed caddisflies to invade non-flowing waters with lower dissolved oxygen levels, because the larvae can create a water current by undulating their bodies within the case, thereby increasing the amount of oxygenated water available[2]. In some species, particularly in the family Limnephilidae, life history adaptations have further contributed to dispersal into seasonally unsuitable habitats including vernal pools.

Figure 21. Large "log-cabin" caddisflies, especially *Limnephilus indivisus,* are readily observed in vernal pools in the spring. (Photograph by Betsy Colburn.)

Three families of caddisflies, Limnephilidae, Phryganeidae, and Polycentropodidae, contain representatives that are common in vernal pools. Members of a fourth family, Leptoceridae, are often found in pools with long hydroperiods.

The caddisflies in the family Limnephilidae include some species whose cases are constructed of vegetation laid in alternating perpendicular courses, as in a log cabin; hence the common name, "log-cabin caddisflies" (Plate 10, Figure 21). One of these, *Limnephilus indivisus*, appears to be preferentially distributed in vernal pools. It can be found throughout the glaciated northeast and is a good indicator species. This caddisfly's distinctive case and the color pattern on the hard, sclerotized plates on its head and thorax make it easy to identify. Its presence can be confirmed even when only empty cases are present, through examination of the larval sclerites left behind in the case. The closely related *L. submonilifer* has a similar pattern on the head and thorax, but it constructs a tubular case from pieces of vegetation laid longitudinally side by side. Other limnephilids incorporate bark, moss, large segments of leaves, snail shells, or other materials into a variety of symmetrical tubes and asymmetrical cases.

The flexibility of many larval limnephilids in their use of case-building materials has been exploited commercially. Some creative jewelers supply caddisfly larvae with colored materials for use in building their cases. The decorative cases are then fashioned into earrings, necklaces, tie clips, and other adornments for naturalists, fly fishermen, and other people interested in jewelry with a "natural history" twist.

The giant-tube-case-makers, family Phryganeidae, are highly visible in some northeastern vernal pools. They have yellow heads brightly striped with black and their cases are constructed from neatly cut pieces of grass or leaves, arranged in cylinders or spirals (Plate 10, Figure 22). Their thin-walled, tubular cases incorporate a variety of carefully cut pieces of vegetation and are often brightly colored. The larvae often select fragments of contrasting colors when building their cases, and the cases can be very pretty, presenting a patterned patchwork appearance. In some species the cases may be 7 to 8 cm (3 or more in) long in the last instar.

Free-living retreat makers (Polycentropodidae) (Figure 23) do not make portable cases, but they construct silken retreats from which they foray in search of prey.

The richness of the caddisfly fauna in vernal pools increases with increasing hydroperiod. Swimming caddisflies (Leptoceridae) are often found in long-cycle and semi-permanent pools, particularly those with emergent vegetation. These caddisflies have long legs and use them to swim swiftly through the water, case and all (Figure 23). Especially when small they can resemble small twigs zipping through the water column.

Caddisflies play important roles within the energy cycle of vernal pools. Many species shred decaying leaves and other detritus into smaller pieces, thereby increasing the surface area of detritus that is available for bacterial and fungal decomposition. Other caddisflies act as predators. *Limnephilus indivisus*, one of the most common species in northeastern vernal pools, preys on early spring *Aedes* mosquito larvae in woodland swamps in Ontario[3], and feeding by members of the family Phryganeidae can produce significant mortality in amphibian egg masses[4]. Glenn B. Wiggins of the University of Toronto has published a summary of the biology of caddisflies in temporary pools, and he and his colleagues have reviewed the life history strategies of vernal pool Trichoptera[5].

Distribution

In general, caddisflies that are common in vernal pools are widely distributed and are not restricted to this habitat. Species in the family Limnephilidae include *Anabolia bimaculata*, *Limnephilus indivisus*, *L. submonilifer*, *L. sericeus*, *Asynarchus batchawanus*, and *Ironoquia punctatissima*, all of which are widespread in southern Canada and the northern United States, and *I. parvula*

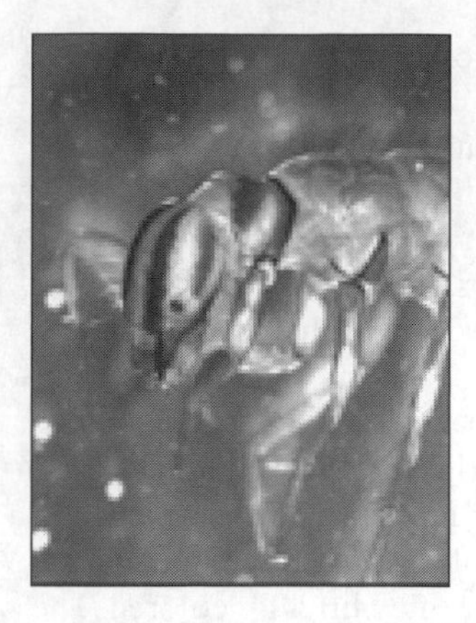

Figure 22. The heads of phryganeid caddisflies are distinctively striped. (Photograph by Holly Jensen-Herrin.)

and *Anabolia sordida*, found at lower elevations and slightly farther south throughout the glaciated northeastern United States[6]. Common in longer-hydroperiod pools are the limnephilid *Nemotaulius hostilis*, found from Canada south to New England and Michigan, and the phryganeids *Banksiola dossuaria* and *B. crotchii*, both of which occur throughout the glaciated northeast[7].

Several species of the family Phryganeidae, including *Ptilostomis ocellifera* and *P. postica*, are also widely distributed through the glaciated northeast in small streams, marshes, and vernal pools[8].

Plectrocnemia crassicornis (formerly *Polycentropus*) is common in vernal pools in Ontario and Massachusetts, and two other species, *P. aureola* and *Holocentropus flavus*, have been collected in short-cycle vernal pools in Ontario[9].

Within the family Leptoceridae, several species of *Oecetis* found in vernal pools, including *O. cinerascens*, *O. inconspicua*, and *O. nocturna*, are widely distributed throughout the glaciated northeast, and *O. ochracea* is found in the western part of the area, with records from Minnesota, Wisconsin, Ohio, and Ontario. *Triaenodes nox* and *T. aba* are also common throughout the glaciated northeast, with *T. nox* particularly common in small, vegetation-rich ponds[10]. None of the leptocerids is specialized for, or typical of, vernal pools, and they probably do not complete development successfully in pools that become dry.

Life History

Adult caddisflies look much like gray or brown moths with hairy, rather than scaly, wings. The life cycles are similar to those of moths and butterflies, although in caddisflies the larvae, and usually the pupae, are aquatic. The larvae hatch from eggs, go through five larval instars, and pupate. Caddisflies produce silk from glands in the head, and they use this to glue materials together in constructing their cases, or to build fixed retreats and nets. In species that construct tubular

Figure 23. Swimming caddisflies (Leptoceridae) such as the one pictured here (left) look like small twigs zipping through the water of long-hydroperiod pools. Polycentropodids like the one shown here (right) are free-living caddisflies that venture out of their silken retreats to seek food. (Photographs by Holly Jensen-Herrin.)

cases, the larvae generally seal the ends of the cases before pupation. Free-living larvae seal themselves in a cocoon-like structure before pupating. The adults emerge from the pupal case and mate, the females deposit eggs, and the cycle is repeated.

Depending on how far adults disperse from vernal pools upon emergence, caddisflies can be classified as migratory breeders that move to pools in fall, or as permanent residents with one or more desiccation-resistant life stages.

In vernal pool caddisflies, two different types of adaptations are seen, both of which contribute to survival in a habitat that is seasonally dry. First, eggs aestivate in the dry pool depression and resist desiccation until the pool fills. Second, in many species, a period of diapause in the adult, larval, or pupal stage allows avoidance of the summer dry period.

Family Limnephilidae

Limnephilids are widely distributed in wetlands and ponds. One basic strategy for avoiding summer drawdown occurs in members of the subfamily Limnephilinae, including the common genera *Limnephilus* and *Anabolia*[11]. In these taxa, both the egg and adult life stages are modified to withstand seasonal pool drying. The eggs, unlike those of most caddisflies, are deposited on moist wood and other debris on the dry pool bottom in fall and are able to resist desiccation. First-instar larvae develop within the egg mass, but they do not emerge from the mass until the pool is inundated. In fall-filling pools, the larvae can sometimes be observed in winter as they move along the pool bottom and under the ice. In spring-filling pools, they remain in the egg mass through the winter. The larvae grow rapidly through the spring and pupation is complete by late spring or early summer. Larvae are ready to emerge at the same time whether in fall- or spring-filling pools. The second life-stage modification is seen in adults, which are reproductively immature upon emergence and spend the summer in a period of reproductive diapause, aestivating in tree holes, caves, and other cool, humid locations. The gonads mature in late summer or early fall, and the adults fly to the dry vernal pool depressions to lay their eggs. This life history pattern is common not only in limnephilines of vernal pools, but also in members of this subfamily from marshes, desert waters, and mountain streams with seasonally warm summer temperatures[12].

The timing of pupation and adult emergence varies among species exhibiting this strategy. In *Limnephilus submonilifer*, pupation is usually complete by early May; in *L. indivisus* and *L. sericeus*, adults emerge in June and July; and in *Anabolia sordida*, a slightly longer hydroperiod appears to be required, with pupation typically occurring in June and July[13]. Pool hydroperiod thus limits the distribution of some vernal pool caddisflies, with more species able to complete their life cycles in pools that retain water later into the spring and early summer.

The life histories of *Asynarchus batchawana* and *Lenarchus crassus* have not been studied in northeastern vernal pools, but based on existing information they are probably similar to *Limnephilus* and *Anabolia*. The larvae of *A. batchawana* are present in vernal pools in June, and adults emerge in July[14].

A different strategy for withstanding summer drawdown is seen in *Ironoquia parvula* and *I. punctatissima*, members of the limnephilid subfamily Dicosmoecinae. As in other limnephilids, larvae hatch from egg masses upon flooding. They grow rapidly, reaching the fourth instar by March. The larvae leave the pools by May, crawl under logs or rotting leaves near the pool edge, and aestivate in unsealed cases through the summer, sealing the cases and starting to pupate in late August. Adults emerge in September and October, mating and leaving eggs in the damp pool depression[15].

Family Phryganeidae

In the most common phryganeid species, *Ptilostomis ocellifera*, females typically lay their eggs on vegetation overhanging water bodies, although Wiggins and his colleagues also collected egg masses from the bottom of a dry pool depression in the fall. It is probable that the life history adaptations of vernal pool phryganeids are similar to those of limnephiline limnephilids, with reproductive diapause in adults and aestivation in the eggs. The eggs can resist desiccation for up to six months, and they hatch when flooded[16].

Family Polycentropididae

In members of the Polycentropodidae in vernal pools, adverse effects of drawdown are avoided through desiccation-resistant eggs, and no other stages experience diapause. Several species of polycentropodids occur in vernal pools, including *Plectrocnemius crassicornis*, *P. aureolus,* and *Holocentropus flavus* (see Appendix). Adults are reproductively mature upon emergence in June and July in Ontario. Females apparently deposit their eggs in the water, and the eggs remain in a state of diapause until the following winter or spring, when the eggs hatch upon flooding. The cues that maintain and terminate diapause are unknown. It may be that, as in some vernal pool crustaceans, the eggs require a period of drying. Alternately, photoperiod might be the controlling factor[17].

Family Leptoceridae

Leptocerids do not appear to be able to survive drawdown. Larvae sometimes occur in short- and long-cycle pools, but they need to recolonize each year from other, permanently flooded water bodies.

Water Beetles (Coleoptera)

Water beetles comprise a highly diverse and important component of the fauna of vernal pools (see Appendix). In a single pool, the sweep of a net is likely to yield adult and larval beetles from several families, including predaceous diving beetles (Dytiscidae), water scavenger beetles (Hydrophilidae), crawling water beetles (Haliplidae), and whirligig beetles (Gyrinidae) (figures 24 to 28). Other families that are often represented include burrowing water beetles (Noteridae), long-toed water beetles (Dryopidae), marsh beetles (Scirtidae), minute bog beetles (Sphaeriidae), snout beetles (Curculionidae), rove beetles (Staphylinidae), straight-snouted beetles (Brentidae), ground beetles (Carabidae), and leaf beetles (Chrysomelidae).

Adult beetles are distinguished by a hardened outer pair of wings, known as elytra, that cover the membranous inner pair used for flight. Depending on the species, beetles vary from tiny individuals of only 1 mm (less than 0.04 in) to giant beetles more than 5 cm (over 2 in) in length. Identification to family is quite easy with a bit of practice, and some species are distinctively colored and marked.

Beetle larvae have hard heads with elongate antennae, and one or two claws on each leg. Many predatory species have sharp, sickle-like mouthparts that have earned their possessors the nickname "water tigers."

Predaceous diving beetles are generally smoothly oval in shape (Figure 25). They have long, linear antennae and fringed hind swimming legs that they move synchronously like a human swimmer doing the frog kick. They carry a bubble of air under their elytra and are often seen as they come to the water's surface to replenish the bubble. Many adult water scavenger beetles look superficially like diving beetles (Figure 26), but they can be told apart by watching how they swim and examining their antennae. They move their swimming legs alternately, as in a scissor kick. Their antennae have a number of segments that are thickened, cupped, scalloped, or otherwise divergent from a simple linear series, and the antennae are

Figure 24. All of these water beetles were found in a small vernal pool on Cape Cod, Massachusetts. (Photograph by Betsy Colburn.)

usually curled around the back of the eyes and are often hard to see. In many water scavenger beetles, the elongate maxillary palps, which are part of the feeding apparatus, project forward and can be mistaken for antennae, and in species whose body shape resembles that of diving beetles this sometimes results in misidentification. Crawling water beetles are often golden with black markings. They are about 0.5 cm (0.2 in) long with long legs and a pair of plates covering the lower half of the underside of the abdomen (Plate 1, Figure 27). They are often visible under ice in fall-filling pools, crawling over leaves in shallow water near the shore. Whirligig beetles school on the water's surface, spinning around in search of prey. They have dual sets of eyes, one pair to see into the air and the other on the underside of the head to see into the water (Figure 28).

The importance of water beetles is seen both in the numbers of species and in the numbers of individuals present in vernal pools. A study in southeastern Pennsylvania found that in fishless ponds, the biomass of water beetles was three times greater than in ponds containing fish[18]. Beetles fill a variety of ecological roles in vernal pools. Adult and larval dytiscids and gyrinids, and larval hydrophilids, can be significant predators on amphibian larvae, crustaceans, and other invertebrates (figures 25, 26, 28). Haliplids and adult hydrophilids feed on algae and decaying organic materials.

Certain aquatic beetles are closely associated with vernal pools (Table 1, Appendix). *Liodessus fuscatus* only occurs in woodland pools, and *Uvarus suburbanus* and *Hygrotus sylvanus* were first identified from woodland pools. Other species typically found in woodland vernal pools include *Agabetes acuductus*, common in the leaf litter on the bottom of highly shaded pools in deciduous forests; *Hygrotus laccophilinus* and its close relatives *H. nubilis* and *H. turbidus*; *Ilybius ignarus*, a species particularly common in acidic waters; *Laccornis difformis*, which prefers rotted leaves in the pools; *Matus bicarinatus*, which also inhabits moist and mossy places in woods; and *Stictotarsus griseostriatus*[19]. The large genus *Hydroporus* has numerous representatives in vernal pools, including *H. despectus*, *H. fuscipennis*, *H. notabilis*, *H. tenebrosus*, and *H. undulatus*. Species of *Hygrotus* and *Hydroporus* are common in temporary

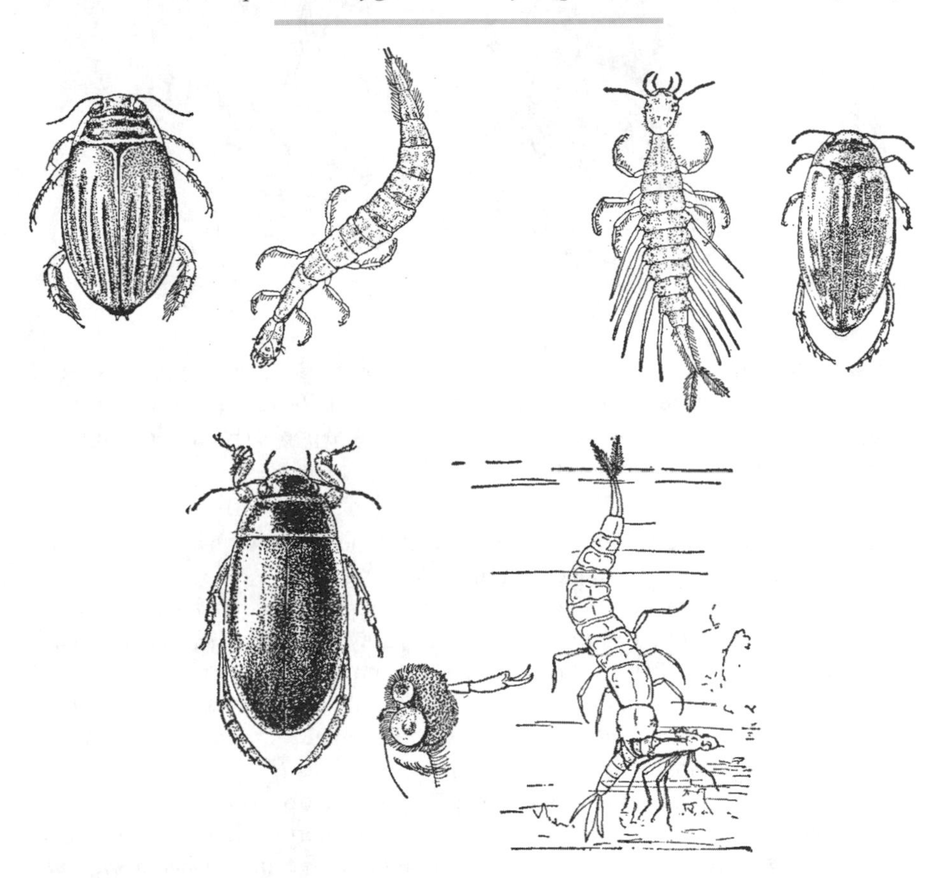

Figure 25. Adult and larval dytiscids, or diving beetles, including species of *Acilius* (upper left), *Coptotomus* (upper right), and *Dytiscus* (bottom), are important predators in vernal pools. (From Morgan 1930, courtesy of the Morgan estate.)

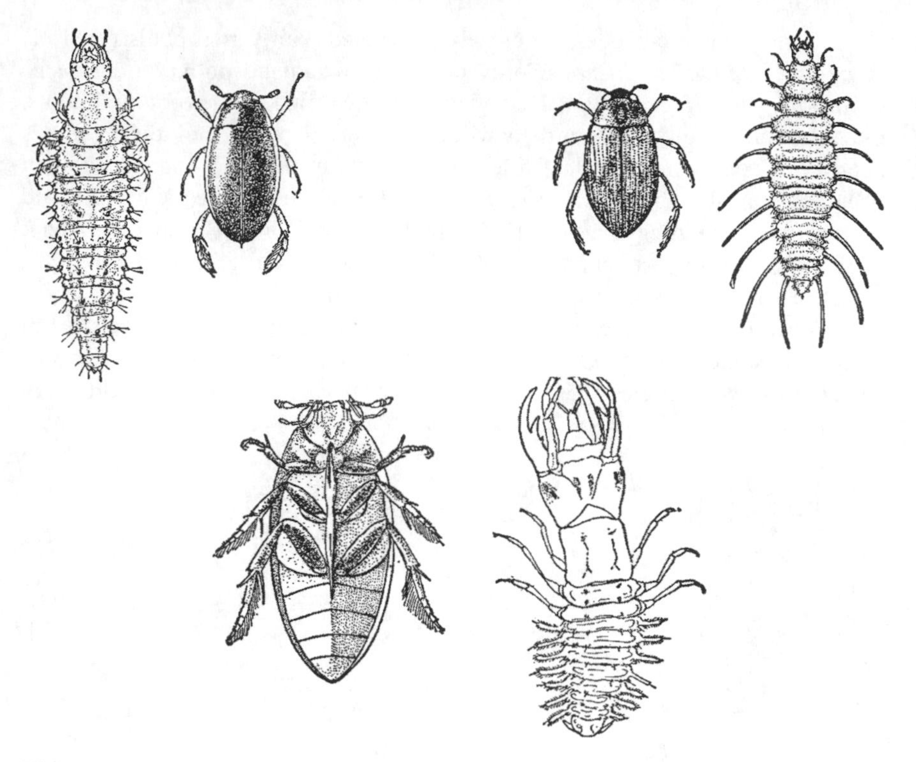

Figure 26. Water scavenger beetles, family Hydrophilidae, including species of *Tropisternus* (upper left), *Berosus* (upper right), and *Hydrophilus* (ventral view, bottom), are among the most common beetles in vernal pools. (From Morgan 1930, courtesy of the Morgan estate.)

pools in Europe, as well as in North America, and the distributions of the different species are associated with the duration of flooding[20].

In small, short-cycle vernal pools with high populations of *Aedes* mosquitoes, snails, and limnephilid caddisflies, water beetles may reach densities of 2.4 animals per square meter (0.22/ft^2). The breeding beetle community in such pools commonly includes *Acilius semisulcatus*, *Agabus erichsoni*, *Agabus falli*, *Colymbetes sculptilis*, *Dytiscus fasciventris*, *Hydroporus tenebrosus*, *Rhantus binotatus,* and *Hydrochara obtusatus*[21]. Some of these beetles, such as *H. obtusatus*, are rarely found in semi-permanent pools and appear to prefer short-cycle pools.

Other species of aquatic Coleoptera found in short-cycle woodland pools, although not necessarily breeding in them, include the diving beetle *Agabus phaeopterus*, water scavenger beetles *Hydrobius fuscipes* and *Tropisternus mixtus*, and whirligig beetles including *Gyrinus affinis* and *G. lecontei*. Often present, although in lower numbers, are *Agabus anthracinus*, *A. bifarius*, *Hydroporus*

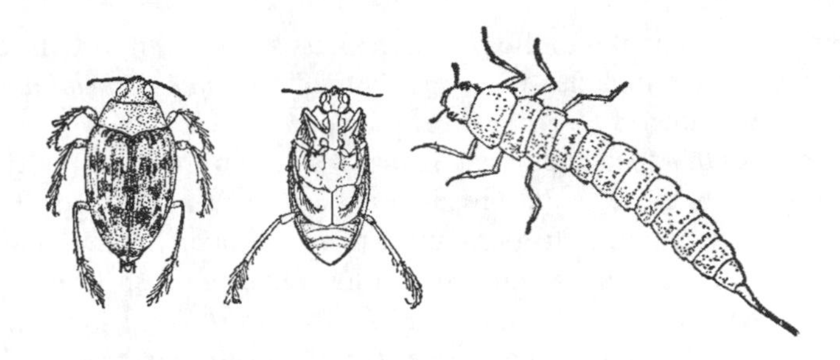

Figure 27. Crawling water beetles, family Haliplidae, are readily identified by their long legs and the large plates covering the underside of their abdomens. This image shows (left to right) the adult in dorsal and ventral views and the larva in dorsal view. (From Morgan 1930, courtesy of the Morgan estate.)

niger, *Hygrotus impressopunctatus*, *Ilybius biguttulus*, and *Tropisternus natator*. *Haliplus blanchardi*, *H. immaculicollis*, *H. longulus*, *H. ohioensis*, *Peltodytes edentulus*, and *P. muticus* are among the crawling water beetles commonly encountered in vernal pools[22].

The beetle community in semi-permanent vernal pools in southern Ontario contains many of the same species as annual vernal pools but is more diverse. Common species include *Acilius semisulcatus*, *Agabus erichsoni*, *Agabus phaeopterus*, *Agabus* spp. adults and larvae, *Colymbetes* sp. larvae, *Dytiscus* sp. larvae, *Gyrinus lecontei*, *Ilybius discedens*, and *Rhantus* sp. larvae. Numbers of *Acilius* are approximately the same as in short-cycle vernal pools. In general, there are more species and more predators in semi-permanent pools than in annually drying pools [23].

Vernal pools in kettle depressions on the outer arm of Cape Cod contain a rich beetle fauna, with species richness increasing with increasing hydroperiod[24].

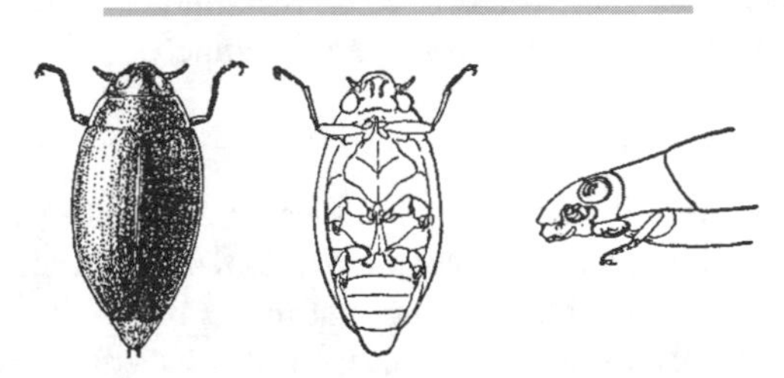

Figure 28. Whirligig beetles, family Gyrinidae, school on the surface of pools and use their divided eyes to seek prey both under and above the surface film. This image shows (left to right) the adult in dorsal and ventral views and the head in lateral view. (From Morgan 1930, courtesy of the Morgan estate.)

In addition to a variety of haliplids, hydrophilids, and small dytiscids, the beetle fauna of these ponds includes the large dysticid *Cybister fimbriolatus* and the strikingly colored noterid *Hydrocanthus iricolor*.

The beetles that feed and breed in annual and semi-permanent vernal pools are not the same as those found breeding in permanent ponds. However, many vernal pool beetles migrate to permanent ponds to overwinter. These include *Acilius semisulcatus*, *Colymbetes sculptilis*, *Dytiscus fasciventris*, *Agabus phaeopterus*, *Hydrochara obtusata,* and *Hydroporus niger*. Permanent pond species not usually found in vernal pools include *Acilius mediatus*, *Agabus ambiguus*, *Coptotomus interrogatus*, *Dytiscus circumcinctus*, *D. harrisii*, *Graphoderus fascicollis*, *Ilybius confusus*, *Rhantus tostus*, and *Tropisternus lateralis nimbatus*[25]. It is evident from these lists that closely related species in the same genus have very different habitat preferences. Often, bioassessments for regulatory purposes identify insects only to family or genus. Such assessments will fail to recognize the unique contributions of annually drying and semi-permanent pools to biodiversity.

In addition to water beetles that are characteristically found in vernal pools, several kinds of beetles show a high degree of plasticity and breed in a variety of habitats. Although these often occur in vernal pools they can be found as well in other sites. *Agabus semipunctatus* is one such widely distributed species. It can be found in both woodland pools and pasture pools. *Dytiscus verticalis* occurs in both temporary and permanent ponds in fields and woods[26].

Several species of *Laccophilus* regularly appear in vernal pools but also occur in permanent waters. *Laccophilus maculosus maculosus* is concentrated in the glaciated northeastern forest region and is found in a wide variety of habitats, although it prefers "shallow, partially shaded ponds that tend to have some water in them all year"[27]. *Laccophilus proximus* is a pioneer species found in many short-lived and recently formed habitats, including pools that develop in discarded tires. In Indiana it was the first of four *Laccophilus* species to reappear in four ponds that had previously dried. *Laccophilus biguttatus* colonizes a wide range of habitats and seems to be more likely to colonize temporary waters in the Canadian prairies than at the edges of its North American range.

Distributions

Water beetles found in vernal pools can be broadly divided into four geographic classes: those whose distributions extend throughout the glaciated northeast; those found north and east but not in the midwestern portions of the region; those found only in northern areas; and those with extremely restricted distributions. Sources on which the summaries below are based are listed at the end of this chapter[28]. Further information on the biology and distributions of vernal pool beetles can be found in these references.

Among the water beetles distributed in vernal pools throughout the glaciated northeast are: *Acilius semisulcatus*, *Agabus bifarius*, *Agabus erichsoni*, *Dytiscus fasciventris*, *Haliplus borealis*, *Haliplus immaculicollis*, *Helophorus grandis*, *Hydrobius fuscipes*, *Hydrochara obtusatus*, *Hygrotus impressopunctatus*, *Hygrotus nubilis*, *Hygrotus laccophilinus*, *Hygrotus turbidus*, *Ilybius biguttulus*, *Laccophilus undatus*, *Laccophilus maculosus maculosus*, *Liodessus fuscatus*, *Peltodytes edentulus*, *Rhantus binotatus*, and *Tropisternus natator*.

Species known from the northern and eastern parts of the region, but not from the midwestern United States, include the following. *Colymbetes sculptilis*, *Gyrinus affinis*, and *Haliplus longulus* occur across Canada and in New England. *Agabus anthracinus*, *Agabus falli*, *Gyrinus lecontei*, *Hydroporus notabilis*, *Hydroporus tenebrosus*, and *Hydroporus undulatus* are found in eastern Canada and New England. *Hydroporus despectus* is known from the Canadian Maritimes and Massachusetts, and *Hydroporus fuscipennis* and *Hydroporus niger* are known from across Canada, one or more New England states, and New York. *Laccornis difformis* is known from Massachusetts, Maine, and New York, and *Uvarus suburbanus* is known from woodland pools in New York and Ontario[29].

Species with northern distributions include *Stictotarsus griseostratus*, known from Canada, Michigan, and the environs of Lake Superior; and *Agabus phaeopterus*, from across Canada, Maine, and New Hampshire.

Hygrotus sylvanus was collected and described in 1917 from a vernal pool in Peekskill, New York. Attempts to collect additional specimens were unsuccessful because the pool had been destroyed, and for many years it appeared that this species had evolved locally in New York and had gone extinct through habitat loss. The species was not collected again and was believed to be extinct until 1979, when populations were reported from temporary snowmelt pools. Records of this rarely collected species are now known from Massachusetts, New York, Minnesota, Wisconsin, and southern Ontario and Quebec[30].

Life History

The life cycle of water beetles includes egg, larval, pupal, and adult stages. The eggs and larvae are aquatic, and the pupal stage is spent on land, usually in damp soil near the edge of the water. Most, but not all adult water beetles are able to fly, and some can travel over long distances, but others have limited dispersal abilities.

Aquatic Coleoptera in vernal pools include permanent residents, migratory breeders, and migratory non-breeders. They use several different strategies for adjusting to periodic drying of the habitat. Some species exhibit highly complex life cycles with several stages of dormancy.

Permanent Residents

Some permanent residents aestivate as adults in the damp or dry sediment of vernal pools, emerging as soon as the pools flood. Several species of *Hydroporus* use this strategy. Wiggins and his colleagues suggest that *Hydrobius fuscipes* and some species of *Haliplus*, *Anacaena*, and *Helophorus* also overwinter as adults in damp pool depressions. Some species of *Agabus* may also aestivate as adults. Adult *Agabus* have been collected from a dry Ontario pool, and *A. disintegratus* have been found to aestivate as adults in dry temporary pool basins in California, although this species migrates to permanent waters from vernal pools in Indiana[31].

Migrants With Simple Life Histories

Many migratory species of beetles move to vernal pools in spring from overwintering sites in permanent waters. These species mate and lay their eggs after reaching the pools, the eggs hatch quickly, and the life cycle is completed before the pool dries. *Tropisternus* lay eggs in a case constructed under leaves or stems of plants in water. The larvae go through several molts and then move onto damp soil above the waterline, where they pupate. The life cycle takes about two months to complete[32].

Migrants With Complex Life Histories

Some of the dytiscid beetles that breed preferentially in vernal pools have highly complex life cycles. For example, in some species of *Agabus* and *Colymbetes*, including *A. erichsoni* and *A. disintegratus*, diapausing eggs rest in the sediment and hatch when the pools flood. The larvae grow rapidly and pupate in late spring or early summer. After the adults emerge, they may feed in the vernal pools for a period of time, but they fly to permanent waters in summer or fall as the pools start to dry. After overwintering in permanent water, they return to their breeding sites late the following spring to mate and lay eggs. These eggs remain dormant on the pool bottom and will not hatch until the pools flood the next year. These species employ a combined life history strategy that incorporates both migration and eggs that hatch on flooding after a period of diapause. There are thus two overwintering periods, the first in the newly emerged adults, and the second in the spring-deposited eggs, and it takes two years for the life cycle to be completed[33].

In vernal pools in Ontario, the larvae of *Dytiscus fasciventris* appear in early spring before the adults arrive, and it appears that this species also experiences two stages of overwintering. The life cycles of *Rhantus* in vernal pools may be similar. Adult *Rhantus binotatus* emigrate from habitats when increasing water temperatures, unsuitable substrate, or high densities create unfavorable conditions[34].

Non-breeding Migrants

Some species of water beetles that are commonly found in vernal pools apparently do not breed in the pools, but they use these temporary waters as feeding areas.

Seasonal Succession in Water Beetle Communities

The composition of the beetle fauna in vernal pools changes over time. Water beetles do not appear in vernal pools at once, but rather they succeed one another in a predictable pattern. Permanent residents such as *Hydroporus* overwinter as adults in the dry pool basin and become active as soon as the pool floods. *Agabus* larvae, which are important mosquito predators, hatch in March and April from eggs laid the previous May and June on moss at edges of the pools, and their appearance coincides quite closely with the hatching of mosquito larvae. Adults of *Agabus* and *Colymbetes* migrate to pools in early spring, and arrive after the eggs deposited in the pool the previous year have hatched. Adult *Acilius*, *Hydrochara obtusata*, and *Gyrinus lecontei* arrive later in the spring, and their larvae appear when mosquitoes are in the pupal stage. *Dytiscus fasciventris* larvae appear in early May, hatching from eggs deposited the previous season, and the migratory adults do not show up until June. The beetle fauna continues to change as long as the pool retains water, with other migratory breeders and non-breeding migrants entering the pools to feed, with newly transformed adults of some species feeding for a time and then dispersing to other waters, and with permanent residents remaining in the pools continuously[35].

Dragonflies and Damselflies (Odonata)

Adult dragonflies and damselflies provide colorful and attractive additions to the pool surroundings as they patrol for insect prey, seek mates, and lay their eggs. Their aquatic nymphs are important predators. Damselfly nymphs are long and slender, with three plate-like gills that extend off the end of the abdomen. Dragonfly nymphs are stouter and heavier, and their abdomens end in three sharp points that surround a chamber into which the nymphs take water, from which they remove oxygen and minerals. Some nymphs can expel the water quickly so that it acts as a jet-propulsion mechanism that shoots the nymphs forward rapidly. Both damselfly and dragonfly nymphs have hinged, extendable mouthparts that let them capture rapidly moving insects, crustaceans, and other prey (figures 29 to 31). Characteristics of the mouthparts are also useful to naturalists in distinguishing the different species.

Worldwide, many odonates require fishless habitats for successful development into adults[36], and this is also true in the glaciated northeast. Although there are no odonates that are restricted to northeastern vernal pools, dragonflies and damselflies are abundant in these habitats and account for a substantial component of the insect species that are present. As with the caddisflies and water beetles, overall species richness increases with increasing hydroperiod, but several species are adapted for life in annually drying pools with short flooding duration.

Many odonates are highly opportunistic in their breeding behavior, with females ovipositing in a variety of habitats, including car hoods and mud puddles. ("Opportunistic" does not necessarily translate into "successful"!) The presence of nymphs of a given species in a vernal pool may reflect this behavior and does not necessarily indicate that the species completes its development successfully in the pool. In the discussion that follows, the emphasis is on dragonflies and damselflies that regularly use vernal pools and are usually able to complete development to the adult stage there.

Among the damselflies (Zygoptera), several representatives of the spread-wing damsels (Lestidae: *Lestes* spp.) and pond damselflies (Coenagrionidae: *Coenagrion resolutum*, *Enallagma* spp., *Ischnura posita*) are common in vernal pools. In the dragonflies (Anisoptera), species common in vernal pools include some members of the darners (Aeshnidae: *Anax junius*, *Epiaeshna heros*) and skimmers (Libellulidae: *Sympetrum* spp.). A pool's hydroperiod is an important determinant of which species are likely to be present, with spread-wing damselflies and some skimmers doing well in annual vernal pools including some short-cycle pools, and with pond damsels and darners more common in long-cycle and semi-permanent pools.

Predation by large dragonfly nymphs plays a major role in structuring amphibian populations in some habitats[37]. Semi-permanent pools support some species that are excluded from annual vernal pools and, in particular, they may harbor large nymphs of species that require several years to complete development. Predator-prey relationships in such waters may be very different from those in pools that dry up annually.

Figure 29. Predatory dragonfly and damselfly nymphs use their hinged mouthparts to capture aquatic insects, crustaceans, and amphibian larvae. (From Morgan 1930, courtesy of the Morgan estate.)

In general, odonates are intolerant of low temperatures during the very early embryonic stages and late nymphal development[38]. Survival in vernal pools requires life history strategies that allow avoidance of cold temperatures by these sensitive stages. Two basic strategies have developed: (1) oviposition in the pools in summer, with obligate egg diapause in late embryonic development, and (2) migratory breeding with early spring oviposition and rapid nymphal growth. A third strategy, employed in fall-filling vernal pools and semi-permanent pools, involves hatching of eggs upon flooding, rapid growth, and overwintering of the nymphal stage. The development of life history strategies and associated physiological tolerances that allow some odonates to withstand an extended period of habitat drying, has allowed representatives of *Lestes*, *Ischnura*, and *Sympetrum* to live in saline and brackish waters, as well as in temporary habitats[39]. Wiggins and his colleagues provide a comprehensive summary of life history strategies in *Lestes* and *Sympetrum* that inhabit spring-filling, short-cycle vernal pools[40].

Spreadwinged Damselflies (Lestidae)

Damselflies of the family Lestidae are known as spreadwings because the adults do not close their wings over their backs like other damselflies, but instead hold them partly open. Certain damselflies of the genus *Lestes* are common in northeastern vernal pools. The nymphs are long and slender, and the extendable lower section of their mouthparts (the mentum) has a long, narrow stalk that provides the nymphs with an extended reach in grasping prey (Figure 30, bottom right). These damselfly nymphs are very active, and they seek prey by climbing on the stems and leaves of submerged vegetation. Rapid growth rates and an obligatory egg diapause have allowed several members of this genus to colonize vernal pools[41].

Distributions

Four species of *Lestes* — the emerald spreadwing (*L. dryas*), the amber-winged spreadwing (*L. eurinus*), the sweetflag spreadwing (*L. forcipatus*), and the lyre-tipped spreadwing (*L. unguiculatus*) — are found preferentially in a variety of seasonally flooded habitats, including vernal pools[42]. Of these, the emerald spreadwing is the most typical vernal pool species. Two other lestids, the spotted spreadwing (*L. congener*) and the slender spreadwing (*L. rectangularis*), prefer permanent or semi-permanent ponds, although both are sometimes found at annual vernal pools. All six of these species occur throughout the glaciated northeast. However, the amber-winged and sweetflag spreadwings tend to vary in abundance. They appear in large numbers in some locations and are relatively uncommon elsewhere.

Of the lestids common in vernal pools, the lyre-tipped spreadwing is typically found associated with temporary waters in open areas, rather than in woodland settings. In contrast, adults of the slender spreadwing prefer heavily shaded areas and are often found near woodland vernal pools[43].

Life History

Adult spreadwing damselflies enter vernal pool basins in summer to lay eggs. They thus exhibit the "overwintering summer recruit" life history strategy of Wiggins and his colleagues. Depending on whether adults disperse long distances or remain in the general vicinity of the pools, they fall into the "migratory breeder" or "permanent resident with desiccation-resistant eggs" category defined herein.

Members of the genus *Lestes* from temperate regions undergo an obligatory egg diapause of up to ten months, an adaptation that has allowed several species to survive in vernal pools[44]. When they are ready to oviposit, females insert their eggs into the stems of emergent plants, especially favoring bulrushes (*Scirpus*), cat-tails (*Typha*), spike-rushes (*Eleocharis*), bentgrasses (*Agrostis*), bur-reeds (*Sparganium*), and docks (*Rumex*). Embryonic development in the eggs proceeds partially — to the blastoderm stage or to the prolarva — but further development and hatching are inhibited and the eggs undergo diapause until the plant stems are wetted by flooding in the spring[45].

Fall-collected eggs of *L. dryas* and *L. unguiculatus* will hatch in the laboratory upon inundation. In pools, however, low temperatures prevent the eggs from hatching until spring, even in those pools that flood in the autumn[46]. In the common spreadwing, *L. disjunctus*, a species that breeds in permanent waters, photoperiod as well as temperature appears to control diapause and prevent the premature hatching of eggs. Possibly, day length helps control hatching in other species, as well. In *L. dryas*, *L. disjunctus disjunctus*, and *L. unguiculatus*, hatching occurs maximally at 10° C (50° F), and the eggs hatch synchronously over a one- to two-week period[47].

Once the eggs have hatched, the nymphs develop quickly. *Lestes* that characteristically are found in vernal pools grow rapidly at low water temperatures and are ready to emerge as adults in as few as fifty to sixty days. Emergence is synchronous over a period of approximately ten days. The adults feed on flying insects for several weeks and begin reproductive activity two to three weeks after emergence[48].

The species that are characteristic of semi-permanent pools typically emerge later in the season than those that are common in annual vernal pools. In western Canada, the spotted spreadwing, which prefers semi-permanent or permanent waters, exhibits the same pattern of egg diapause, rapid nymphal growth, and early adult emergence as the lyre-tipped and emerald spreadwings, but in the glaciated northeast, adult spotted spreadwings are rarely found before July and they emerge most commonly in August[49]. Other species that are typical of permanent and semi-permanent ponds, such as the slender spreadwing, also grow fairly slowly and typically do not emerge before July or August[50]. It may be these species are usually unable to reproduce successfully in short-cycle pools because the water does not persist there long enough for the nymphs to develop into adults.

Figure 30. Damselfly nymphs such as this young pond damsel (Coenagrionidae) (top) have slender bodies with three plate-like gills at the end of the abdomen. The hinged mouthparts of pond damselflies (bottom left) are more-or-less rectangular; those of spread-wing nymphs (bottom right) are shaped somewhat like a long-handled spoon. (Photographs by Holly Jensen-Herrin.)

Pond Damselflies (Coenagrionidae)

Pond damselflies (Coenagrionidae) are abundant in lakes, ponds, marshes, and semi-permanent vernal pools, and they do not appear to be able to tolerate pool drying. The slender, clasping nymphs are typically associated with aquatic vegetation, and they are especially common in open-canopy, semi-permanent pools dominated by sedges, rushes, and grasses (Figure 30). Pond damselfly nymphs are often found in long-cycle vernal pools, and sometimes they occur in short-cycle pools, as well, although they may not be able to complete development to the adult stage in these habitats. Because members of this family lack the ability to resist desiccation, the presence of coenagrionid nymphs in a pool in early spring is a good indication that the pool is either a semi-permanent pool or a fall-filling vernal pool[51].

Distribution

Coenagrionids reported from annual and semi-permanent vernal pools include the taiga bluet (*Coenagrion resolutum*), the northern bluet (*Enallagma boreale*), Hagen's bluet (*E. hageni*), the marsh bluet (*E. ebrium*), the familiar bluet (*E. civile*), and the fragile forktail (*Ischnura posita*) (see Appendix). With the exception of the taiga bluet, a northern species that does not extend south of northern New York, northern New England, and Michigan, the distribution of all of these species extends broadly throughout the glaciated northeast, although many are extremely localized in occurrence. All have fairly wide habitat plasticity and occur in a variety of habitats with submerged and/or emergent vegetation, although the northern bluet is largely restricted to fishless waters, the marsh bluet does not tolerate acidic conditions, Hagen's bluet is found in more acidic marshes and ponds, and the familiar bluet colonizes small and temporary sites including those altered by human activity[52]. *Ischnura posita* is often found in shaded waters as well as the more open situations typical of the bluets.

Life History

Coenagrionids exhibit the "spring-summer migrant" strategy defined herein and can reproduce successfully only if water persists after the eggs have been deposited. These damselflies hatch in summer from eggs laid in the water, and they overwinter as nymphs. Some members of this group may be able to reproduce successfully in fall-filling vernal pools and in semi-permanent pools. In western Canada, the taiga bluet hatches within three weeks of egg deposition, and the young nymphs grow rapidly until falling temperatures slow down their growth. Growth resumes in spring, and the adults emerge synchronously in May and are reproductively mature a week later[53]. This pattern of rapid nymphal growth and more or less synchronous development is common in many species of Odonata[54].

Skimmers (Libellulidae)

Dragonflies in the family Libellulidae, the skimmers, are widely distributed in vernal pools. Several members of the genus *Sympetrum* (meadowhawks) are characteristic of vernal pools throughout the glaciated northeast. Like spreadwing damselflies, they are particularly well adapted for life in seasonally drying habitats. Libellulid nymphs are often squat, with long spidery legs, and their mouthparts are in the form of a bowl that cups the lower part of the head (Figure 31, top).

Other libellulids are common in a wide variety of semi-permanent and permanent waters, with several species regularly found in semi-permanent vernal pools. The species richness in this group increases markedly with increasing hydroperiod. Among the skimmers commonly reported from semi-permanent vernal pools are the whiteface (*Leucorrhinia intacta*), the twelve-spotted skimmer (*Libellula pulchella*) and the four-spotted skimmer (*L. quadrimaculata*), the blue dasher (*Pachydiplax longipennis*), the wandering glider (*Pantala flavescens*), the eastern pondhawk (*Erythemis simplicicollis*), the Carolina saddlebags (*Tramea carolina*), and the black saddlebags (*T. lacerata*). Some libellulids may require more than one year to complete development, depending on water temperatures and food availability, and large nymphs that are present in early spring can play an important role in structuring the aquatic community in semi-permanent vernal pools.

Distribution

The white-faced meadowhawk (*Sympetrum obtrusum*) and the ruby meadowhawk (*S. rubicundulum*) occur in short-cycle, spring-filling vernal pools in Ontario and Massachusetts. The blue-faced meadowhawk (*S. ambiguum*) also breeds in temporary ponds, but little is known of its biology in the glaciated northeast. The cherry-faced meadowhawk (*S. internum*) is known from a fall-filling vernal pool in Ontario. Other *Sympetrum*, including the black meadowhawk (*S. danae*) and the saffron-winged meadowhawk (*S. costiferum*), have been reported from semi-permanent vernal pools. Two others, the band-winged meadowhawk (*S. semicinctum*) and the yellow-legged meadowhawk (*S. vicinum*), appear to prefer permanent waters (see Appendix). All of these species occur throughout the glaciated northeast, although *S. rubicundulum* may not extend into Wisconsin and Minnesota[55].

Of the other libellulid genera that sometimes occur in semi-permanent vernal pools, *Pantala*, *Leucorrhinia*, and *Libellula* occur widely throughout the glaciated northeast. The geographic range of *Pachydiplax* encompasses the northeastern United States and extends into southernmost Canada. *Erythemis* and both species of *Tramea* are more southern in distribution, the northward edge of their ranges extending into southern Ontario and Quebec but not northern New England or the Canadian Maritime Provinces. As with other odonates found in vernal pools, none of these species is restricted to this habitat. Rather, they all occur in a variety of fresh waters[56].

Life History

Two life history patterns are seen in libellulids in vernal pools. The white-faced and ruby meadowhawks survive the period from drawdown until spring in short-cycle vernal pools as diapausing embryos within desiccation-resistant eggs. The females deposit their eggs directly onto the dry surface of vernal pools in the fall. As in *Lestes*, the eggs do not hatch even if flooded in fall or winter, suggesting that a combination of temperature and day length control diapause. Once the eggs have hatched, the nymphs grow rapidly and the adults emerge in late spring or early summer. Adults of both these species and the blue-faced meadowhawk fly from June into September or October, and they are regularly seen as late as early

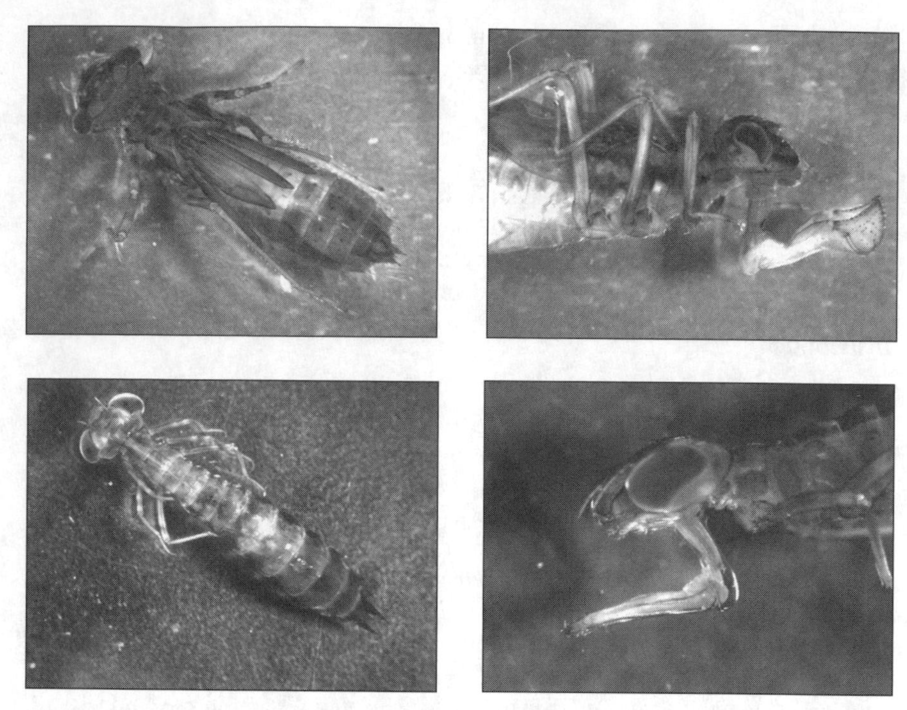

Figure 31. Skimmers (Family Libellulidae) (top) and darners (Family Aeshnidae) (bottom), shown here with entire body in dorsal view and head and abdomen in lateral view, are the common families of dragonflies in vernal pools. Differences in the shape of the hinged mouthparts are useful in identifying the families. (Photographs by Holly Jensen-Herrin.)

November in southeastern New England[57]. The pattern of diapausing eggs also appears to hold for *Sympetrum* in semi-permanent and permanent waters, although the nymphs may grow more slowly and the adults may emerge later than in species found in vernal pools.

Other libellulids apparently lack a mechanism for withstanding drying in vernal pools. In the Carolina and black saddlebags, as well as the wandering glider, migratory adults fly to vernal pools in spring or early summer and deposit their eggs in the water[58]. It seems likely that other libellulid species colonize vernal pools in a similar manner. For most of these species, vernal pools may dry before the nymphs complete development, but the flooding duration in some long-cycle and semi-permanent vernal pools may be long enough to allow adults to emerge successfully, and even pools that normally dry early in the summer may be able to support development in these dragonflies during unusually wet years.

Darners (Aeshnidae)

The darners (Aeshnidae) are a large group of dragonflies, only a few of which are found in northeastern vernal pools. They are often streamlined in body outline, and their hinged mouthparts are flat (Figure 31, bottom). Among our largest dragonflies, a few species find sufficient food in small vernal pools to complete development in just a few months. None of these species is restricted to vernal pools. The giant swamp darner (*Epiaeshna heros*) has been collected from long-cycle vernal pools in Massachusetts. The common green darner (*Anax junius*) has been found in short-cycle vernal pools as far north as southern Ontario, and although it does not appear to complete development successfully in short-cycle pools it probably can do so in long-cycle pools that remain flooded into August or September. The fawn darner (*Boyeria vinosa*) has been found in small woodland pools and large, open, semi-permanent pools[59].

Several species of darners are commonly encountered in semi-permanent pools, including the black-tipped darner (*Aeshna tuberculifera*), the variable darner (*A. interrupta*), and the spatterdock darner (*A. mutata*) which, while typical of permanent ponds, is also found in semi-permanent vernal pools on Cape Cod.

Distribution

Most darners that can be found in vernal pools have a wide geographic distribution throughout the glaciated northeast, although their abundance varies considerably, depending on the presence of appropriate habitat. The spatterdock darner occurs in the southern part of the glaciated northeast and in Massachusetts, New York, Michigan, and southern Ontario, but it does not extend into northern New England, the Maritime Provinces, or Quebec. The variable darner is a northern species whose range does not extend into the southern part of the glaciated northeast[60].

Life History

In general, darners appear to lack mechanisms for resisting drying in vernal pools. Some species of *Aeshna* are known to deposit desiccation-resistant eggs[61], but this strategy has not been documented in members of this genus in vernal pools. I have observed a female *Epiaeshna heros* ovipositing in a dead tree trunk well above the water line of a vernal pool in early summer, and the presence of this species in long-cycle pools suggests that the eggs may undergo a diapause and hatch upon flooding in fall or the following spring.

Many female darners probably move facultatively into vernal pools in spring and summer to deposit eggs, but their nymphs are unlikely to survive the winter in pools that dry during the summer or fall. An exception is seen in the common green darner, a species in which migrants from farther south move into northeastern vernal pools in early spring and lay eggs which hatch rapidly and develop over the course of two to three months, with adult emergence taking place in mid July to late

August[62]. Interestingly, the green darner also maintains resident populations in the glaciated northeast, but in these populations eggs are deposited later in the season. The nymphs of resident populations require up to eleven months to complete their development and, therefore, they typically must overwinter and do not survive in annual vernal pools.

Water Bugs (Hemiptera: Heteroptera)

As with many of the other aquatic insect taxa found in vernal pools, there do not seem to be any water bugs that are restricted to temporary habitats. Few hemipterans are to be found in vernal pools during the winter and early spring phases of the vernal pool seasonal cycle (see Chapter 13), but the numbers of species and individuals increase from late spring into summer. In long-cycle pools, during the summer low-water period when pools are drawn down but not yet dry, water boatmen (Corixidae) and backswimmers (Notonectidae) are often the most abundant macroinvertebrates present (Figure 32).

Most aquatic hemipterans are predators, feeding on crustaceans and aquatic insects. They play an important role in regulating populations of other aquatic insects, including mosquito larvae[63]. Larger species, including backswimmers and giant water bugs (Belostomatidae), will take tadpoles. Corixids are omnivorous, taking animal prey and scraping algae, detritus, protozoans, and other small organisms from surfaces.

Distribution

All of the major families of aquatic Hemiptera are represented in vernal pools. Water boatmen (Corixidae) and backswimmers (Notonectidae) are the most abundant and diverse of the vernal pool Hemiptera (see Appendix). Four of the seventeen genera of corixids are commonly found in annual vernal pools, as is one of the three notonectid genera. A second genus of notonectid occurs in semi-permanent pools. Other families that occur regularly in annual vernal pools include the giant water bugs (Belostomatidae), water striders (Gerridae, Veliidae, Hebridae, Mesoveliidae), water scorpions (Nepidae), and marsh treaders (Hydrometridae).

Species of water bugs reported from vernal pools tend to be widely distributed with broad habitat plasticity. Examples include water striders (*Gerris marginatus*, *G. remigis*) (Figure 33), giant water bugs (*Lethocerus americanus*, *Belostoma flumineum*), water scorpions (*Nepa aciculata*, *Ranatra fusca*), and water boatmen (*Hesperocorixa vulgaris*, *Callicorixa alaskensis*, *C. audeni*, and *Sigara alternata*). A backswimmer, the ubiquitous *Notonecta undulata*, is the most common vernal pool hemipteran, and it occurs in vernal pools throughout the glaciated northeast[64].

Life History

All vernal pool hemipterans are migrants that overwinter in permanent waters and fly into vernal pools in spring or early summer to feed and breed. Some species use vernal pools solely as feeding areas and breed elsewhere. The concentrated life in the drying pools serves these species as a ready source of food. Of the water bugs that do breed in vernal pools, most apparently have only one generation per year, with the exception of some water striders of the genus *Gerris*, which are able to complete two generations[65]. Taxa that are sometimes apterous or brachypterous — with no or reduced wings, and thus unable to fly — in permanent waters, are macropterous — with fully developed wings — in vernal pools, a necessary adaptation to allow migration to permanently flooded overwintering sites.

Wiggins and his colleagues provide a more detailed discussion of life history strategies of vernal pool hemipterans[66]. In their sampling of Ontario vernal pools, they found evidence that, at least in some years, some hemipterans may not be able to complete their life cycles in short-cycle pools. Adult water scorpions, *Ranatra fusca*, appeared in May, and two species of giant water bugs, *Belostoma flumineum* and *Lethocerus americanus*, migrated into short-cycle pools in early June. In both species, the immature nymphs were still present in early July, shortly before pool drying, and probably did not complete development. Adults of a

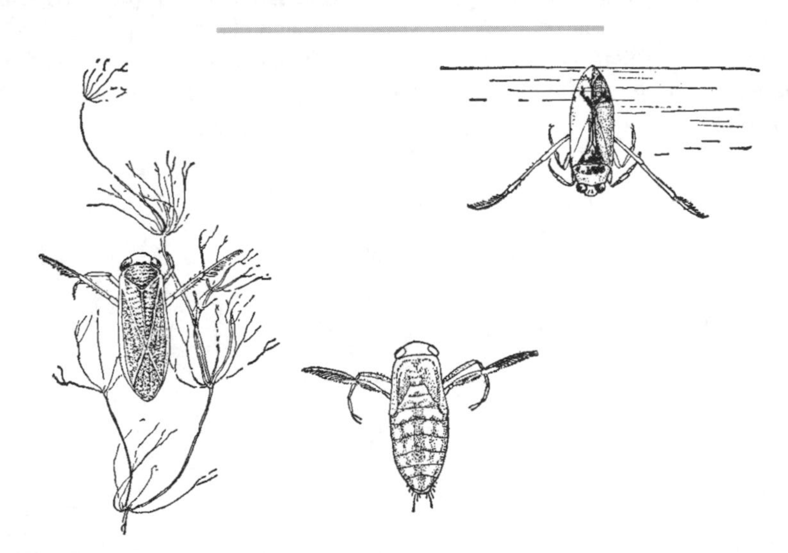

Figure 32. Water bugs such as predatory backswimmers, family Notonectidae (top right), and omnivorous water boatmen, family Corixidae (left, adult; bottom center, nymph), are among the many migratory breeders that colonize vernal pools in spring from permanent waters where they overwinter. (From Morgan 1930, courtesy of the Morgan estate.)

backswimmer, *Notonecta undulata*, arrived in short-cycle pools as early as mid April, but they did not produce young before mid June and these nymphs, similarly, were still immature in early July. These taxa did complete development in nearby long-cycle and semi-permanent pools. Corixids, in contrast, arrived from mid April to mid May, and their nymphs were halfway through development by the beginning of June and had completed growth before pool drying. Similar patterns of distribution in corixids, notonectids, and gerrids were observed in a short-cycle Ontario vernal pool[67].

Mosquitoes (Diptera: Culicidae)

Vernal pools support a wide variety of mosquito species (Figure 34). None of the major carriers of human or domestic animal diseases is typical of vernal pools, although some have been found occasionally in vernal pools. Some of the species commonly occurring in vernal pools are nuisances, and some can occur in sufficient numbers to be serious pests. In Massachusetts, 8% of mosquito problems are associated with seasonally flooded basins and vernal pools[68].

Some of the important pest species occur in shallow depressions that flood in early spring or following rainstorms but are too small and short-lived to meet the definition of a vernal pool. Often, adult mosquitoes encountered near vernal pools spent their larval lives elsewhere. C. Brooke Worth studied mosquito larvae in a vernal pool along the eastern shore of New Jersey and observed that the majority of the mosquitoes that were biting him were salt marsh species that had flown inland and treehole species from the surrounding forest, not species from the vernal pool[69].

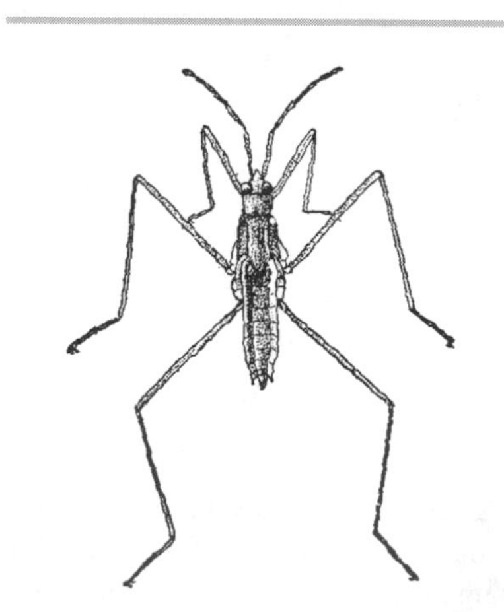

Figure 33. Water striders, including *Gerris,* skate on the surface tension of the water in search of prey. (From Morgan 1930, courtesy of the Morgan estate.)

In some vernal pools, mosquito larvae reach high densities and can be the most abundant macroinvertebrates present. Densities of more than 500 *Aedes* larvae per square meter (more than 46/ft^2) have been observed in Ontario pools[70]. Mosquito larvae are particularly common, and are likely to be dominant members of the community, in small, shallow, short-cycle vernal pools. Their rapid growth rates and high turnover suggest that they play an important role in energy flow and nutrient cycling. Most mosquitoes in vernal pools are filter-feeders or grazers, feeding on small particles of detritus, algae, and microorganisms in the water. *Psorophora* larvae are predaceous on other mosquito larvae and on tadpoles; this genus is generally poorly represented in vernal pools, but it may be a significant predator in some circumstances.

Fewer than 1% of mosquito eggs mature into adults. The rest are eaten by a variety of predators or fall victim to parasites[71]. Mosquito larvae are fed upon by aquatic beetles, snails, water bugs, dragonfly and damselfly nymphs, newts, and salamander larvae. Several species of water beetles prey heavily on mosquito larvae in Ontario vernal pools, and evidence from other habitats suggests that some predaceous beetle species selectively colonize habitats with high densities of mosquitoes. The caddisfly *Limnephilus indivisus*, a common vernal pool species, is another important predator on mosquito larvae in Ontario, and the phantom midge *Mochlonyx* sp. preys on *Aedes communis* in Massachusetts vernal pools. Cyclopoid copepods are such effective predators that they have been used to control species of *Aedes* and *Anopheles* that transmit dengue fever in the tropics and the southeastern United States. The degree to which these predators control mosquito populations in northeastern vernal pools, and the importance of the mosquito larvae to the growth and development of the predators, are unknown.

Distribution

Species of *Aedes* that are typical of northeastern woodland vernal pools include *A. canadensis*, *A. communis*, *A. diantaeus*, *A. grossbecki*, *A. intrudens*, *A. punctor*, and *A. stimulans*.

Certain species and associations of species are characteristic of particular types of vernal pool habitats. Often several species occur in the same pool, usually with one dominant and the others present in relatively low numbers[72]. For example, *Aedes abserratus*, *A. punctor*, *A. canadensis*, and *A. cinereus* commonly co-occur in boggy habitats in Canada. *A. punctor* tends to be the dominant species in such pools in coniferous forests, where it is often found with *A. abserratus*. *Aedes communis*, *A. diantaeus*, and *A. intrudens* commonly co-occur in temporary woodland pools in hardwood forests. Of the three, *A. communis* tends to be the most abundant in these associations. In small vernal pools that contained water for three-and-a-half months in southern Ontario, *A. stimulans* and *A. provocans* (as *A. trichurus*) co-occurred with small numbers of *A. excrucians*. The same three species were present along with *A. euedes* (as *A. barri*), *A. communis*, *A. fitchii*, *A.*

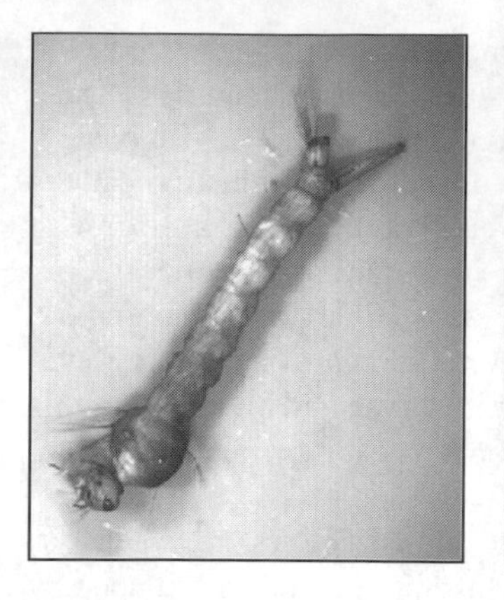

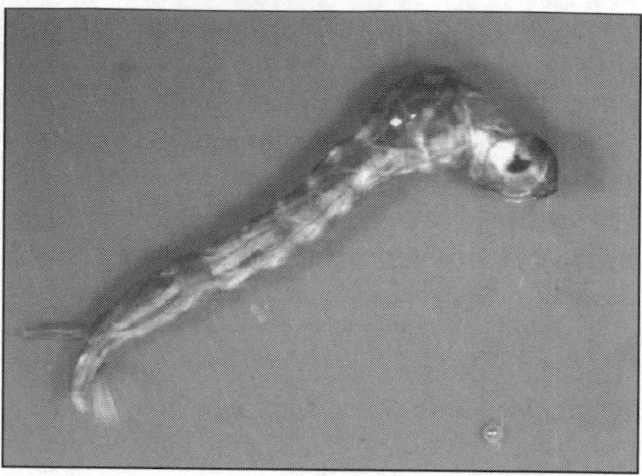

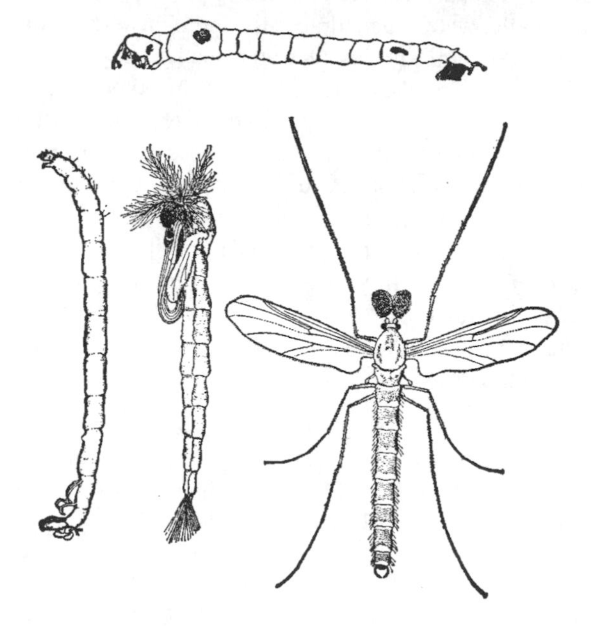

Figure 34. The larvae of many kinds of true flies (Diptera) occur in vernal pools. Mosquito larvae, including many species of *Aedes* (upper left), hang from the surface film and filter bacteria and small particles with their brush-like mouthparts. Phantom midges, including the mosquito-like *Mochlonyx* (upper right) and the transparent *Chaoborus* (center), capture water fleas and mosquito larvae with their grasping antennae. Chironomid midges (bottom), live in the mud and on the surface of plants and detritus, and form swarms of non-biting adults when they emerge. All images shown here are of larval stages except the chironomid pupa at bottom center and the adult chironomid at bottom right. (Photographs by Holly Jensen-Herrin. Line drawings from Morgan 1930, courtesy of the Morgan estate.)

diantaeus, and *Culiseta morsitans* in semi-permanent pools of similar sizes. *A. canadensis* and *A. grossbecki* were present in small vernal pools in southern New Jersey. References containing additional information on mosquitoes of northeastern vernal pools are listed at the end of this chapter[73].

Life History Strategies

Mosquitoes exhibit several general life history strategies in vernal pools. Although in some species the adults remain near the pool after emerging, in others they disperse over long distances to new pools. Mosquitoes thus include both migratory breeders and permanent residents and employ a variety of mechanisms to withstand pool drying. In general, most mosquitoes spend the dry period as desiccation-resistant eggs. A few species are early-spring migrants that overwinter as adults and move to flooded pools to breed. Five broad types of life history patterns can be identified for mosquitoes in vernal pools.

(1) Species that breed in spring-filling, short-cycle pools that contain water for only a few months overwinter as eggs that hatch in March upon flooding. The larvae grow rapidly and the adults emerge, feed, and lay their eggs in May or June in moist soil surrounding the drying pools. The eggs remain in the sediment until the following spring. *Culiseta morsitans* and most species of *Aedes* from vernal pools fall in this group. When *Aedes stimulans* and *A. provocans* lay their eggs in vernal pools as the water levels recede, the eggs tend to be distributed in a relatively narrow band at a particular elevation in the pool. This behavior, as with marbled salamanders, ensures that the eggs will hatch after the pool has been flooded with enough water to provide adequate food and habitat for larval development[74].

(2) Mosquitoes typical of fall-filling and semi-permanent vernal pools overwinter as eggs that hatch, not when first exposed to water, but rather, in spring after a period of cold-conditioning. The larvae develop in a few weeks, the spring-emerging adults mate, and the females deposit eggs that hatch within days. Successive generations follow until day lengths and temperatures decrease and stimulate the production of diapausing winter eggs. *Anopheles walkeri* exhibits this life history pattern.

(3) Species of floodplain pools and other depressions that flood with summer rainstorms are thermally-cued hatchers. Examples include *Aedes sticticus*, *A. trivittatus*, *A. vexans*, and *Psorophora* spp. They overwinter as eggs that hatch in late spring or summer in response to warm rains. The adults are active in summer, and the females deposit their eggs on moist pool bottoms or floodplains. These eggs may hatch in response to later flooding to produce additional generations in a given year, or they may remain in the sediment for several years if appropriate flooding conditions do not occur. Many of the pools inhabited by these species may not meet the definition of a vernal pool.

(4) Several mosquito species found in a wide variety of habitats overwinter as adult females that emerge to feed and deposit eggs in spring. These species

usually have several generations in a year. *Anopheles quadrimaculatus*, *Aedes punctatus*, *Aedes earlei*, *Culex* spp., most *Culiseta* spp., and *Urotaenia sapphirina* are species in this group.

(5) *Culiseta melanura* has a unique life history strategy. It overwinters as a larva and completes its development and emerges in the spring. When this species occurs in vernal pools it is restricted to fall-filling or semi-permanent pools.

Phantom Midges (Diptera: Chaoboridae and Corethrellidae)

Phantom midge larvae are also sometimes known as glassworms, because species of *Chaoborus* are transparent and almost invisible as they float motionless in the water. These insects are closely related to mosquitoes, and some genera that are common in vernal pools can easily be mistaken for mosquito larvae (Figure 34). Like mosquito larvae, they are most abundant in vernal pools in early spring. The larvae are fierce predators. They capture zooplankton and small insects with their antennae, which are modified into grasping structures. Adult phantom midges do not bite.

Three species of *Chaoborus* have been reported from vernal pools in the glaciated northeast, and all of them are found in a variety of freshwater habitats including deep lakes (see Appendix). The larvae are truly phantoms of the aquatic world and blend in perfectly with the water. They become visible only when they move suddenly or when the light reflects off the two air sacs, one at each end of the body, which distinguish members of this genus from other phantom midges (Figure 34). The larvae can adjust their buoyancy by altering the amount of air in the sacs. In vernal pools they often float horizontally near the surface at a constant depth, suddenly striking out with their prehensile antennae at water fleas or other small prey that pass nearby. In some vernal pools, *Chaoborus* larvae can be important predators on mosquito larvae. The larvae may migrate vertically within the water column as light, temperature, or other conditions change. In lakes, the diurnal vertical migration of *Chaoborus* larvae is pronounced, with some species spending the day on the lake bottom and migrating tens of meters to the surface at night to feed; the migrating layer of larvae is often so dense that it is visible on sonar surveys of the water column. Less is known about vertical migration patterns in vernal pools. In permanent waters, *Chaoborus* overwinter as fourth instar larvae, and they pupate and emerge after water temperatures warm up in spring. The adults are thought to fly to vernal pools and lay eggs on the water surface after emerging from permanent waters. There is some evidence from the literature, however, that some *Chaoborus* may overwinter in vernal pool basins as eggs or larvae, and the eggs of one western species, *C. cooki*, have been shown to aestivate in vernal pools in Alberta[75].

Larvae of the three species of *Mochlonyx* are also common in vernal pools throughout the glaciated northeast (see Appendix). Two of them, *M. velutinus* and *M. cinctipes,* are difficult to distinguish in the larval stage. Unlike the transparent

Chaoborus larvae, members of this genus lack air sacs, and they look somewhat like mosquito larvae with gray bodies, swollen thoraxes, and a respiratory siphon on the last abdominal segment, but their jointed prehensile antennae distinguish them as phantom midges (Figure 34). *Mochlonyx* is common in short-cycle and semi-permanent pools in southern Ontario, where it is an important predator on mosquito larvae. All three species of *Mochlonyx* aestivate in dry pool basins as eggs that are deposited on the damp substrate in late spring and develop partially before entering diapause when the pool draws down. The eggs hatch upon flooding, and the larvae develop rapidly, feeding on zooplankton and insect larvae. Wiggins and his colleagues found that eggs from a spring-filling, short-cycle pool hatched in the laboratory when they were flooded in autumn, suggesting that their diapause is ended by inundation and does not require a prolonged period of cold-conditioning[76].

Eucorethra underwoodi is widely distributed in pools of northern coniferous forests, and it has been found in pools as far south as northern New Jersey. Larvae of *Eucorethra* have prehensile antennae and a swollen thorax, and they lack hydrostatic air sacs and respiratory siphons. Unlike other phantom midges, which feed on aquatic insects and crustaceans, *Eucorethra* preys on terrestrial insects and other arthropods that fall onto the surface of the water[77].

Another phantom midge, *Corethrella brakeleyi* (Corethrellidae), has been collected from some semi-permanent vernal pools in the southern portions of the glaciated northeast[78]. Like *Mochlonyx*, *Corethrella* has a respiratory siphon and looks much like a mosquito larva, but it has jointed prehensile antennae that lie close to each other on the top of its head. This insect occurs in a variety of habitats including cold springs and seeps.

Non-Biting Midges (Diptera: Chironomidae)

A variety of other kinds of true flies representing several different families occurs in vernal pools, but, overall, dipterans have been inadequately studied in these habitats. There is little information available about which species occur in pools, whether any of them are particularly indicative of these habitats, how different taxa survive pool drying, and how fly larvae, pupae, and adults contribute to the food webs and overall energetics of vernal pools and their environs.

Among the fly larvae that are very abundant in the sediments of vernal pools are non-biting midges in the family Chironomidae (Figure 34). This is a large and diverse family with adults that range in size from tiny gnats only 1 to 2 mm (0.04 to 0.08 in) long to large midges the size of adult mosquitoes. The larvae of many species have hemoglobin in their blood, which both gives them a red color — and the popular name of "bloodworms" — and helps them to capture oxygen from the water. This allows the midges to survive in the sediments of pools and other wetland habitats where the levels of dissolved oxygen are so low that other insect species are excluded. The larvae often construct soft dwelling tubes from fine sediments on the bottom. Many larval chironomids ingest sediment and are important processors of organic oozes in lakes, wetlands, and streams. Others are predators. In appropriate habitats, midge larvae can occur at densities of several thousand larvae per square meter[79]. Large mating swarms that emerge from some kinds of wetlands are important sources of food for nesting tree swallows and other birds[80], and it is possible that chironomid midges are among the insects being preyed upon by bats that forage above vernal pools. These flies can be an important vehicle for the transfer of energy from pools to the terrestrial ecosystem. Some midges in vernal pools survive the drawdown period as drought-resistant larvae. Others overwinter as adults and fly to pools in spring to deposit their eggs[81].

Craneflies (Diptera: Tipulidae)

Cranefly larvae, family Tipulidae, are also found in vernal pools. This is another extremely diverse group of insects, with several thousand species in North America alone. The adults look like oversized, long-legged mosquitoes, but they do not bite. The larvae have heavy cylindrical bodies with a series of projections or tentacles surrounding a flattened disk-like area at the hind end. Depending on the species, they feed on detritus, sediment, algae, and plant materials, or are predaceous on other invertebrates. Some craneflies are shredders and can play a role in preparing detritus for uptake by filter-feeders such as fairy shrimp and cladocerans. They pupate in damp sand or mud near the edge of pools. Some cranefly larvae are not aquatic and live in damp moss or mud. The genera of craneflies found in vernal pools are common in wet organic sediments and detritus in forests and floodplains. Although the details of their life history strategies in vernal pools have not been described, it seems likely that the larvae tolerate drying[82].

Other True Flies (Diptera)

Marsh fly larvae, family Sciomyzidae, look much like cranefly larvae. They feed on fingernail clams and snails and can be very important regulators of the densities of their prey. In some parts of the world, sciomyzids have been used as biological controls of snails that serve as intermediate hosts for liver flukes and schistosomiasis[83].

Horseflies and deerflies, family Tabanidae, are also represented in vernal pools. Their long (up to 6 cm or 2.4 in), cylindrical larvae are pointed at each end and the segments are encircled by fleshy bands that help the animals burrow through the sediments in the bottom of pools and other wetland habitats. Like some craneflies, some horsefly larvae are semi-terrestrial and occur in damp sediments rather than in water. Many are fierce predators that consume a wide variety of prey including invertebrates and amphibian larvae, and some are detritivores. Some of the species that occur in vernal pools survive drawdown as drought-tolerant larvae that aestivate on the pool bottom until the pool floods in spring[84]. Pupation occurs on land, and the adults can inflict a painful bite.

Adult females of one genus in the family Ceratopogonidae, *Culicoides*, inflict a painful bite, and common names for the family include biting midges, punkies, and no-see-ums. *Culicoides* have not been reported from vernal pools. The many other genera, including those that occur in vernal pools, feed on the blood of other insects and generally do not bite mammals (see Appendix). The larvae are small and cylindrical with tapered heads and tails, and many of them are transparent and difficult to see. Some larvae prey on other insects, and some species feed on detritus in the pools.

The Appendix includes a list of dipteran taxa that have been reported from vernal pools in the literature, and Wiggins and his colleagues provide an excellent summary of life history strategies in some of these insects[85].

Other Kinds of Insects

Mayflies (Ephemeroptera) and stoneflies (Plecoptera) are two major orders of aquatic insects that are generally depauperate in vernal pools, although members of both are well represented in the faunas of temporary streams[86]. Both of these orders are most widely distributed in flowing waters, and stoneflies, in particular, are uncommon in ponds and quiet water habitats. Several species of mayflies in the family Baetidae have been reported from woodland vernal pools. Wiggins and his colleagues suggest that adult *Callibaetis ferrugineus* alternate generations between vernal pools and permanent waters: adults fly to vernal pools from permanent waters in the spring, a generation is completed in the pools before they dry, and the next generation is completed in permanent waters (Figure 35)[87]. Members of the mayfly families Siphlonuridae and Leptophlebiidae and of the stonefly families Leuctridae and Nemouridae are often found in vernal pools with intermittent stream inlets or in floodplain pools, but none of these taxa is characteristic of vernal pools. Some leuctrids and nemourids are characteristic of spring seeps and headwater streams that dry seasonally. Adaptations to these habitats presumably allow survival in vernal pools fed by surface flows, as well.

Larval fishflies (Megaloptera, Corydalidae, genus *Chauliodes*) are often collected in vernal pools. The larvae are known commonly as hellgrammites and are popular as bait for fishermen. They are found in a variety of aquatic habitats including both short-cycle woodland pools and semi-permanent vernal pools. The predaceous larvae have tubular filaments that extend off the sides of each of the abdominal segments, a pair of elongated tubular gills on the end of the abdomen, and large hooked mandibles that give them a quite formidable appearance that is superficially similar to some beetle larvae. Late in larval development they reach a length of 4 cm (over 1.5 in). Female fishflies deposit their eggs on vegetation, woody debris, and rocks overhanging the pool bottom, and when the eggs hatch the larvae fall into the water. Most species of Megaloptera require two or three years to complete their life cycles, and the details of the life history of *Chauliodes* in vernal pools, including strategies for surviving pool drying, are unknown. They pupate on shore, tranforming into adults over a two-to-three-week period, and the adults live only a few days[88].

Some Remaining Questions about Aquatic Insects

- What species occur in vernal pools? Where are they found? What are the habitat characteristics associated with different species? Are there species that clearly serve as vernal pool specialists and can be used as indicator species?
- What are the life histories of vernal pool insect species? How are they related to seasonal patterns of temperature, hydrology, and day length? Do they differ geographically, or with different pool habitat conditions?
- How do vernal pool insects cope with seasonal drawdown?
- For species that spend the dry period as eggs or cysts, what environmental cues stimulate hatching? How long can eggs or cysts resist drying? Is there an egg bank?
- What are the characteristics of the overwintering habitats used by species that breed in vernal pools and migrate seasonally to permanent waters? How far are they from the pools?
- For species with weak dispersal abilities, how much genetic differentiation is seen in vernal pool populations? Are there species that are clearly restricted to vernal pools that should be the focus of greater conservation attention?
- Is there homing in vernal pool insects? To what extent are vernal pools repopulated by new colonists each year? How rapidly are new pool habitats colonized by aquatic insects? Are there successional patterns over time in the insect community, as newly created pools mature?
- How do vernal pool insects contribute to decomposition and energy flow? Do they serve as important prey for salamander larvae? How much energy is

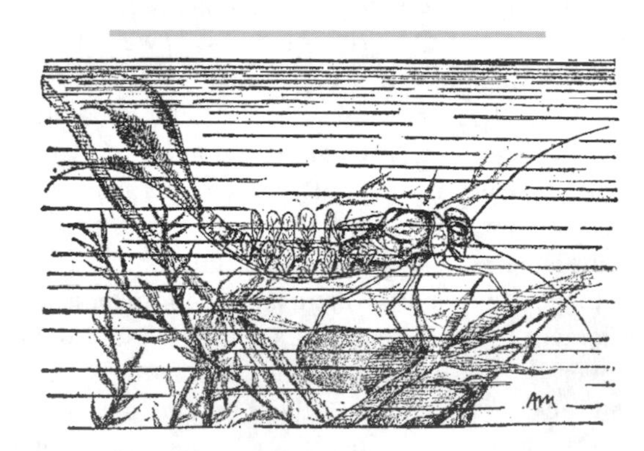

Figure 35. Species of the mayfly genus *Callibaetis* can often be found in vernal pools in the spring. (From Morgan 1930, courtesy of the Morgan estate.)

transferred to the terrestrial ecosystem by the emergence of adult midges, mosquitoes, caddisflies, and other insects from pools? How important are insects from vernal pools as food for forest-breeding birds, bats, or other animals?

- How do closely related species interact when they co-occur in vernal pools?
- What is the role of insect predators in the economy of vernal pools? To what extent do long-hydroperiod pools support long-lived predators, such as large dragonfly nymphs? How do such predators structure the invertebrate and amphibian communities of the pools?

References for Chapter 9

[1] Larson et al. 2000; Merritt and Cummins 1996a; Wiggins 1977, 1996, 1998. [2] Ross 1944, Wiggins 1977. [3] Collins and Washino 1985. [4] Rowe et al. 1994, Stout and Stout 1992. [5] Wiggins 1973, Wiggins et al. 1980. [6] Flint 1960, Flannagan and MacDonald 1987, Ross 1944, Ruiter 1995. [7] Wiggins 1977, 1998. [8] Ross 1944. [9] Colburn unpublished data , Flannagan and MacDonald 1987, Wiggins 1973. [10] Floyd 1995, Glover 1996. [11] Flint 1960, Richardson and Mackay 1984, Wiggins 1973, Wiggins et al. 1980. [12] Bouvet 1971, Colburn 1984, Novak and Sehnal 1963, Sutcliffe 1960. [13] Flannagan and MacDonald 1987, Flint 1960. [14] Colburn unpublished data, Flannagan and MacDonald 1987, Nimmo 1971, Ross and Merkley 1952. [15] Flint 1960, Wiggins 1973, Wiggins et al. 1980. [16] Wiggins et al. 1980. [17] Wiggins 1973, Wiggins et al. 1980. [18] Fairchild et al. 2000. [19] Alarie et al. 2002, Kenk 1949, Larson 1987b, Leonard 1928, Wiggins et al.1980. [20] Cuppen 1983, Eyre et al. 1992, Galewski 1971. [21] James 1961, 1969; Larson 1996. [22] Colburn unpublished data, Kenk 1949, Wiggins et al. 1980. [23] James 1966. [24] Colburn and Garretson 1997. [25] James 1967. [26] Formanowicz and Brodie 1981, Leonard 1928. [27] Roughley and Nilsson 1994, Zimmerman 1970. [28] Boobar et al. 1998, Bousquet 1991, Formanowicz and Brodie 1981, Larson et al. 2000, Leonard 1928, Smetana 1988, Wiggins et al. 1980, Zimmerman 1970. [29] James 1961, Leonard 1928, Museum of Comparative Zoology Collections, Zimmerman 1970. [30] Anderson 1976, Bousquet 1991, Daussin 1979, Hilsenhoff 1994, Larson et al. 2000. [31] Garcia and Hagen 1987, Wiggins et al. 1980. [32] Hosseinie 1976. [33] Garcia and Hagen 1987, James 1961. [34] James 1961, Smith 1973, Wiggins et al. 1980. [35] Higgins and Merritt 1998, James 1961. [36] Corbet 1999, Johnson and Crowley 1980, McPeek 1990. [37] Brockelman 1969; Smith 1983; Bryan Windmiller, Hyla Ecological Services, personal communication. [38] Corbet 1999. [39] Cannings et al. 1980, Corbet 1999, Mantilacci et al. 1976, Westfall and May 1996. [40] Wiggins et al. 1980. [41] Corbet 1999. [42] Walker 1953, Westfall and May 1996, Wiggins et al. 1980. [43] Walker 1953, Westfall and May 1996. [44] Fischer 1966; Sawchyn and Gillott 1974a, 1974b. [45] Gower and Kormondy 1963, Walker 1953. [46] Needham 1903. [47] Sawchyn and Gillott 1974b. [48] Sawchyn and Gillott 1974a, 1974b. [49] Christopher Leahy, Massachusetts Audubon Society, personal communication; Sawchyn and Gillot 1974a; Walker 1953. [50] Gower and Kormondy 1963, Walker 1953. [51] Wiggins et al. 1980. [52] Westfall and May 1996. [53] Sawchyn and Gillott 1975. [54] Benke and Benke 1975, Corbet 1999, Weir 1974. [55] Colburn unpublished data, Walker and Corbet 1975, Wiggins et al. 1980. [56] Walker and Corbet 1975. [57] Christopher Leahy, Massachusetts Audubon Society, personal communication; Walker and Corbet 1975; Wiggins et al. 1980. [58] Corbet 1962, Walker and

Corbet 1975, Wiggins et al. 1980. [59] Maria Aliberti, New England Aquarium, personal communication; Colburn unpublished data; Trottier 1966; Walker and Corbet 1975; Wiggins et al. 1980. [60] Needham and Westfall 1954, Walker and Corbet 1975. [61] Corbet 1999. [62] Trottier 1966, Wiggins et al. 1980. [63] Hilsenhoff 1991. [64] Hungerford 1953, Wiggins et al. 1980. [65] Calabrese 1979. [66] Wiggins et al. 1980. [67] Williams 1983. [68] Edman and Clark 1988. [69] Worth 1972. [70] James 1961. [71] Baldwin et al. 1955; Collins and Washino 1985; Edman and Clark 1988; James 1961, 1965, 1966a, 1966b, 1967; New Jersey Agricultural Experimental Station 1960. [72] Horsfall 1963; James 1961, 1966; Marten 1999; Wood et al. 1979; Worth 1972. [73] Edman and Clark 1988; Horsfall 1963; James 1961, 1966a, 1966b, 1967, 1969; Massachusetts Department of Public Health 1940; Spencer 1967; Wiggins et al. 1980; Williams 1983; Wood et al. 1979; Worth 1972. [74] James 1966b. [75] Borkent 1979, 1980, 1981; Wiggins et al. 1980. [76] Borkent 1981; James 1957, 1966a; Wiggins et al. 1980. [77] Borkent 1981. [78] Colburn unpublished data. [79] Ali 1980, Coffman and Ferrington 1996, Epler 2001. [80] Kiviat and McDonald 2002, Kraus 1989. [81] Wiggins et al. 1980. [82] Byers 1978. [83] Hilsenhoff 1991, Pennak 1978. [84] Wiggins et al. 1980. [85] Wiggins et al. 1980. [86] Williams 1987, Williams and Hynes 1976. [87] Wiggins et al. 1980. [88] Pennak 1978, 1989.

Chapter 10

Water Mites and Miscellaneous Other Arthropods

The phylum Arthropoda is an enormously diverse group, and a number of its members besides crustaceans and insects occurs in vernal pools (see Appendix). Of these, water mites — of which dozens of species are found in temporary ponds in the glaciated northeast — are most likely to be observed by both casual observers and researchers studying pools. Water mites have fascinating life cycles and can play important roles in regulating the populations of other pool inhabitants. Spiders and springtails are also found in vernal pools, but they are not sufficiently well studied in this habitat to allow for the inclusion of a detailed discussion here.

Water Mites (Hydrachnidia)

Water mites look somewhat like tiny spiders, with a round body on which there are two eyes near the front, eight legs, and a tiny head-like structure — the gnathostoma — that bears the mouth and feeding appendages (Figure 36). Those in vernal pools include migratory species and species that remain in the dry pool basin in a dormant condition. Possibly a hundred or more species of water mites occur in vernal pools in the glaciated northeast — there are sixty species in fourteen genera and eight families in temporary waters of southern Ontario and Quebec alone[1]! The accounts below provide a general summary of the life cycles and ecology of water mites in vernal pools. The chapter on water mites by Ian Smith et al. in the second edition of *Ecology and Classification of North American Freshwater Invertebrates,* edited by James Thorp and Alan Covich, is an invaluable resource for those who wish to learn more about this very interesting group of animals[2].

Water mites are dramatically colored, coming in bright reds, blues, and forest greens, as well as yellows and browns. Adults of some species reach 5 mm (about 0.25 in) in diameter. Identification requires the use of a microscope and a good key[3]. Some species are most easily identified as larvae, others as adults. All water mites are both parasites and predators, adopting different feeding modes during

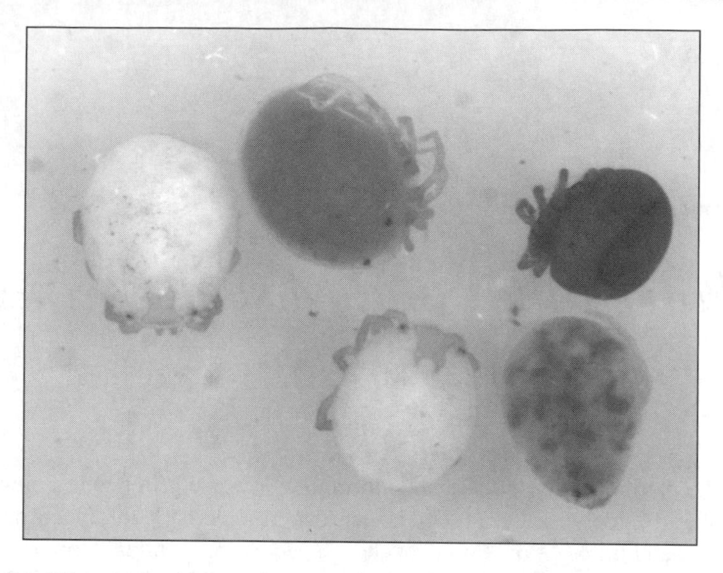

Figure 36. Water mites are predators and parasites during different stages of their life cycles, and they play important roles in regulating populations of insects and crustaceans in vernal pools. (Photograph by Holly Jensen-Herrin.)

different stages of their lives. Although small, these animals play an important role in regulating the populations of aquatic insects and some crustaceans in vernal pools[4].

The life cycles of water mites are complex with several stages including egg, larva, nymphochrysalis, deutonymph, imagochrysalis, and adult. Adult females deposit their eggs individually or in masses of up to 400 onto plants and other materials in the pool. The larvae hatch out of the eggs after one to several weeks. They are parasitic on aquatic insects and most species have highly specific host requirements. After the larvae hatch, they seek hosts, attach to the exoskeleton, insert their mouthparts through the cuticle, and feed on the host's body fluids. The larvae of different species remain associated with their hosts for varying lengths of time, from a few days to several months. After a larva has engorged, it drops back into the water and enters a non-active stage known as the nymphochrysalis, which is somewhat analogous to the pupal stage in insects. Over a period of a few days to six months, depending on the species, the nymphochrysalis undergoes structural and physiological reorganization and molts into a deutonymph, which like the larva is reproductively immature, and which is actively predaceous. Depending on the species, the deutonymph stage may last only a few days or several months before entering another transformative, pupa-like stage, the imagochrysalis or teliochrysalis. After further reorganization the imagochrysalis molts into an adult. Adults are also predators and feed on the same prey as the deutonymphs. They mate, the females lay eggs, and the cycle starts anew[5].

Quite a wide variety of life history strategies for avoiding or withstanding seasonal drawdown can be seen in this group. Water mites in vernal pools include migrants and permanent residents. Migratory species overwinter in permanent water bodies as larvae, attached to hosts that are themselves migrants. Permanent residents overwinter as larvae attached to hosts that remain in the dry vernal pool basin or as drought-resistant deutonymphs, nymphochrysalises, or adults.

The larvae of some species of water mites are parasitic on aquatic insects that breathe air and carry air bubbles under their elytra or elsewhere. These hosts include adult giant water bugs (Belostomatidae), water boatmen (Corixidae), backswimmers (Notonectidae), predaceous diving beetles (Dytiscidae), crawling water beetles (Haliplidae), and water scavenger beetles (Hydrophilidae)[6]. The larvae attach to the host in the air bubble and feed while the host is submerged. Other species parasitize adult insects during the hosts' aerial phase. Many mites are parasitic on adult dipterans, especially mosquitoes and chironomid midges. After the larvae of these mites hatch, they usually attach themselves to larvae or late-stage pupae of particular dipteran species or families, but they do not feed until the adults emerge from the pupa and become airborne. At that time, the mite larvae insert their piercing mouthparts through the host's cuticle and engorge on its blood while the host is flying over the water, and then drop back into the pool.

Family Eylaidae

About fifteen members of the Eylaidae, genus *Eylais*, are specialized for life in vernal pools. The adults are large and bright red. They are good swimmers and are highly visible as they move through the water. Most eylaids are migrants. They colonize the habitat annually, being carried in as larvae attached to migratory water bugs, water boatmen, predaceous diving beetles, and other water beetles that overwinter in permanent waters and enter vernal pools in early spring to feed and/or breed. After entering the vernal pools, the larval eylaids mature into deutonymphs, which drop off their hosts and feed on ostracods. They transform into imagochrysalises and then into adults, mating and laying eggs that hatch into larvae. The newly hatched larvae break through the water surface and crawl on the surface film until they attach to members of the Belostomatidae, Corixidae, Dytiscidae, Haliplidae, and Hydrophilidae. These hosts carry them to permanent waters, where they overwinter. Some mites have broad host plasticity, while others appear to have narrow preferences and to parasitize only one host species[7]. The following spring, the cycle is repeated.

The larvae of at least one eylaid, *E. harmani*, attach to beetles of the genus *Hydroporus*, several species of which overwinter in dry vernal pool basins[8]. Thus, *E. harmani* may be a permanent resident rather than a migrant.

Family Piersigiidae

Members of the Piersigiidae, genus *Piersigia*, are red, crawling mites reported to occur in temperate and boreal temporary pools, marshes, and swamps. The larvae of *Piersigia* parasitize water beetles, and the deutonymphs and adults prey on ostracods. These mites are closely related to the eylaids and are similar to them in their biology as larval parasites and adult predators. The life history strategies that allow piersigiids to survive drawdown in vernal pools are not yet known[9].

Family Hydrachnidae

The family Hydrachnidae, genus *Hydrachna*, is another large group, with about fifteen species that are common in vernal pools. Adult hydrachnids are large and slow-moving and can reach 8 mm (0.3 in) in diameter. They are parasitic as larvae on water bugs and aquatic beetles, and their hosts include backswimmers and water scorpions (Notonectidae and Nepidae) as well as the families listed above for eylaids. The adults and deutonymphs prey on the eggs of aquatic insects. Like eylaids, hydrachnids probably overwinter as larvae with their hosts in permanent water[10].

Family Arrenuridae

The family Arrenuridae, genus *Arrenurus*, contains a large number of mites that occur in vernal pools (see Appendix). As larvae, members of the subgenus *Arrenurus* are parasitic on dragonflies or damselflies. Other arrenurid subgenera parasitize adult chironomid midges or mosquitoes. *Arrenurus (Arrenurus) planus* is widely distributed and has a unique adaptation to drying: the nymphochrysalis develops a leathery covering and resists desiccation[11]. All other arrenurids that have been studied spend the drawdown period as adults in the pool basin[12].

Families Hydryphantidae, Limnesiidae, and Pionidae

Among other water mites common in vernal pools are members of several genera of the families Hydryphantidae, including *Hydryphantes*, *Euthyas*, *Thyas*, *Thyasides*, and *Zschokkea*; Limnesiidae, with one genus, *Limnesia*; and Pionidae, including *Piona* and *Tiphys*. The larvae in all of these families are parasitic on dipteran adults. Adult hydryphantids are large, red, crawling mites that feed on aquatic insect eggs and early instar insect larvae. Limnesiid adults are strong swimmers and are colored various shades of blue, gray, or brown. They prey on insect eggs and the larvae of chironomid midges. Pionids are dark red, and both the adults and the deutonymphs feed primarily on cladocerans and chironomid larvae.

Members of these families are permanent residents in vernal pools, overwintering as deutonymphs or adults in the damp sediment. Pionid deutonymphs embed their mouthparts into mosses on the pool bottom, and remain quiescent until the pool floods. Adults and deutonymphs in these groups show no obvious adaptations for surviving desiccation, and the overwintering stage apparently passively tolerates drawdown[13].

Host-Mite Interactions

Water mites are important predators and parasites, and they can affect the populations of prey and hosts. Adult mites influence population densities by preying on eggs and early instar larvae of aquatic insects, and on small crustaceans and other pool invertebrates. Larvae of several species may attach to the same adult host insects. Different mite species attach to the hosts at different locations, with some attaching to the wings, and others to the abdominal segments. The body burden of attached larvae can amount to a significant proportion of the host's weight[14]. The larvae of many species attach to immature insects, transferring to the next instar each time the host larva or nymph molts. They feed only when the host has reached the final adult stage. When immature hosts carry large numbers of mite larvae, their locomotion may be affected, and they could potentially be more visible to predators.

Mite larvae affect not only their hosts, but also one another. When different species parasitize the same host, they distribute themselves differently on the host's body than when they are the sole parasite. Multiple parasites show density-dependent growth and development — the size a mite reaches before it drops off the host is negatively correlated with the number of parasites the host is carrying[15].

Oribatid Mites (Oribatidae)

Members of another group of mites, family Oribatidae, only distantly related to the water mites discussed above, also occur in vernal pools. They are tiny, with a maximum diameter of 1 mm (less than 0.04 in), and to the naked eye they look like microscopic flecks of black pepper in the water. Because of their small size, they are difficult to collect without a medicine dropper or a fine-meshed plankton net. Oribatids are extremely important in breaking down leaves and other organic debris in forest soils. They can reach densities of hundreds of thousands per square meter. Most aquatic oribatids feed on algae, fungi, and decaying organic materials, and they play a role in decomposition and energy flow. The role of oribatid mites in vernal pools has not been studied.

Spiders (Araneae)

Spiders are commonly encountered in vernal pools, especially in grassy pools and other pools with extensive emergent vegetation. Fishing spiders in the genus *Dolomedes* have been shown to prey on mosquitoes in short-cycle woodland pools in Ontario[16]. No comprehensive information on either species distributions or the ecology of spiders in vernal pools is available.

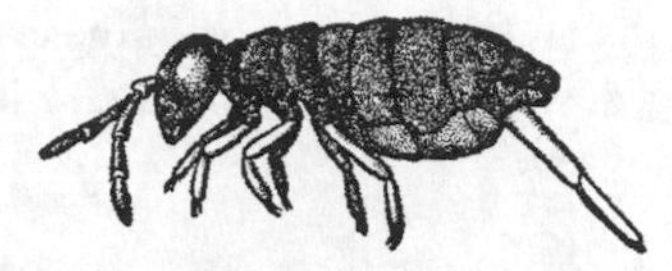

Figure 37. Hardly larger than specks of black pepper, springtails often occur in large numbers on the water surface along the edges of pools where they feed on decaying organic materials. (From Morgan 1930, courtesy of the Morgan estate.)

Springtails (Collembola)

Springtails are abundant in damp situations and in a variety of freshwater habitats (Figure 37). In fresh waters, they feed on algae and organic materials in the surface film[17]. Because of the difficulty in capturing and identifying them, members of this group are poorly known. Species in the families Isotomidae, Sminthuridae, and Poduridae have been collected from vernal pools[18], but it is not known whether any species are particularly characteristic of these waters. Several general works on freshwater invertebrates as well as specialized keys provide additional information on freshwater collembolans of North America[19].

Some Remaining Questions About Other Arthropods

- What species of water mites, spiders, springtails, and other non-insect and non-crustacean arthropods occur in vernal pools?
- What are the habitat preferences, distributions in space and time, and life cycles of these animals? Are any of these species restricted to vernal pools as habitats?
- What are the host preferences of the parasitic forms of water mites that occur in vernal pools? How specific are they?
- How important are water mites in regulating the populations of hosts and prey in vernal pools?
- To what extent do spiders affect populations of insects, crustaceans, and amphibians in vernal pools?
- What are the relationships between springtails and vernal pools?

References for Chapter 10

[1] Smith 1997. [2] Smith et al. 2001. [3] Good keys to subfamilies of water mite larvae and adults, as well as additional taxonomic references are found in Smith and Cook 1991 and Smith et al. 2001. Keys in Pecharsky et al. 1990 are also helpful. [4] Lanciani and Boyett 1980, Smith and Cook 1991, Smith et al. 2001. [5] Pennak 1978, 1989; Smith and Cook 1991; Smith et al. 2001. [6] Lanciani 1970, 1971; Smith 1997. [7] Lanciani 1970, 1971; Smith 1997. [8] James 1969, Lanciani 1971, Wiggins et al. 1980, Williams 1983. [9] Ian M. Smith,

Biosystematics Research Centre, Agriculture Canada, personal communication; Smith 1997; Smith and Cook 1991; Smith et al. 2001; Wiggins et al. 1980. [10] Smith and Cook 1991, Smith et al. 2001, Wiggins et al. 1980. [11] Wiggins et al. 1980. [12] Ian M. Smith, Biosystematics Research Centre, Agriculture Canada, personal communication. [13] Kenk 1949, Newell 1959, Smith 1997, Smith and Cook 1991, Wiggins et al. 1980, Williams 1983. [14] Lanciani 1970, 1975; Reilly and McCarthy 1993; Smith 1997. [15] Lanciani 1970, 1971, 1976. [16] James 1961. [17] Hilsenhoff 1991. [18] Williams 1983. [19] Pecharsky et al. 1990; Pennak 1978, 1989; Waltz and McCafferty 1979.

Chapter 11

Amphibians

Much of the recent interest in northeastern vernal pools has focused on the importance of these small wetlands as breeding habitat for certain amphibians, specifically mole salamanders (Ambystomatidae) and the wood frog (*Rana sylvatica*). To maintain their populations, these amphibians require breeding ponds that lack predatory fish[1]. Several other amphibians also use vernal pools as breeding sites, although they breed successfully in other types of wetlands and in permanent waters. In some cases, if permanent waters or other more typical breeding habitats are not available, local populations of some of these species depend on vernal pools.

This chapter begins with an overview of the life history strategies of amphibians in vernal pools. It then presents detailed syntheses of the literature on life history and habitat for mole salamanders and wood frogs and provides more abbreviated discussions of the biology of other amphibians that frequently occur in pools. Plates 1 and 11 to 13 illustrate some of the more common species.

The amphibians that breed in vernal pools are forest dwellers that require large areas of upland forest as habitat. Relatively little is known about the ecological role of these animals in the forest, other than that they feed on terrestrial insects and other invertebrates. They are known to travel substantial distances from vernal pools. The median distances traveled by adults of all species in their migrations from their upland habitats to vernal pools where they breed exceed 100 m (330 ft). Juveniles may travel even farther when they leave their natal pool for the first time (Table 14). As discussed in the individual species accounts, the degree of connection between woodlands and vernal pools, and the distances to other vernal pools, influence the suitability of a vernal pool as breeding habitat for amphibians.

Features of the Life History Strategies of Vernal Pool Amphibians

Vernal pool amphibians are migratory breeders that move into pools to reproduce but spend most of their lives elsewhere. Unlike many other amphibian species, they are aquatic only as eggs and larvae, and during short periods of reproductive activity during adulthood. Their life cycles are characterized by

Table 14. Summary of Amphibian Home Ranges and Migration Distances Traveled between Uplands and Vernal Pools

Wood Frog (*Rana sylvatica*)

- home ranges are small, from < 45–64.5 m^2 (<484–694 ft^2)
- genetic neighborhood is ca. 5,000 ha (12,355 ac)
- populations need to have other pools within 1,000 m (ca. 3,300 ft) to maintain gene exchange
- adults commonly migrate 400–800 m (1,300–1,600 ft) from uplands to breeding pools
- first-time breeders that disperse to new pools travel an average of 1,100–1,200 m (3,600–3,900 ft) (this represents about 15–20% of population — the rest home with high fidelity to the natal pool)

(Data from Bellis 1961, 1965; Berven and Grudzien 1990; Heatwole 1961; Windmiller 1996)

Spotted Salamander (*Ambystoma maculatum*)

- home ranges are relatively small:
  - 38.6 m^2 (415 ft^2) (males, Indiana)
  - 1–8.5 m^2 (10–91.5 ft^2) (females, Indiana)
  - 10 m^2 (107.6 ft^2) (all animals, Michigan)
  - 25–100 m^2 (270–1075 ft^2) (adults within 200 m/650 ft of breeding pond, Massachusetts)
  - 5–33 m^2 (54–355 ft^2) (juveniles within 55 m/180 ft of pool, Massachusetts)
- adult migratory travel distances differ in different studies:
  - range 6–220 m, mean 150 m (20–720 ft, mean 490 ft) (Kentucky)
  - range 18–823 m (59–2700 ft) (North Carolina)
  - range 157–249 m, mean 192 m (515–815 ft, mean 630 ft) (Michigan)
  - range 0–125 m, mean 64 m (0–410 ft, mean 210 ft) (Indiana)
  - range 35–100 m (115–305 ft) (Massachusetts)
- in eastern MA during non-breeding season, distances from breeding pools are:
  - > 100 m (330 ft) for 50% of adult population
  - > 200 m (650 ft) for 14–28% of adults
  - > 29 m (95 ft) for 50% of newly transformed, early season metamorphs
  - > 47 m (154 ft) for newly transformed, late-season metamorphs
  - > 100 m (330 ft) for 12–14% of young-of-the-year in the months following transformation
- juveniles have been found 75 m (245 ft) from breeding pools in New York, and 300 m (984 ft) from a breeding pool in Michigan

(Data from Douglas and Monroe 1981; Gordon 1968; Kleeberger and Werner 1983; Shoop 1965; Wacasey 1961; Williams 1973; Wilson 1976; Windmiller 1996)

(continued)

Table 14. (continued)

Marbled Salamander (*Ambystoma opacum*)

- adult migratory distance to breeding pools ranges from 0–450 m, mean 194 m (0–1475 ft, mean 636 ft)

(Data from Williams 1973)

Jefferson's and Blue-Spotted Salamanders
(*Ambystoma jeffersonianum, A. laterale,* and hybrids)

- home ranges vary widely
  - 1–181 m^2 (11–1,950 ft^2) (males)
  - 9.3–114 m^2 (100–1,227 ft^2) (females)
- mean adult migratory travel distances from different studies:
  - range 20–625 m, mean 252 m (66–2050 ft, mean 826 ft) (Indiana)
  - mean 250 m (820 ft) (Kentucky)
  - no more than 152 m (500 ft) (Michigan)
  - 21, 27, 107 m (69, 89, 351 ft) (Ohio)
- mean juvenile dispersal distance:
  - range 3–247 m, mean 92 m (10–810 ft, mean 302 ft) (Indiana)

(Data from Bishop 1941; Douglas and Monroe 1981; Downs 1989; Wacasey 1961; Williams 1973)

breeding early in the growing season, rapid embryonic and larval development, high larval mortality, dispersal of juveniles to upland woods, several years of terrestrial growth before sexual maturity, and repeat breeding with high site fidelity. General characteristics are summarized below, with further details provided in the accounts of the individual species.

Early Egg Deposition

Many vernal-pool-dependent amphibians are "explosive breeders," with mating and egg deposition restricted to a short period of time in early spring (most mole salamanders, wood frog, chorus frog) or fall (marbled salamander). This strategy ensures that eggs are in vernal pools early and that embryos will start to develop when water levels are high, increasing the probability that the young amphibians will hatch just as food becomes plentiful and will have grown large enough to leave the pools before drawdown.

Rapid Development

Embryonic and larval development in vernal pool amphibians is rapid. There is a minimum size threshold that larvae must reach before they can undergo metamorphosis; once they have reached this threshold, individuals can transform into juveniles, leave the pools, and move into the upland woods. Juveniles leave individual pools earlier and at smaller sizes in years when the pool dries early in the season; in any one year larvae remain longer and transform at larger sizes in pools with longer hydroperiods. The size at metamorphosis is positively correlated with the size and age of first reproduction.

High Mortality

Mortality during the aquatic phase of the life cycle is high, with only a small proportion of the deposited eggs surviving through larval development to metamorphosis. In some years, there may be 100% mortality of eggs and larvae and no recruitment of juveniles into the population.

Dispersal to Upland Woods

Upon leaving the pools, juveniles disperse into uplands where they feed on invertebrates for two to five or more years before reaching sexual maturity.

Repeat Breeding and Site Fidelity

Many vernal pool amphibians breed in more than one year. In the anurans, repeat breeding is documented in wood frogs, spring peepers, chorus frogs, and toads. Adult mole salamanders have life spans of one to two decades, and they may breed as often as every year or as rarely as every five or six years. A high degree of site fidelity, with adults returning to their natal pools to breed and with consistent repeat breeding in the same pool, has been demonstrated for many species.

The accounts that follow emphasize the characteristic vernal pool species. Brief accounts are provided for other species that commonly breed in both vernal pools and other waters.

Wood Frog

Of the true frogs (Ranidae), the wood frog (*Rana sylvatica*) is the most indicative of vernal pools. The wood frog is a distinctively marked terrestrial frog that reaches a maximum body length of 5.5 cm (just over 2 in) and has a brownish or pinkish skin and a dark mask over its eyes (plates 1, 11).

Distribution

The wood frog has the widest distribution of all North American frogs, occurring in wooded areas as far south as Arkansas, Tennessee, and South Carolina, and north to the Canadian Maritime Provinces and Alaska. Some populations can be found above the Arctic Circle. It is very common in the glaciated northeast. Wood frogs are the most common vernal pool amphibians in northern forests, and they have been found breeding in 72 to 77% of randomly sampled vernal pools in eastern Massachusetts and western Rhode Island[2].

Life History

Wood frogs are generally considered to be dependent on temporary habitats for reproduction. Their life history encapsulates the main features associated with reproduction in seasonal waters.

Breeding

This species breeds as soon as the ice in vernal pools starts to melt, usually when there is still considerable snow cover on the ground. Wood frogs are explosive breeders, breeding all together during a very short time period. Adults migrate en masse to vernal pools during evening showers in late winter or early spring — as early as January in the southern part of its range, early February in Rhode Island, mid March in Massachusetts, late March to mid April in southern Michigan, New York, Minnesota, and Ontario, and as late as July in the far north[3]. The date at which breeding begins varies from year to year; over a twenty-one-year period in southern Ontario, the date of first calling varied from 25 March to 16 April, and the date of last calling from 15 April to 6 May[4]. The size of the breeding population varies more than ten-fold among pools and, within pools, from year to year[5].

Breeding adults home with high fidelity to their natal pools. Based on the results of a seven-year study in Virginia and a similar unpublished study in Maryland, 15 to 20% of juveniles disperse from the natal pool and breed elsewhere in their first year of reproduction; the remaining 80 to 85% return to their natal pool to breed. Regardless of the pool of natal origin, once adults have selected a breeding pool they return to that pool with 100% fidelity in successive breeding seasons (based on 11,195 marked adults)[6]. Other studies provide additional evidence of high site fidelity in wood frogs. At a site with seven vernal pools in central Massachusetts, genetic analysis found greater genetic differences between wood frogs from pools separated by only a few meters than between frogs from more distant pools[7]. These results suggest that high site fidelity and pool-specific selection pressures, and not just the proximity of breeding pools, may be important for maintaining genetic diversity among subpopulations. A recent study in southern Michigan recaptured a number of marked animals at new locations, confirming that wood frogs may move long distances from their breeding pools, but fewer than 1% of breeding individuals were recaptured in pools other than those where they were first marked[8].

When it is time to breed, males arrive at the pools first. They call, challenging other males and attracting females. Calling males sound much like ducks with their quack-like vocalizations; for this reason, some people call wood frogs "the quacking frogs." Females actively select pools for breeding and avoid sites that contain fish even if males are present[9]. Males eagerly seize females or anything else that moves

— there is a special male vocalization that warns an over-eager suitor "Let go, I am not a female!" Amplexed pairs swim to parts of the pool with appropriate substrate for oviposition. There, females extrude their eggs and males release their sperm into the water, fertilizing the eggs externally. Adults remain in the vernal pools for only a few days, usually no longer than two weeks, completing their breeding activities rapidly and then returning to the surrounding forest.

Eggs

Estimates of the number of eggs deposited by individual female wood frogs vary from a few hundred to several thousand[10]. The ovaries of breeding female wood frogs characteristically contain 1,000 to 3,000 eggs. The eggs are deposited in globose masses containing from 200 to more than 895 eggs. The discrepancy between reported ovarian egg counts and the numbers of eggs in masses may result from overestimates of egg production, greater variability in production of eggs by individual females than has been recognized, or the deposition of more than one egg mass by breeding females.

The eggs in each mass are loosely aggregated without a surrounding matrix of jelly (plates 1, 11). The dark poles of the eggs face upward, and the paler halves point down toward the bottom of the pool. The egg masses are laid in communal clusters, with sometimes more than 100 females ovipositing in the same location[11]. The clusters are usually quite close to the surface in water of shallow or intermediate depths, and they are typically associated with submerged branches which help shelter the eggs from water movements. Females oviposit selectively where the water temperatures are slightly warmer than elsewhere in the pools[12]. Absorption of sunlight by the dark eggs in the large aggregations contributes to local warming of the masses, and the temperatures in the center of the communal egg cluster are warmer than at the periphery. Embryonic development is temperature-dependent, and the thermal benefits of communal oviposition help speed up hatching. Clusters are usually in the same locations in vernal pools from one year to the next. Counts of egg masses have been shown to be a good measure of variations in the breeding population from year to year[13].

Unless wood frog eggs freeze, dry, are deposited in very acidic water (pH below 4.2), or are preyed on by leeches, caddisflies, or turtles, their hatching success is high (80 to 96%)[14]. Hatching usually occurs within three weeks of egg deposition. Algae grow on the surface of the egg masses, and toward the end of embryonic development and for a period after hatching the egg masses look like large blobs of algae floating on the water's surface.

Tadpoles

Unlike the larvae of many other anuran species, wood frog tadpoles are omnivorous, feeding on algae, detritus, and dead and living animals[15]. They sometimes feed on *Ambystoma* egg masses and include toad eggs and tadpoles in their diet if the opportunity arises. Consequently, American toads preferentially avoid breeding in vernal pools containing wood frog tadpoles. The tadpoles grow rapidly, developing limbs and starting to resorb their tails by late spring.

Mortality during the larval stage is high[16]. Mark-recapture studies following tadpoles from hatching to metamorphosis in four vernal pools in Alaska showed that, on average, fewer than 4% of eggs survived to metamorphosis. Similarly, on average, only 2% of eggs deposited in a Pennsylvania population completed development into metamorphosing juveniles. Most of the mortality occurs in small tadpoles. Mortality rates are highest in the month immediately following hatching when tadpoles are particularly vulnerable to predation and decline as larval size increases and densities decrease. Predation by dytiscid beetle larvae is a significant source of mortality in young tadpoles. There was relatively low mortality (65.7%) in mark-recapture studies of month-old tadpoles in a vernal pool in New Jersey's Great Swamp during their last month of larval life.

Juveniles and Adults

Newly metamorphosed wood frogs are usually ready to leave the pools and move into the upland by early summer. The timing of metamorphosis varies with the hydrologic cycle of the pool. Animals leave earlier in short-cycle pools than in pools that retain water for longer periods of time. Transformation occurs as early as late May in Delaware, the first week of June in New Jersey, mid June in Rhode Island, and the beginning of July in Michigan and Minnesota[17]. Wood frogs are thus able to breed successfully in many spring-filling, short-cycle vernal pools. The number of juveniles leaving a pool may vary by more than three orders of magnitude from one year to the next and from pool to pool[18].

Young wood frogs take several years to reach reproductive maturity[19]. In Minnesota, males are mature two years after hatching and females after three years; three years is average for both in the Shenandoah National Park and George Washington National Forest in Virginia. The growth of juveniles is rapid following their initial dispersal, but it slows as air temperatures decrease later in the summer. The following year, growth again is rapid at first, but it decreases as individuals get larger. Food availability appears to affect both growth rates and the size of territories, with productive areas supporting more rapid growth and allowing animals to maintain smaller territories than areas with limited food.

Given the relatively low survival from eggs to metamorphosis, it has been calculated that a survival rate of at least 0.25 per year is needed for juvenile wood frogs to reach maturity and replace their parents[20]. This means that 25% or more of the population of juveniles needs to survive each year. However, this calculation does not necessarily reflect the actual amount of mortality that can be or is sustained in wild populations, because it fails to account for the amount of year-to-year variability in vernal pools and the associated variations in wood frog breeding success. It may be that in some years when there is high recruitment of young wood frogs from vernal pools, a lower juvenile survival rate would still allow adults to be replaced by offspring, and even for the population to experience an overall increase in numbers. In contrast, in years when conditions are unfavorable for the development of wood frogs in pools so that there is little or no recruitment of juveniles, only a very high juvenile survival rate would allow adults to replace themselves. In addition, adult life spans are three to five years, and some adults breed in more than one year. This multiple-breeding strategy also needs to be considered, because if an animal breeds more than once it is less critical that it successfully produce offspring in a given year. In fact, as noted earlier, the size of the breeding population of wood frogs varies by orders of magnitude from year to year, and this probably reflects variability in survival both in vernal pools and in the surrounding uplands.

Overwintering

Wood frogs overwinter under leaf litter or in shallow burrows near the surface of the ground[21]. Adults are occasionally heard calling in fall. Some males may move toward breeding pools in the fall and overwinter near the pools, and this behavior may provide these individuals with a selective advantage by providing access to females in populations where sex ratios are highly skewed and males greatly outnumber females. Adults come out of hibernacula and migrate to the breeding pools in response to warm rains associated with the spring thaw.

Unlike other anurans that breed in early spring, wood frogs do not anticipate freezing by sequestering antifreeze compounds in their blood and tissues. Instead, they produce cryoprotectants only as needed[22]. Within five minutes of the start of freezing, wood frogs accumulate high levels of glucose in the liver and leg muscles, subsequently releasing the glucose into the blood and other tissues, where it functions as an antifreeze. By preventing tissues from freezing and the associated cellular damage, this mechanism protects the animals if hibernacula are insufficiently insulated from freezing temperatures, as well as from the effects of freeze-thaw cycles during the breeding period.

Terrestrial Habitats and Home Ranges

Wood frogs are terrestrial throughout their lives, except during their larval phase and during the brief breeding period in the spring. Adult wood frogs can be found in deciduous and coniferous forests, as well as in wooded wetlands and bogs and along the moist edges of ponds[23]. In Minnesota, wood frogs are particularly abundant in tamarack, spruce, fir-ash, and basswood-maple communities, and they are the only amphibians that breed consistently in vernal pools in the northern forests of the state. They are likewise abundant in all wooded habitats except spruce swamps and Jack pine stands in Michigan. They occur widely in red maple and northern hardwood forests in New Hampshire and are common, but less abundant (29% of captures), in balsam fir forest stands. In maple and hardwood stands they are more abundant in streamside than upland habitats (61.5% and 38.5% of captures, respectively). In central Maine, this species is the most commonly encountered amphibian, accounting for 70 to 92% of drift-fence captures in lagg, forested bog, moss-*Chamaedaphne*, and shrub heath habitats. Habitat use varies seasonally; in a Massachusetts population, adults were found in upland oak forests during fall and winter, up to the spring migration, but not during summer sampling.

Newly metamorphosed juveniles typically remain near the pools for at least a week or until the pools are dry[24]. Humidity may be an important factor controlling their dispersal. Once they move into the uplands they use the same habitat as adults.

The terrestrial habitat of wood frogs may be a considerable distance from vernal pools. Individuals commonly migrate 400 to 800 m (1,310 to 2,625 ft) or more from uplands to their breeding pools (Table 14).

In the uplands, individuals establish home ranges within which they forage and overwinter. Home ranges vary in size (Table 14)[25]. In a Michigan population they average 64.5 m^2 (704 ft^2). Based on the numbers of breeding adults caught in pitfall trapping at a breeding pool, home ranges of adult wood frogs in a Massachusetts woodland are no larger than 45 m^2 (485 ft^2). Individual frogs can be found in the same territories two years in succession.

Wood frogs are present in large numbers in northeastern woodlands[26]. In a single breeding season, 4,684 individual adult wood frogs were captured at a large (0.5 ha, 1.3 ac), semi-permanent vernal pool in Concord, Massachusetts, and the estimated biomass of these animals in the adjacent 21-ha (52-ac) forest exceeds 68 kg (150 lbs). These estimates do not include individuals breeding at four other, smaller pools adjacent to the forest. The wood frog is also the most commonly encountered amphibian in New Hampshire forests; 1,200 individuals caught in pitfall traps over a two-year period represent 57% of all amphibians captured.

Despite their abundance, little is known about the ecological role of these frogs within northeastern woodlands[27]. They are active during the day and feed on ground beetles, crickets, bugs, caterpillars, other small insects, earthworms, snails, and spiders. In the far north, where there may be as few as forty frost-free days per year, adults are often found engorged with food, especially beetles and other insects.

Breeding Habitats

Small woodland pools are the preferred sites for breeding[28]. The type and amount of vegetative cover in wood frog breeding pools are variable. This species does well in densely shaded pools with heavy canopy closure. In some years, up to half of the pools that are used by wood frogs dry before the tadpoles can transform successfully into juvenile frogs. Although most reported breeding locations are small pools, calling adults or egg masses have also been reported from roadside ditches, bogs, marshes, ponds, lake edges, ruts made by heavy equipment, red maple swamps, shrub swamps, and kettle hole depressions. There are no reports in the literature as to whether wood frogs breeding in ditches and other anthropogenic sites successfully produce juveniles and establish self-sustaining reproductive populations. Data from anthropogenic pools used by wood frogs in industrial forests in Maine suggest that hydrologic variability in such pools may result in premature drying and mortality of tadpoles, and that such pools may serve as population sinks[29].

Wood frogs breed successfully in pools of varying chemistry, including acidic vernal pools with pH as low as 4.0[30]. Populations in areas with acidic soils and water appear to have developed genetic tolerance to low pH. Reduced hatching success is correlated with excessively low pH and high aluminum levels.

It is unclear whether there is a relationship between a pool's size and the number of wood frogs breeding there[31]. In a population of wood frogs in Concord, Massachusetts, most females deposit their eggs in a single, relatively large, 5,000-m^2 (3.7-ac) pool. Smaller pools nearby are used by fewer breeding individuals. A similar direct relationship between pool surface area and the number of wood frog egg masses is seen in eight Pennsylvania vernal pools ranging from 250 to 1,000 m^2 (2,690 to 10,764 ft^2) . In contrast, there is no such relationship in fifteen vernal pools ranging from 40 to 520 m^2 (430 to 5,598 ft^2) in Quebec or in 304 vernal pools in Maine. In a cluster of fourteen pools on Cape Cod, breeding by wood frogs is correlated with short hydroperiod, not pool size. The relative importance of pool size in terms of long-term population stability is unknown.

The 15 to 20% of young wood frogs that do not return to their natal pools when ready to breed travel an average of 1,140 m (males) and 1,236 m (females) (3,740 ft and 4,055 ft) to new pools[32]. The average genetic neighborhood is about 5,000 ha (12,350 ac), and genetic exchange among pools within 1,000 m (3,300 ft) of one another appears to be sufficient to prevent differentiation of the populations.

A North Carolina study casts some light on the selection of breeding pools in wood frogs that are dispersing to new breeding sites[33]. Sixteen depressions were excavated to depths below the water table by backhoe, and half were stocked with fish. Breeding by wood frogs, stocked as juveniles in the nearby woodland two to three years earlier, was observed over the following two seasons. Adult males were observed in all pools, although they were more abundant in the fishless pools, and several males were observed moving between pools. Females deposited eggs only in pools without fish. These results, in conjunction with other studies on amphibian breeding site selection, such as avoidance of pools with wood frogs by American toads and breeding site fidelity in wood frogs, suggest that, in the portion of the population that disperses, site selection is not random and the choice of breeding sites by females may be the determining factor in establishing new populations[34]. From a population perspective, this makes excellent sense, as successful breeding in wood frogs appears to be restricted to fishless pools.

Other Ranids Sometimes Found in Vernal Pools

Other true frogs that periodically occur in vernal pools are the green frog, the leopard frog, and the bullfrog. None of these is typical of vernal pools, and their biology is described only briefly.

Green Frog

Green frogs (*Rana clamitans*) typically breed in permanent water bodies, but they often breed in semi-permanent vernal pools and, occasionally, during unusually wet years, in annual vernal pools. They are greenish to brownish frogs, often with mottling and blotches, and with a body length of 6 to 9 cm (2.4 to 3.5 in) (Plate 12). Distinct dorsolateral ridges extend along both sides of the back. This species is widely distributed in water bodies and damp areas from the Canadian Maritimes to Minnesota and south to the Carolinas and Oklahoma[35]. In the midwestern United States it is absent from most of Illinois.

Adult green frogs become active in spring, and they may appear at vernal pools early in the season and feed on invertebrates including fairy shrimp. Males call over an extended period throughout the warm months of the year. Breeding may begin as early as the end of March in Rhode Island. In south-central Ontario, over a twenty-one-year period, the first calling occurred as early as 22 May and as late as 3 July; calling ended as early as 20 July or as late as 15 August[36]. In most of the glaciated northeast, this species requires one to two years to develop, depending on food availability and water temperatures. Tadpoles overwinter and transform into young frogs throughout the following summer. Water must be present continuously for at least sixteen months for green frogs to reproduce successfully[37]. Successful recruitment is thus limited to pools that remain flooded through one summer and well into the next.

Juvenile green frogs disperse widely after emerging, and they often can be found around the margins of vernal pools in summer and autumn. Adults occupy home ranges in spring and fall, and travel up to 300 m (1,000 ft) to breeding areas. They are more common in streamside habitats than in upland forest areas in New Hampshire[38].

Leopard Frog

Leopard frogs (*Rana pipiens*) are occasionally reported from vernal pools, but they are not typical vernal pool breeders, preferring temporary waters in open habitats adjacent to meadows and other grassy areas. They range in body length from 5 to 9 cm (2 to 3.5 in) and are brown or greenish with two or three rows of distinctive, rounded spots with pale borders. Like green frogs, leopard frogs have strong dorsolateral ridges running down the back.

The distribution of the leopard frog includes the entire glaciated northeast except portions of New York, Pennsylvania, New Jersey, and Connecticut, although they are not widespread within much of their range[39]. Adults range freely from water in summer and are common in grassy fields, meadows, and pastures.

Leopard frogs breed at about the same time as wood frogs in much of their range and metamorphose by early to mid summer[40]. Leopard frog tadpoles interact competitively with wood frog tadpoles when they co-occur. It may be that habitat segregation, with wood frogs typically occurring in wooded sites and leopard frogs in open ones, allows these two similar species to minimize competition and maximize survival.

Bullfrog

The bullfrog (*Rana catesbeiana*) is our most aquatic frog and, unlike the species discussed thus far, it is rarely found far from water. It generally prefers permanent water bodies with submerged and emergent aquatic vegetation for breeding and feeding. During wet periods, or in areas where it is expanding its range and other, more suitable habitat is unavailable, this species may move into semi-permanent pools to breed. This is the largest North American frog, ranging from 9 to 20 cm (3.5 to 8 in) in length. The upper surfaces of this frog are green or green mottled with gray or brown, and the underside is whitish or with a yellow tinge. It is readily distinguished from the green frog by both its size and the absence of distinct dorsolateral ridges running down the sides of the back (Figure 38).

The range of the bullfrog includes most of the glaciated northeastern United States, with the exception of western Wisconsin and most of Minnesota, and extends into the Canadian Maritime Provinces, southern Quebec, and southeastern Ontario[41].

Breeding occurs over an extended period once air and water temperatures have warmed[42]. Males establish territories and begin calling at breeding pools in late May or early June in Michigan, and they remain active into July. Females arrive

Figure 38. The bullfrog occurs in permanent and semi-permanent waters with abundant vegetation. (From Morgan 1930, courtesy of the Morgan estate.)

throughout the breeding period and most of them are reproductively active for only one night, although older and larger females may produce a second clutch of eggs. Territory characteristics, including susceptibility to leech predation, affect embryonic survival, and females appear to select males based on the quality of their territories. Most breeding males are three to four years old, although some larger and older individuals as well as some two-year-old males are active breeders.

Like green frogs, bullfrogs usually require several seasons to complete development[43], and their tadpoles typically need to overwinter once or twice before they are ready to metamorphose. However, development from eggs to transformed juveniles in less than three months has been reported for one population in a productive Missouri pond.

Bullfrogs are highly effective predators[44]. Bullfrog tadpoles have been observed to feed on hatchling wood frog tadpoles emerging from eggs, and recently transformed bullfrogs have been documented to prey on other anuran tadpoles.

Treefrogs (Hylidae)

Treefrogs are distinguished by swollen toes that act as suction devices and allow the animals to climb onto trees, shrubs, and other vertical surfaces. Among the treefrogs in northeastern vernal pools are the western chorus frog (*Pseudacris triseriata*), the spring peeper (*Pseudacris crucifer*), and the gray treefrog (*Hyla versicolor*). These species breed in a variety of habitats and are not restricted to vernal pools. A brief summary of major life history characteristics is presented below. Sources of additional information are listed at the end of this chapter.

Western Chorus Frog

Chorus frogs (*Pseudacris* spp.) are tiny (<3.8 cm, 1.5 in). They have eye stripes and some people get them confused with wood frogs. In fact, the two types of frogs are quite easy to tell apart. Adult chorus frogs are much smaller than wood frogs and they have the swollen toes characteristic of treefrogs. Their lateral stripes, unlike the simple face mask of the wood frog, run from the nostril along the shoulder and sides of the body. Chorus frogs are also distinguished by several broken or continuous longitudinal stripes down the back; wood frogs lack such stripes.

Chorus frogs are more typical of grassland areas than of woodlands and are generally absent from forested areas, but they can be found in sparsely wooded areas as well as in agricultural, suburban, and other habitats. The western chorus frog, *Pseudacris triseriata*, also known as the midland chorus frog and the swamp treefrog, is the most abundant frog in much of the Midwest, and it is often found in vernal pools in or near open areas in the upper Midwest, southern Ontario, and southwestern Quebec[45]. The ranges of two closely related species, the New Jersey chorus frog, *Pseudacris kalmi*, and the upland chorus frog, *P. feriarum*, barely extend into the glaciated northeast, in northern New Jersey and southeastern Pennsylvania. The boreal chorus frog, *P. maculata*, overlaps with the western chorus frog in northern Wisconsin and most of Minnesota. Chorus frogs are absent from New England, the Maritime Provinces, and most of New York and Pennsylvania.

Western chorus frogs are among the earliest vernal pool breeders and often migrate to their breeding sites at the same time as wood frogs[46]. They breed in vernal pools, swales, ditches, and other shallow waters, usually where emergent or submerged vegetation is present. Eggs are laid as early as February in Maryland, and in March and April in Indiana and Michigan. In southern Ontario, breeding choruses start as early as mid March or as late as the third week in April; depending on the year, breeding may be complete by the end of April or it may continue into the first week of June.

Females lay from 150 to 1,500 eggs, depositing several small masses of 20 to 100 eggs attached to grasses or other materials in shallow water[47]. As in wood frogs, embryonic development is temperature-dependent. On average, the eggs hatch in about two weeks. The tadpoles maintain neutral buoyancy and take up midwater positions in the pool. They feed on algae and plant materials. After seven to nine weeks, tadpoles transform into juveniles and leave the pools. Chorus frogs mature rapidly, and the juveniles may be ready to breed the year after they emerge.

Outside the breeding season, chorus frogs are found in a variety of habitats including open areas such as prairies and grasslands as well as wet woods, marshes, swamps, grassy swales, and dry uplands[48]. In winter, they secrete glucose into their blood, and this protection from freezing allows them to overwinter at the soil surface under logs, leaves, moss, and snow[49]. They are thus similar in their overwintering behavior to the wood frog.

Spring Peeper

Spring peepers (*Pseudacris crucifer*, formerly *Hyla crucifer*) are small treefrogs (<3.6 cm, <1.5 in) with a distinctive brown X-shaped mark on their light pinkish-brown backs (Plate 1). They are best known for their loud peeping or trilling breeding calls in early spring. They occur throughout the glaciated northeast.

Spring peepers use many kinds of flooded wetlands as breeding areas[50]. In vernal pools, they are particularly likely to breed in long-cycle pools with emergent shrubs in or bordering the basin.

This species mobilizes glucose from its liver into the blood as a cryoprotectant, thereby preventing freezing during winter and allowing individuals to be active early in spring[51]. Adults migrate to breeding pools with, or a week or two later than, wood frogs and chorus frogs; typically, this breeding activity starts near the beginning of March in Rhode Island, and in mid April in Ontario [52]. The breeding period lasts several weeks. The onset and duration of breeding are highly variable from year to year: first calling over a twenty-one-year period in Ontario ranged from 30 March to 22 April; last calling from 22 May to 17 June.

Males arrive at breeding pools first and establish the chorus. Peak calling activity of spring peepers coincides with warmer weather and is usually several weeks later than that of chorus frogs in areas where the ranges overlap. The females arrive in waves, starting soon after chorusing begins and continuing over a five-to-eight-week period. After mating and depositing up to 1,000 eggs, the females leave the pools. Breeding individuals are two to four years in age[53].

The eggs hatch in about a week, and the tadpoles remain in the pools for five to seven weeks, filter-feeding on small suspended particles in the water column. Metamorphosis may be complete by June, and transformed juveniles continue to leave the pool over a period of up to eight weeks[54]. Transformation occurs earlier, and at a smaller size than usual, if pond drying is early. If the pools dry too early, the tadpoles may be stranded and die. Newly transformed juveniles remain near the pools for a period of time before dispersing into the uplands.

Spring peepers are found on the ground and in shrubs and trees in a variety of wooded upland habitats[55]. They occur in red maple, northern hardwood, and balsam fir stands in New Hampshire forests and in tamarack, spruce, and fir-ash stands in Minnesota.

Gray Treefrog

The gray treefrog comprises two similar species[56], *Hyla chrysoscelis* (diploid, with the normal two sets of chromosomes) and *H. versicolor* (tetraploid, with four sets of chromosomes, twice as many as is usual). Like all other treefrogs, they have swollen toes that allow them to climb, but they are somewhat larger than chorus frogs and spring peepers, reaching adult lengths of up to 5 cm (2 in). Both are mottled gray with a patch of white below the eye and yellow or orange on the underside of the rear legs. Morphologically indistinguishable, they have different calls. These two species are allopatric in most of their range, with *H. chrysoscelis* becoming dominant in the south. In the glaciated northeast, most populations consist of *H. versicolor*.

Gray treefrogs breed later in the season than spring peepers, after weather has warmed[57]. In the Ottawa Valley of Ontario breeding characteristically begins in mid to late May and continues into July. In Rhode Island, adults migrate to pools in late April or early May and leave by the beginning of July. Gray treefrogs secrete glycerol derived from liver glycogen into their blood as a cryoprotectant to prevent winter freezing, and they retain this freeze-tolerance into the spring[58].

The preferred breeding habitat for gray treefrogs is permanent and semi-permanent pools[59], usually where there are flooded shrubs. In Tennessee, adults migrate 100 to 630 m (330 to 2,080 ft) to breeding ponds, and females breed at the same ponds in successive seasons. A small proportion (5%) of adults appears to breed in more than one pool in a given year[60]. Females select breeding pools and can distinguish between pools with and without snails that serve as hosts for trematodes, flukes whose free-living larvae attack treefrog tadpoles and cause significant mortality[61].The eggs hatch in less than a week, and the larvae are neutrally buoyant and take up mid-level positions in the water column, where they filter-feed on small particles. Metamorphosis usually occurs in two to three months[62].

Adult gray treefrogs are arboreal and are rarely seen outside the breeding season, although they are often heard calling from trees.

Toads (Bufonidae)

American toads (*Bufo americanus*) and Fowler's toads (*B. woodhousei fowleri*) breed in a variety of aquatic habitats throughout much of the glaciated northeast, including vernal pools. Both are brown with warty skin and reach a maximum body length of about 5 to 9 cm (2 to 3.5 in), with adult females generally 2 to 3 cm (about 1 in) larger than males (Plate 12). They are most readily distinguished from one another by the number of warts within the larger brown spots or patches on the back: one or two of these small warts per patch are characteristic of the American toad, and up to six or seven are typical of the Fowler's toad.

These species are sympatric through much of their ranges.

American Toad

The American toad occurs throughout the glaciated northeast, with the exception of outer Cape Cod. It breeds in a variety of habitats, including streamside pools smaller than 2 m^2 (20 ft^2) in Kentucky, isolated upland pools in Minnesota, and a wide range of shallow-water habitats in Ontario[63]. Female toads actively avoid breeding in pools that contain wood frog tadpoles.

Breeding takes place in late spring, starting any time from late March in the south to late April in the north, and continuing into June[64]. Adults migrate to breeding pools during warm spring rainstorms, and the males call, trilling loudly to attract females. Breeding males range from two to five years in age[65].

Toad eggs are deposited in long, double strings containing thousands of eggs. They hatch in a few days. The larvae are negatively buoyant and feed on benthic and suspended algae until metamorphosis[66]. In experimental populations of American toads, increased larval densities lead to decreased numbers and reduced sized of metamorphs, suggesting that a few animals grow well at the expense of the rest in high-density populations[67]. Emergence of metamorphs may start as early as late June and continue into October[68]. Mass emergences of hundreds or thousands of juvenile toadlets from breeding pools are not uncommon in summer.

American toads disperse for considerable distances from breeding pools[69]. Individuals have been recorded as moving more than 594 m (1,930 ft). On land, American toads feed on a variety of insects and other invertebrates. They establish small territories and may be found repeatedly within them. In a Minnesota bog, the average distance between captures of marked individuals was 13.2 m and the median distance 9.2 m (43 and 30 ft)[70]. The American toad is the second-most abundant amphibian in New Hampshire forests[71]. The greatest numbers are found in northern hardwood stands, but toads also occur in red maple and balsam fir forests.

Toads do not secrete cryoprotectants into their blood. They are thus less well protected from freezing than species such as wood frogs and spring peepers, and therefore adults burrow approximately 1 m (3.3 ft) below the ground surface in winter[72]. They remain inactive throughout the winter and do not emerge until air and soil temperatures have warmed in spring[73].

Fowler's Toad

Fowler's toad is common on the Atlantic Coastal Plain and occurs irregularly through much of the interior of eastern North America. It is absent from Minnesota, Wisconsin, and eastern Michigan, as well as from most of Pennsylvania, New York, northern New England, and Canada[74]. This species appears to prefer sandier substrates than the American toad.

Fowler's toad breeds slightly later than the American toad, starting in late May or June in southern New England and continuing well into July. Like the American toad, Fowler's toad breeds over an extended period of several weeks although, in any location, breeding may be completed within a period of only a few days[75].

Adults breed in shallow floodplain pools, isolated woodland vernal pools, and semi-permanent and permanent pools[76]. In the New Jersey Pine Barrens, where natural pools are acidic and highly colored, this species breeds in permanent *Sphagnum* bogs, seeps, and small ponds. Amplexed pairs and tadpoles are not found in pools with pH below 4.25, and there is high embryonic mortality in pools with high tannin content and low pH[77].

The eggs are deposited in long strings, often in double rows. Embryonic development is rapid, and the eggs hatch in three to four days. The larvae develop over a one to two month period. Because breeding occurs over an extended period of time, tadpoles often occur within a pool at several stages of development[78]. After metamorphosis, the young toads double their length in the first summer, and by the end of the third summer they have reached adult size[79].

On Monomoy Island, Massachusetts, a coastal barrier island, Fowler's toads are active nocturnally on sand with sparse dune grass cover[80]. They burrow into loose sand during the day and for hibernation. In other parts of southern New England, Fowler's toads arc commonly found in wooded areas, meadows, gardens, fields, and pastures.

Spadefoot Toad (Pelobatidae)[81]

Spadefoot toads (*Scaphiopus holbrooki holbrooki*) represent a special case of vernal pool amphibian, because they breed both in vernal pools and in ephemeral pools that do not retain water long enough to support a characteristic vernal pool fauna. Most typically, they breed in highly ephemeral waters, but in some locations they occur in vernal pools. They are small (4.5 to 5.7 cm, 1.8 to 2.25 in), granular-skinned toads, with a sharp callus (the "spade") on the hind foot which allows the animals to dig rapidly into sandy soils. The upper surface is yellowish, light brown, or greenish, and the underside is pale. The eyes have vertical pupils and a golden iris.

Within the glaciated northeast, spadefoot toads are restricted to southern New England, southeastern New York, northern New Jersey, and southeastern Pennsylvania. They are typically found in areas with sandy soils.

This species is an explosive breeder but, unlike the wood frog, breeding may occur any time from spring to fall following heavy rains. The toads spend their lives underground, emerging at night in wet weather to feed. They emerge in large numbers during intense rainstorms and migrate to shallow pools to breed. Males have a deep, explosive grunt, groan, or wail that carries long distances over the sound of heavy rain, and the inflated white air sacs in their throats reflect flashes of lightning. Amplexed pairs move to grasses or other egg deposition sites within the pools, and the females release strings of up to 2,500 eggs, attaching them to grass stems or other vegetation. The eggs hatch rapidly — within days — and the tadpoles feed on suspended materials, algae, zooplankton, and a variety of dead and living organisms. Depending on food availability and water temperatures, tadpoles may complete their development into juveniles in only a couple of weeks. This species thus is able to breed both in vernal pools and in shallow, ephemeral waters that do not meet the definition of vernal pools.

Spotted Salamander

Spotted salamanders (*Ambystoma maculatum*) are among the best known of the species that are considered to be vernal pool indicators. They are large animals, reaching an adult length of 15 to 20 cm (6 to 8 in). Their appearance, with distinctive orange or yellow spots on a dark background (plates 1, 13), is striking. Individual animals can be identified by the spot pattern[82]. The books on salamanders by Norman Bishop and James Petranka[83] provide thorough reviews of the literature on this species.

Distribution

The spotted salamander is the most widely distributed of the North American mole salamanders. With the exception of Minnesota, northwestern Wisconsin, and most of Illinois, it is found throughout the glaciated northeast, including the Canadian Maritime Provinces, southern Quebec, and Ontario. Farther south, this species can be found throughout the eastern United States, with the exception of parts of Delaware and most of Florida.

Life History

Breeding

Spotted salamanders breed in early spring in short-cycle, long-cycle, and semi-permanent vernal pools throughout their range. Adults migrate to their breeding pools during nighttime rainstorms, engage in courtship activities, deposit eggs, and return to the upland forest.

Breeding migrations occur at night in response to warm spring rains. The onset of breeding varies from year to year[84]. Migration may begin as early as the first week in March or as late as the first week in April in Ohio; as early as 13 March or as late as 13 April in Ithaca, New York; and as early as 18 March or as late as 8 April in Albany, New York. In an eastern Massachusetts population, some individuals arrive five to six weeks later than the peak migration, and in Rhode Island, individuals may arrive over a seven-week period beginning the second week in February. In Maine, eggs may be deposited as late as May. Depending on local conditions from one year to the next, the breeding period may be compressed into as few as nine days or may last as long as twenty-nine days[85].

Spotted salamanders exhibit high fidelity to their breeding pools[86]. It is presumed that migrating adults return to their natal pools to breed; studies documenting the proportion of a population that disperses to other pools are needed. Translocated individuals return to breeding pools and do not stray to other, more convenient pools encountered during the return. Individuals marked at a breeding pool in a Michigan woodlot were not encountered at any of ten other nearby pools during a five-year study, despite the suitability of four of these pools for breeding as evidenced by the presence of breeding populations.

Individual spotted salamanders tend to enter and leave breeding pools along the same general compass heading, and to use the same route year after year[87]. In a population in an eastern Massachusetts pool, most individuals enter and leave the pool to the north, the direction in which most of the suitable upland habitat is located. In contrast, in a North Carolina population, there is no particular directional bias in individual migratory routes.

Males usually arrive first at the breeding pools. In Ohio, males arrive within twenty-four hours of the start of rainfall when air temperatures have reached 10° C (50° F). Females have a higher temperature threshold for starting their migration, arriving within eighteen hours after the start of rainfall, at average temperatures of 11.8° C (53.2° F)[88]. Other cues may override thermal thresholds, and hard and fast rules about minimum air temperatures needed for migration do not appear to be valid.

The number of individuals that breed varies from year to year[89]. Individuals do not necessarily breed annually. In a Michigan population, an estimated 36% of males and 32% of females breed in a given year, and several individuals skipped three years during a five-year study. In a Rhode Island population, most individuals breed annually. In a Massachusetts population, the proportion that bred varied from 20 to 55% over a four-year study.

Males usually outnumber females[90]. Interestingly, in a study in Connecticut, males outnumbered females in a temporary pool but the sex ratio was balanced in a permanent pond. Spotted salamanders engage in elaborate courtship behavior, and a busy breeding pool may sport hundreds of animals swimming in sinuous

Table 15. Numbers of Eggs in Spotted Salamander Egg Masses

Location	Number of Eggs	Reference
Massachusetts	93	Portnoy 1990
Massachusetts	88	Stangel 1988
Massachusetts	100–200, mean 155	Talentino and Landre 1991
Massachusetts	51–170, mean 101	Windmiller 1996
New York	12–256, mean around 100	Bishop 1941
Ohio	55–88	Brodman 1995

whorls around one another. In the course of this mating dance, which is known as "congressing," males deposit spermatophores, which are small packets of sperm wrapped in jelly (Plate 1). Having chosen a male, a female will pick up his spermatophore by straddling it and taking it into her cloacal opening, or vent. The eggs are fertilized internally. Extra spermatophores litter sections of the pool bottom for several days. Resembling bird droppings or small pieces of bread, they provide a good indication that salamanders are breeding in the pool[91].

Eggs

Female spotted salamanders deposit one to several globose egg masses, 5 to 15 cm (2 to 6 in) across. The numbers of eggs in masses vary from a dozen to several hundred, averaging around one hundred (Table 15)[92]. In some areas (e.g., northeastern Massachusetts, parts of Maine), the number of spotted salamander egg masses in a pool is correlated with the surface area of the pool; in other areas (e.g., parts of Maine, Cape Cod), there is no apparent relationship between pool area and the number of egg masses. In one study in Connecticut, ovarian egg counts were greater, and egg diameters were smaller, in females breeding in a permanent pond than in those breeding in a temporary pond.

The egg masses are commonly attached to twigs, the stems of grasses or rushes, or other supports[93]. Depending on the conditions within a particular pool, egg masses may be laid in clusters within open water moats beneath overhanging shrubs at the edges of heavily vegetated pools, on branches in deep water, or in a variety of locations throughout the pool. The distribution of eggs within pools is non-random, with respect to both the surfaces to which the egg masses are attached and the depth of deposition. In Ohio, eggs are deposited non-randomly on logs and emergent vegetation, 10 to 25 cm (4 to 10 in) below the water surface. In the Connecticut Valley, Massachusetts, twigs with a mean diameter of 3 mm (0.12 in) are used preferentially for egg deposition, and eggs are laid 13 cm (5 in) below the water surface. In vernal pools in Rockport and Cape Cod, Massachusetts, 81% of egg masses are attached to vegetation and 19% are unattached on the pond bottom. Similarly, most of the eggs in an eastern Massachusetts pool without suitable attachment sites were on the bottom. In New York pools, egg masses tend to be in portions of pools that are exposed to sunlight during most of the day, usually 10 cm (4 in) below the surface and never deeper than 1 m (3.3 ft).

Egg masses show color polymorphism, with some eggs developing a milky translucence or a white opacity while others remain clear (Plate 1)[94]. This color polymorphism is associated with differences in the proteins in the gelatinous mass: hydrophobic protein provides opacity, while a water-soluble protein results in a clear egg mass. Masses with hydrophobic and water-soluble gel proteins do not differ in time to hatching, hatching success, or size of larvae at hatching. There is also no effect on the eggs' susceptibility to predation by caddisfly larvae. Predation by wood frog tadpoles on egg masses is reduced in opaque masses.

A symbiotic alga, *Oophila amblystomatis*, develops in the egg membrane within the egg masses[95]. Detailed studies show that the alga is specific to *Ambystoma* eggs and that its presence apparently provides benefits to the developing larvae. The alga enters the eggs from the pond water. In experimental chambers, eggs containing algae and incubated in the dark develop more slowly and have lower hatching success than eggs incubated in light. In addition, embryos in algae-free masses develop more slowly than embryos in eggs containing the alga. These data suggest that the alga actively benefits the embryos. Field data also support this suggestion. In vernal pools, egg masses deposited at greater depths have less of the alga present than eggs closer to the surface; the combination of reduced algal presence and lower incubation temperatures are implicated in the lower hatching success in eggs deposited in deeper water. It is probable that the nutrients and carbon dioxide released by the developing eggs are used by the alga in photosynthesis, and that oxygen produced by the alga benefits the embryos.

The time from the deposition of eggs to the start of hatching is variable, ranging from twenty-seven to thirty-one days after oviposition in Michigan, thirty to sixty days in Rhode Island, twenty-three to forty-eight days in Massachusetts, and forty-two to forty-nine days in New Jersey[96]. Over a four-year period, the peak of hatching in a semi-permanent pool in eastern Massachusetts always occurred in the first two weeks of May, regardless of the date of the breeding migration.

The time to hatching and hatching success are correlated with incubation temperature[97]. Laboratory incubation of eggs from a North Carolina population at constant low temperatures of 2° C (36° F) results in complete hatching failure; fluctuating temperatures averaging 10° C and 15° C (50° and 59° F) lead to high hatching success with the size of hatchlings decreasing as temperature increases.

Estimates of hatching success and larval survival vary. This variation reflects natural differences among populations, years, and breeding pools, but it also reflects differences in the ways investigators estimate the numbers of eggs deposited[98]. Such estimates are obtained by counts of eggs within masses in pools, by measurement of total egg mass volume divided by mean volume per egg, and by ovarian egg counts. All three methods have limitations. It is difficult to count eggs within egg masses, and in turbid or highly colored pools many egg masses may be missed. In anurans, the accuracy of egg estimates based on weight or volume displacement increases with increased numbers of samples. Ovarian counts are valid only if females deposit all of the eggs in their ovaries. Some gravid females migrate to pools and leave again without ovipositing[99].

Hatching success is variable, ranging from 0 to 100% (Table 16)[100]. Spotted salamander eggs hatch successfully over a wide range of pH and other chemical conditions[101]. Hatching success decreases in highly acidic pools, and tannins and aluminum appear to be correlated with decreases in embryonic survival and increases in embryonic deformity. In the Connecticut Valley of Massachusetts, hatching success is not correlated with water chemistry. The incidence of deformed larvae is low, 0 to 0.67%, and when deformities do occur they are correlated with aluminum in the pool water. Hatching success is also independent of pool chemistry in vernal pools in Pennsylvania and on Cape Cod, with the exception of highly acid pools with pH near 4.2. Hatching success is also reduced in highly acid ponds in New York, Ontario, and Maryland. A combination of laboratory and field investigations suggests that populations in areas with naturally acidic soils and water chemistry are relatively tolerant of low pH within the range found in vernal pools, but that below a threshold low pH, organic acids and aluminum interact with pH to affect embryonic development and larval survival. This threshold varies geographically.

Embryonic mortality from predation by turtles, leeches, and caddisfly larvae may be severe[102]. Spotted turtles feeding in vernal pools consume egg masses. Predation by larvae of the phryganeid caddisfly *Ptilostomis postica* results in up to 100% mortality in egg masses in Pennsylvania vernal pools and accounts for 39% of overall spotted salamander mortality in laboratory experiments (Plate 10). In a Virginia pond, 10% of embryos were killed by the related phryganeid *Banksiola dossuaria*; eight of eighty egg masses were completely macerated by the caddisfly larvae.

Larvae

The average size of spotted salamander hatchlings is correlated with ovum size and with incubation temperature[103]. Larvae are potentially vulnerable to predation from immature dragonflies and damselflies, backswimmers, phryganeid caddisfly larvae, wood frog tadpoles, and other *Ambystoma* larvae[104].

Spotted salamanders remain near the bottom of their pool during most of their larval lives[105]. They tend to congregate in shallow water, 0.5 to 1 m (1.6 to 3.3 ft) deep. In some populations the larvae stratify late in development, floating near the water surface. Experimental populations in Michigan prefer vegetated areas.

Salamander larvae are voracious predators[106]. Newly hatched larvae start to feed immediately, preying on microcrustaceans including copepods and cladocerans. Chironomid and chaoborid midges are also important food items. As larvae grow, they take successively larger prey, adding isopods, mayfly nymphs, fairy shrimp, odonate nymphs, caddisfly larvae, and larval dytiscids (*Agabus* and *Hydroporus*). Late instar larvae fed on larval newts in a New Jersey pool.

Reports differ on the numbers of spotted salamander larvae that can be found in vernal pools[107]. The densities of larvae in a Massachusetts vernal pool in mid July are constant from year to year, regardless of the number of breeding females or eggs deposited. In contrast, densities in North Carolina pools vary from year to year and from pond to pond. Growth rates and mortality appear to be density dependent.

Overall, only a small proportion of eggs survives through larval development and completes metamorphosis (Table 16). The length of the larval period and the timing of metamorphosis are variable both within and among pools and years[108]. A minimum larval size must be attained before metamorphosis is possible; if the pool

Table 16. Hatching Success and Survival to Metamorphosis in Spotted Salamanders

Location and Conditions	Hatching Success (Percent)	Reference
Massachusetts: 12 pools over 3 years	33.9–98	Jackson and Griffin 1991
Massachusetts: 1 pool over 4 years	consistently 97	Stangel 1988
Massachusetts: 13 pools	70–98	Cook 1983
Massachusetts: 1 pool	over 90	Windmiller 1996
Ohio: 1 pool over 4 years	consistently 65	Brodman 1995
Ontario: 16 pools	0–99	Clark 1986
North Carolina: 7 pools 1 year	0–100	Stenhouse 1987
and 8 pools the next year	60–100	Stenhouse 1987
Virginia: 1 permanent pond	49.3	Ireland 1989
Location	**Survival to Metamorphosis (Percent)**	**Reference**
Massachusetts	1–12.6	Shoop 1974
Massachusetts	0–3	Stangel 1988
Massachusetts	3.3–10.3	Windmiller 1996
North Carolina	0.11	Stenhouse 1987
Virginia	0.4–1.9	Ireland 1989

dries before this minimum size is reached, the larvae are stranded within the pool. Larvae from small, early-drying pools transform earlier and at a smaller size than larvae from long-cycle, semi-permanent, or permanent pools, and the number of juveniles recruited per female may be lower. In New York, the larval stage lasts sixty to one hundred days, with metamorphosis taking place from mid July through October. In New Jersey, metamorphosis may occur as soon as six to seven weeks after hatching, and it is complete by the end of June in short-cycle pools that start to draw down in June and are almost dry by July. In Michigan, larval development averages ninety-four days. In a series of semi-permanent Missouri pools, metamorphosis may extend from June to September in some years; in other years, no animals transform before November and all either die or overwinter. Similarly, metamorphosis in Massachusetts extends over several months, with late-metamorphosing animals attaining larger sizes than those that transform early.

In pools that do not dry completely, it is common for some larvae to overwinter instead of metamorphosing in summer or fall[109]. Spotted salamander larvae from one year were still present in a vernal pool on Cape Ann, Massachusetts, in July of the following year — fourteen months after they hatched. In general, spotted salamander larvae remain in vernal pools longer than larvae of other species of *Ambystoma*[110]. Spotted and tiger salamanders are the only northeastern, spring-breeding mole salamanders for which larval overwintering within breeding pools has been reported.

Juveniles and Adults

After juveniles leave the pools, they disperse during rainfall[111]. Little is known about the juvenile stage. Newly metamorphosed juveniles migrate preferentially toward upland forests and tend to follow the same orientation as the post-breeding adults. Males take two to three years to mature in Michigan and females three to five years. In an eastern Massachusetts population, preliminary data suggest that males mature the second year after metamorphosis and females a year later.

Spotted salamanders are long-lived, with adult life spans exceeding eighteen years. They continue to grow throughout their adult lives, although the rates of growth slow dramatically in older animals[112]. Adult survival over a five-year period is estimated at 79 to 94% for males and 63 to 80% for females in a Michigan population.

Terrestrial Habitats and Home Ranges

The spotted salamander is found in a variety of forested habitats[113]. Adult habitats in the glaciated northeast have been described as open field and woodland, mixed hardwood and pine forest, well-drained woodlots, thick oak-hickory forest, deciduous forest, hardwood and softwood forests, beech-maple forest, sumac-sycamore-tulip tree forest, oak forest, red maple forest, northern hardwoods, and balsam fir forest. In general, cleared areas are avoided by this species. Males crossing pavement have been found to have higher corticosterone levels than individuals migrating through forest, suggesting that habitat fragmentation and degradation may be a source of stress for this species[114].

Deciduous forest with vernal pools appears to be the optimal habitat for spotted salamanders[115]. These animals prefer dry, well-drained soils, slopes of around 10% or more, considerable leaf litter and woody debris, and an abundance of shrew tunnels[116]. The preference for well-drained soils appears consistently in studies across the spotted salamander's range, including Indiana, New York, Michigan, and Ohio.

Adults are found under logs and leaf litter or in small mammal burrows. Most of the time is spent underground: 72% of observations of six isotope-tagged individuals in Michigan found the animals underground, as did 76.8% of observations of eighteen radio-tagged individuals in Massachusetts. Members of a New York population were found equally in blind vertical burrows and small mammal burrows. Adults prefer shrew burrows to other habitats in Indiana and Massachusetts. When not in burrows, individuals in the Michigan study were under logs 21% of the time and under leaf litter 7% of the time; in Massachusetts, individuals were under leaf litter less than 0.5 m (1.6 ft) from the burrow entrance 15.2% of the time, and under litter more than 0.5 m from the burrow 8% of the time. In Massachusetts, 63% of all thirty locations in which radio-tagged individuals were found were under logs larger than 45 mm (1.8 in) diameter.

Upland habitats used by adult and juvenile spotted salamanders may extend some distance from breeding pools (Table 14)[117]. Reports of distances traveled by individuals vary and probably reflect differences in habitat quality and population size. Studies of migration by breeding adults report travel distances of from 6 to 250 m (20 to 820 ft). A significant proportion of the adult population may be more than 300 m (984 ft) from the breeding pool during most of the year.

In the forest, spotted salamanders apparently remain within home ranges or territories where they feed on earthworms, snails, slugs, spiders, crickets, millipedes, beetles, ants, other adult insects and insect larvae[118]. There is some evidence that males have larger home ranges than females (Table 14).

Breeding Habitats

Spotted salamanders breed in ditches, bogs, marshes, ponds, lakes, red maple swamps, kettle depressions, and vernal pools[119]. Long-cycle pools and semi-permanent pools with shrubs, emergent herbaceous vegetation, or twigs and branches that can serve as oviposition sites appear to be the preferred habitats, although in some locations small, short-cycle pools have high reproductive effort and high recruitment.

Marbled Salamander

The marbled salamander (*Ambystoma opacum*) is another characteristic vernal pool species. It is readily distinguished from other mole salamanders by its unique black-and-white mottled or marbled pattern. Individuals can be identified on the basis of this patterning[120]. Smaller than many other mole salamanders, marbled salamanders reach an adult length of 10 to 12 cm (4 to 5 in).

Distribution

The marbled salamander is at the northern edge of its range in southern parts of the glaciated northeast. Its distribution extends from New England to Florida and west to eastern Texas and the lower Midwest and includes southeastern New York, the southern half of Pennsylvania, southern Ohio, all of Indiana but the northeast, eastern Illinois, and a small portion of southeastern Wisconsin. This species is uncommon in central New England and northwest Ohio and absent farther north.

Life History

The marbled salamander exhibits a different life history strategy than other northeastern *Ambystoma*. The breeding migration takes place on rainy nights in late summer or early autumn. Most adults home to their natal pools to breed, with about 7% of individuals dispersing to other pools. Individual salamanders tend to follow the same routes in entering and exiting the pool basin[121]. Courtship takes place in a dry pool. Females excavate nests under logs or leaf litter, or they may use crayfish burrows or other hiding places as nests. The eggs, which can resist drying for up to several months, are deposited in the nests, and the mothers remain with their eggs, protecting them from predators, until the pool fills. Guarded eggs have greater survival than abandoned ones[122].

Marbled salamanders select nest sites at intermediate elevations within vernal pool basins[123]. This minimizes the chance that the eggs will hatch prematurely in response to small flooding events that do not provide enough water for larvae to complete development. Hatching success is greater in fall-filling pools than in spring-filling pools, but, in South Carolina, some eggs can survive the winter in the dry basin, and in New Jersey, eggs incubate 120 days or more in dry pool basins before hatching when the pool floods[124].

In a North Carolina population, egg mortality prior to hatching was 64%, and individuals surviving to metamorphosis amounted to only 8.7% of the total eggs that were deposited[125]. In Alabama, the dates of hatching and the length of the larval period vary between pools. The duration of embryonic and larval development increases from south to north.

Larval growth is slow in the first several months, but the growth rate increases in April as water temperatures warm[126]. Larval behavior varies with the stage of development and, apparently, with geographic location[127]. During the day, marbled salamander larvae tend to congregate in submerged vegetation in shallow water near the pond edge. In an Alabama population, larvae remain on the pool bottom during the day and stratify in the water column at night. As they near metamorphosis, large larvae do not stratify. In contrast, in a New Jersey population, the young larvae do not stratify, but they swim near the bottom at night, switching their behavior to float at the surface at night as they get older. In April, as the growth rate increases, the larvae again change their nighttime behavior, swimming 15 cm (6 in) below the water surface, above vegetation.

Newly hatched larvae feed on zooplankton, especially copepods[128]. Prey size increases as the larvae get larger. Larval *A. opacum* are opportunistic predators and feed both day and night. They are important predators on spotted salamander larvae and strongly influence the reproductive success of other species of *Ambystoma* in pools in which they co-occur. Zooplankton remain an important food throughout larval life, comprising up to 84% of the diet of metamorphosing larvae. Larval density influences size at metamorphosis, and size at metamorphosis in turn influences size and age at first reproduction.

Metamorphosis takes place in June in New Jersey. In South Carolina, approximately 50% of juveniles survive their first year[129]. The maximum life span is nine to ten years.

Terrestrial Habitats and Home Ranges

Adult marbled salamanders migrate an average of 194 m (636 ft) (range 0 to 450 m, or 0 to 1,476 ft) between upland habitats and breeding pools in Indiana[130]. They occur in oak forests, sumac-sycamore-tulip tree forests, deciduous woodlands, floodplain forests, mature and second-growth hardwood forests, mixed-deciduous woods, oak-maple woods, oak-hickory woods, and pine barrens[131]. In southern New England, the optimal habitat is provided by ecotone forest edges, especially associated with deciduous forest, on well drained, undisturbed slopes. In the forest they are usually found in small mammal burrows, often under thick leaf litter, or under logs. There is some evidence that individuals establish home ranges or territories and return to them each year[132]. Some individuals can be found in the same burrows two years in a row. In breeding migrations, certain animals tend to arrive each year before the others, suggesting that they travel the same distance to the pools[133].

Breeding Habitats

Marbled salamanders breed in woodland pools that fill in fall, stream-fed temporary ponds, permanent ponds altered by human activity, and spring-filling vernal pools[134]. Fall-filling pools with abundant leaf litter, moss, or other substrate under which eggs may be deposited appear to be the preferred habitats. Many sites lack emergent vegetation, but others are well vegetated. Adults home to their natal pools, and no more than 7% of juveniles migrate away from pools and ultimately breed in pools other than those they came from.

Hybridization in *Ambystoma*

Four other species of vernal-pool-dependent *Ambystoma* occur in the glaciated northeast: the Jefferson's (*Ambystoma jeffersonianum*), blue-spotted (*A. laterale*), small-mouthed (*A. texanum*), and tiger salamanders (*A. tigrinum*). In places where their ranges overlap, these species have hybridized[135]. It was thought for many years that the blue-spotted salamander had a northern distribution and the Jefferson's salamander a more southerly one, and that two types of triploid hybrids occurred in the intermediate zone of overlap. With improved cytogenetic techniques and more intensive sampling, it is now recognized that there is a hybrid complex that includes all four of these species, with numerous diploid, triploid, and tetraploid unisexual hybrids produced between them in areas of sympatry. James Petranka's treatise on salamanders provides a detailed review of the literature on the unisexual hybrids in this group[136].

Hybrids

In referring to hybrids of these four species, the following shorthand notation will be used to designate parental species: J = *A. jeffersonianum*, L = *A. laterale*, T = *A. texanum*, Ti = *A. tigrinum*.

Hybrids between *A. jeffersonianum* and *A. laterale* are most common, while hybrids involving *A. texanum* or *A. tigrinum* are relatively rare. In the summaries presented below, Jefferson's and blue-spotted salamanders and their hybrids will be discussed first, followed by separate discussions of the small-mouthed and tiger salamanders and their hybrids.

Jefferson's Salamander and Blue-Spotted Salamander

The Jefferson's salamander (*Ambystoma jeffersonianum*) is a slender salamander with a body length of 11 to 18 cm (4 to 7 in) and long toes. Its background color is grayish brown, and there are small bluish, grayish, or white flecks throughout. The blue-spotted salamander (*Ambystoma laterale*) is usually slightly smaller, 10 to 13 cm (4 to 5 in) long, chunkier, and marked on the back, legs, head, tail, and sides with distinctive flecks of bright blue to gray and white on a bluish-black background (Plate 1). The flecking is usually larger and denser on the blue-spotted salamander than in Jefferson's salamanders and somewhat resembles the pattern on old-fashioned enameled cookware. Hybrids between these two species may look like either of the parents or have characteristics intermediate between them.

Distribution[137]

The Jefferson's salamander has a limited distribution in the glaciated northeast. It occurs largely south of the greatest extent of Wisconsin glaciation, with some populations that extend northward into glaciated parts of Indiana, Ohio, western Pennsylvania, southwestern New York, western New Jersey, and western New England. Populations are also found in southern Ontario.

The blue-spotted salamander occurs in northern and east-central Minnesota, Wisconsin, Michigan, northern Illinois and Indiana, northwestern Ohio, northern and central New England, the Canadian Maritime Provinces, southern Quebec, and Ontario. It is absent from southern sections of Indiana, Illinois, Ohio, and New York, as well as from Pennsylvania, Rhode Island, and southeastern Connecticut. An isolated population occurs in east-central Iowa.

The best-known hybrids are triploids between *A. jeffersonianum* and *A. laterale*. These have the widest distributions of the hybrids in the complex. J x J x L is a unisexual (all-female) hybrid, formerly known as the silvery salamander, *A. platineum*; it is known from Michigan, Indiana, Ohio, Massachusetts, northwestern New Jersey, and Ontario. J x L x L is also a unisexual (all-female) hybrid, formerly known as Tremblay's salamander, *A. tremblayi*; it is known from Wisconsin, Michigan, Indiana, Ohio, northwestern New Jersey, Massachusetts, Maine, the Maritime Provinces, Quebec, and Ontario. *A. jeffersonianum* and *A. laterale* rarely co-occur, but *A. platineum* and *A. jeffersonianum* co-occur, as do *A. tremblayi* and *A. laterale*.

Life History

In the field, it is difficult to distinguish hybrids from the dominant diploid parent. Due to difficulties in identifying individuals to species and long-standing confusion about the taxonomy of this group, the literature is confusing about the courtship behavior, egg masses, oviposition sites, and larvae of blue-spotted and Jefferson's salamanders and their hybrids[138]. It is not at all clear, especially in older literature, which species is (or are) involved. Further, many life history studies on blue-spotted or Jefferson's salamanders have been carried out in locations where hybrids also occur. Some of these studies probably involve a mixture of diploid and triploid populations, whether expressly recognized as such or not.

Breeding

Members of the Jefferson's/blue-spotted salamander complex are explosive breeders, typically concentrating their breeding effort into a period of several days. Breeding dates may coincide with those of spotted salamanders, or they may be offset in time[139]. In much of its range, the blue-spotted salamander breeds approximately a week earlier than the spotted salamander, and in New Jersey pools where the two species co-occur the peaks of migration are offset. Similarly, the breeding migration of *A. jeffersonianum* starts several days earlier than that of *A. maculatum* in Ontario and Ohio, although in Ontario the peaks of egg deposition in the two species coincide. In a Massachusetts pool with sympatric populations, in contrast, *A. laterale* and *A. maculatum* migrate and deposit their eggs at the same time, with egg deposition occurring over a two-to-three-day period.

Breeding migrations take place at night during the first warm spring rains[140]. In a Kentucky population of *A. jeffersonianum*, both males and females migrate when daytime temperatures exceed 7.2° C (45° F), nighttime temperatures exceed 4.4° C (40° F), and precipitation occurs between 2200 and 0400 hours. These temperature thresholds are lower than those found for *A. maculatum*. Females have higher thresholds for movement than males and are less likely to migrate if temperatures are low or in the absence of precipitation; it may be adaptive for males to arrive at breeding pools first in populations where males outnumber females. In some populations, larger individuals migrate earlier than small ones[141].

Breeding occurs in early spring[142]. In an eastern Massachusetts vernal pool, breeding in blue-spotted salamanders took place the last week in March or the first week in April in each of three successive years. In Ontario, it usually occurs in early to mid April, and in southeastern Michigan the migration occurs any time from the third week in March to mid April. In an Ohio population of *A. jeffersonianum*, breeding over a four-year period took place between 6 and 19 March, depending on the year.

Breeding populations often contain a higher percentage of females than males, reflecting the presence of all-female hybrids[143]. Males court females by mounting and clasping them, and rubbing them with their snout and hind legs; once the female is responsive the male steps in front of her and deposits one to several spermatophores which the female may take up into her cloaca. The spermatophores of *A. jeffersonianum* are approximately twice the size of *A. laterale* spermatophores.

The diploid forms reproduce sexually. The all-female triploid hybrids compete with diploid females for spermatophores; the sperm activate the triploid eggs but the sperm genome is rejected and the eggs develop gynogenetically, expressing only the maternal genes[144]. Males do not court triploid females as energetically as diploid females, and they produce fewer spermatophores per female when courting triploids. As in the spotted salamander, individuals do not breed every year, with females more likely than males to skip a breeding season[145].

Eggs

The egg masses differ from those of spotted salamanders in size, number of eggs, and, usually, location[146]. In the *A. laterale/A. jeffersonianum* complex, the egg masses are smaller than in *A. maculatum*. Diploid females have greater numbers of eggs than triploids, with both J x J x L and J x L x L producing on average two-thirds as many eggs as diploids. In *A. laterale*, the eggs are usually deposited singly or in pairs on leaf litter at the edge of vernal pools. Triploid *A. tremblayi* (J x L x L) also typically deposit their eggs singly or in small clumps on twigs, leaves, and other surfaces. In *A. jeffersonianum* and *A. platineum* (J x J x L), the egg masses are larger, averaging fifteen to twenty eggs, and they are attached to twigs or the stems of vegetation in the breeding pool.

Hatching success differs depending on whether eggs are diploid or triploid[147]. A high proportion of triploid eggs is infertile. In a Michigan population, 61% of 101 egg clumps were inviable, and low egg viability has been observed for populations in Ohio, Massachusetts, and Ontario. Hatching success in diploids is comparable to that in spotted salamanders: embryonic development was completed successfully in 69 to 96% of eggs in six pools in the Connecticut Valley of Massachusetts, 65% of eggs in an Ohio population, and 90% of eggs in southern Indiana.

The sensitivity of embryonic development to pool chemistry varies geographically[148]. Hatching success is not correlated with pool acidity in populations in Massachusetts. Deformities were found in only 0.13 to 2.7% of larvae and when they did occur they were correlated with aluminum content of the water. In seven acidic Pennsylvania vernal pools (pH near or less than 4.5), mortality in naturally deposited *A. jeffersonianum* eggs was 20% in one pool, 80% in a second, and 100% in the remaining five. In general, members of the Jefferson's/blue-spotted salamander complex are more sensitive to low pH than the spotted salamander.

Reproductive effort in a given pool may be high[149]. Estimated densities of *A. jeffersonianum* eggs are 21 to 350/m^2 (mean 123 eggs/m^2) (2 to 35, mean 12.3 eggs/ft^2) for an Indiana vernal pool, and 57/m^2 (5.7/ft^2) for an Ohio pool that also supports *A. maculatum*. Typical estimates of the numbers of Jefferson's/blue-spotted salamander hatchlings in vernal pools range from 30,000 to 84,000.

The eggs develop rapidly, averaging twenty-eight days to first hatching in New Jersey, fifteen to eighteen days in Michigan, and thirty days in eastern Massachusetts[150]. The rate of embryonic development increases with increasing water temperatures. Hatchlings are larger when incubated at low temperatures.

Larvae

Newly-hatched larvae stratify in the pools and hover in the water column above vegetation[151]. This behavior is maintained throughout the larval stage. The larvae feed on microcrustaceans, including cladocerans, copepods, and ostracods, and on mosquito larvae. As they grow they also feed on isopods, chironomid midges, phantom midges, beetle larvae (*Hydroporus* and *Agabus* spp.), and water boatmen. In a Michigan population in four small woodland pools, larvae feed on crustaceans in spring and switch to chironomids and oligochaete worms as these prey become abundant in summer. The leech *Batracobdella picta* commonly parasitizes larvae of *A. laterale* in June in Michigan vernal pools[152].

The larvae grow rapidly[153]. Diploids develop faster than triploids, typically completing metamorphosis a week earlier. Larvae that develop more slowly become larger than fast-growing larvae, and larval size is inversely related to water temperature. The larvae of both diploids and triploids grow more rapidly, have greater survival, and reach a larger size at lower densities, suggesting that food may be limiting. Triploids produce only female offspring, and triploids leaving the pools are larger than diploid juveniles.

Members of the Jefferson's/blue-spotted salamander complex do not appear to overwinter as larvae[154]. Individuals complete metamorphosis and leave the pool seventy to seventy-eight days after hatching in Massachusetts, after sixty-six to eighty days in southern Indiana, and after eighty-eight to ninety-four days in Michigan. Larval mortality is high in the weeks preceding metamorphosis. The survival to metamorphosis is comparable to that in the spotted salamander: an estimated 0.07 to 0.6% of the eggs that were laid over a two-year period produced juveniles in Indiana.

Juveniles and Adults

Following transformation, juveniles leave their natal pools and move into the upland, where they feed for several years until they reach sexual maturity[155]. It has been estimated that blue-spotted salamanders mature at two years of age. In members of the Jefferson's/blue-spotted salamander complex in Ontario, males mature in two years, and tetraploids apparently mature later and at a larger size than triploids. Based on studies of growth rates in *A. jeffersonianum* and drift-fence sampling of breeding adults over a four-year period, the estimated maturation period in a population in Indiana is three years for males and four years for females.

Terrestrial Habitats and Home Ranges

Like spotted and marbled salamanders, members of the Jefferson's/blue-spotted salamander complex occur in a variety of forested conditions but appear to be preferentially distributed in forested areas with well-drained soils containing burrows and surficial woody debris[156]. In New Jersey, adult blue-spotted salamanders occur in hardwood forests near ridges, and adults and juveniles are found under logs near ponds in densely wooded areas. In New York, these salamanders occur in burrows along roots of deciduous trees, in deciduous woods, and in mixed beech-hemlock-oak-maple forests. Indiana populations are found in oak forests and beech-maple forests that are fifty to seventy years old, with distribution apparently independent of forest type. Adults occur in both hardwood and softwood forests in the Connecticut Valley, Massachusetts, and in second-growth deciduous forests in southwestern New England. Individuals often overwinter near vernal pools. Jefferson's and blue-spotted salamanders establish home ranges similar to other salamanders (Table 14). Some animals may return to the same location two years in a row. Metamorphs of *A. jeffersonianum* prefer soils with higher pH. Laboratory studies on Pennsylvania populations show that body water and sodium content decrease over time in acidic soils[157].

The distances traveled by dispersing juveniles of the Jefferson's/blue-spotted salamander complex from breeding pools average 92 m (302 ft) in southern Indiana (range 3 to 247 m, 10 to 810 ft)[158]. Some adults remain near breeding pools, while others may travel more than a mile (Table 14).

Breeding Habitats

Members of the Jefferson's/blue-spotted salamander complex breed in small, temporary woodland pools with heavy canopy cover, temporary ponds, small shallow woodland pools, grassy ditches, pasture ponds, red maple swamps, roadside ditches, limestone sinkholes, abandoned gravel pits, farm ponds, and semi-permanent woodland pools[159]. Small pools with forest canopy cover and bordering forest appear to be the preferred habitats, particularly for Jefferson's salamanders. In Michigan, *A. jeffersonianum* greatly outnumbers *A. maculatum* in small woodland pools. In Ontario, *A. laterale/A. tremblayi* are observed more often in grassy field ponds and roadside ditches than *A. jeffersonianum/A. platineum*, which prefer woodland ponds. Blue-spotted salamanders in southwestern New England and southeastern New York appear to breed preferentially in floodplains rather than in isolated woodland pools[160].

Short-cycle pools do not always retain water long enough for these salamanders to complete their life cycles[161]. Breeding can be successful in long-cycle pools that dry by August, and many breeding pools are semi-permanent or permanent. In some cases, the longer hydroperiod is the result of human alteration of the habitat and may not reflect the preferred habitat for the population.

Jefferson's and blue-spotted salamanders are more sensitive than spotted salamanders to low pH and cannot reproduce successfully in highly acidic waters[162]. Studies of large numbers of vernal pools in Pennsylvania show that these species do not breed in pools with low pH and high aluminum, sulfate, and zinc, and laboratory studies of embryonic survival in relation to these water quality parameters predict the field distribution.

Small-Mouthed Salamander

The small-mouthed salamander (*Ambystoma texanum*) is medium sized, with an average body length of 13 to 15 cm (5 to 6 in), and with a slender head and a small mouth. The background color of the back, legs, tail, and head is dark brown to black with lighter grayish blotches, especially on the sides. The underside is lighter brown to yellowish.

Distribution

The small-mouthed salamander's range is largely southern and western[163]. This species does not occur throughout most of the glaciated northeast, but populations are known from southwestern Ohio, southeastern Michigan, Illinois, southern Indiana, and Pelee Island, Ontario. It sometimes co-occurs with other species of *Ambystoma*. In Michigan it co-occurs in a pool with *A. laterale*, *A. tremblayi*, and *A. maculatum*; *A. tigrinum* is also occasionally present but there is no evidence that the latter breeds in the pool. On Pelee Island, Ontario, the small-mouthed salamander co-occurs with the blue-spotted salamander, and hybrids between these two species are also present.

As noted above, small-mouthed salamanders have hybridized with both Jefferson's and blue-spotted salamanders in areas of geographic overlap[164]. Diploid hybrids L x T and triploid hybrids L x L x T have been identified in Ontario, and triploid hybrids J x L x T have been identified in Ohio and Michigan.

Life History

The small-mouthed salamander has two forms with very different breeding strategies[165]. One form breeds in streams, gluing its eggs individually to the undersides of rocks. This form is now known as *Ambystoma barbouri.* The other form breeds in shallow, temporary woodland pools and ditches, depositing eggs in masses on the surfaces of vegetation or detritus. It is thought that the stream form evolved from the pond form in parts of the species' range where geologic conditions do not support ponds with hydroperiods long enough to allow for the successful development of the larvae.

Breeding in vernal pool populations parallels that of other spring-migrating salamanders[166]. It begins as early as February in Ohio, with eggs deposited singly or in small masses attached to sticks or vegetation. The eggs hatch in three to five weeks. Small larvae stratify and are actively swimming predators; large larvae spend more of their time on the bottom as crawling predators. In Ohio and Indiana, the larvae are ready to transform by June.

Habitats[167]

In the northern part of its range, the small-mouthed salamander is common in hilly areas under logs and debris. In spring, individuals can be found under logs and debris near ponds and swamps and in river bottoms. Indirect evidence suggests that extensive loss of uplands surrounding breeding pools leads to population declines in this species.

Tiger Salamander

The eastern tiger salamander (*Ambystoma tigrinum tigrinum*) is the largest of the northeastern mole salamanders, reaching an adult length of 25 cm (10 in) or more. This is a stout, chunky salamander with a deep brown or dull black background color overlain with marbling of pale golden, olive, or yellow-brown blotches or spots. These blotches often form continuous bands along the sides and tail. They are irregularly shaped, in contrast to the rounded or oval spots of the spotted salamander.

Distribution

The tiger salamander (*Ambystoma tigrinum*) evolved in the American southwest and moved north in the early Pliocene[168]. In addition to the eastern subspecies (*A. t. tigrinum*), there are several western races. In the glaciated northeast, the tiger salamander occurs from central Ohio west through Indiana and Illinois, north to the southern half of Michigan, across most of Wisconsin and all but the northeastern part of Minnesota, and in a few locations in southern Ontario. In northern Ohio and southern Michigan it is sympatric with other species of *Ambystoma* in an association that is believed to have been stable for 4,000 to 5,000 years. Populations are known from central Long Island, New York, and southern New Jersey. The tiger salamander is absent from the Appalachian highlands and New England, and it is probably extirpated from Pennsylvania[169].

Triploid hybrids J x L x Ti have been identified in Ohio and Michigan, and tetraploid hybrids J x J x L x Ti have been identified in Indiana[170].

Life History

Tiger salamanders show great plasticity in their life cycles, depending on the geographic location and habitat conditions. In populations that use vernal pools, adults migrate to breeding pools in late winter or early spring, engage in courtship activity, and deposit eggs in masses of fifteen to fifty eggs attached to sticks, plants, or the pool bottom[171]. The onset of breeding varies with latitude. In southern New Jersey, breeding begins in January, and eggs are deposited from January to early March, while in the northern part of the range, breeding starts in March. Females lay several egg masses with approximately forty-four eggs each; the total number of eggs per female is approximately 250 to 350. The egg masses differ from those of spotted salamanders in having a thicker gelatinous envelope and, on average, fewer eggs per mass. The eggs are typically deposited 15 to 35 cm (6 to 14 in) below the water surface in water depths of 35 to 100 cm (14 to 40 in).

Females select favorable oviposition sites, and they may also select favorable ponds[172]. At a New Jersey site with twelve breeding pools, almost all of the eggs are deposited in only three of the available pools, with more than half of the eggs in a single pool; the rest are distributed sparsely among the nine other pools. Embryonic development is rapid, with some eggs hatching in as few as twenty-four to thirty days after oviposition, typically when water temperatures have increased to 9° C (48° F) or above.

As in other species, a high proportion of eggs fails to hatch, and most of the larvae die before metamorphosis[173]. Average survival from eggs to late larvae is 3.3%. Sources of embryonic mortality include feeding on egg masses by caddisflies, low pH, and freezing of egg masses in ice at the pond surface. Two major periods of egg mortality have been identified, one associated with the gastrula and neurula stages and probably exacerbated by low temperatures, and the second associated with hatching — about one-third of the mortality in the pre-larval phase is associated with hatching. Large masses experience the least mortality, small masses the most. Newly hatched larvae may be eaten by marbled salamander larvae or larger tiger salamander larvae, if they are present, and predation by red-spotted newts may also have significant effects on populations of tiger salamanders. No data are available on the percentage of late larvae surviving through metamorphosis.

New Jersey researchers[174] suggest that mortality at different stages of development may select for different genetic characteristics. According to the model proposed by these investigators, early embryonic mortality removes unfit genotypes, mortality of gastrula and postgastrula selects for appropriate responses to local climatic conditions, mortality during late embryo and hatching stages selects for adaptations to specific environmental factors in ponds, hatchling losses are associated with density-dependent biotic factors in ponds, larval mortality selects for species characters associated with niche and behavior, and mortality at metamorphosis responds to physiological stresses and selects for adult-favoring characters.

Tiger salamander larvae are nocturnal and are fierce predators. They feed on dragonfly nymphs, mosquito and phantom midge larvae, wood frog tadpoles, and larvae of other *Ambystoma* salamanders[175]. The larval stage typically lasts for three to five months, and metamorphosis can be completed as early as June in some populations[176]. Juveniles disperse from pools from July through September. In permanent waters, larvae may overwinter, metamorphosing the next year at very large sizes. Juveniles dispersing from semi-permanent or permanent pools are larger than larvae from pools that dry in mid summer and may reach sexual maturity later[177]. Males typically mature in two years, females in three to five years. Under favorable conditions, some individuals may mature the spring following dispersal from the pool[178].

Habitats[179]

The tiger salamander breeds in temporary and permanent ponds near or within mixed hardwood or coniferous forests. Permanent ponds used for breeding typically are fishless. This species sometimes co-occurs with wood frogs and other vernal pool ambystomatids, including *A. maculatum*, *A. laterale*, *A. jeffersonianum*, and hybrids of the *laterale/jeffersonianum* complex, but it is also often the only amphibian breeding in some vernal pools. Adults can be found in mole burrows, under logs, and in tunnels they dig themselves.

Red-Spotted Newt[180]

Red-spotted or eastern newts (*Notophthalmus viridescens*) (Salamandridae) typically breed in permanent waters, but individuals of some populations breed in vernal pools. Adult newts reach lengths of 8 to 12 cm (3 to 5 in). The head and back are dark olive, and a row of small, red or orange circles ringed with black extends down either side. The underside is pale yellow. Juveniles are terrestrial and bright red-orange with the same lateral row of circles as the adults.

The red-spotted newt is found throughout the glaciated northeast. Unlike the mole salamanders, this species is aquatic in both the adult and larval stages. There is an intermediate juvenile stage, the red eft, which is fully terrestrial and which may move long distances during a two-to-seven-year period of maturation, after which it develops a keeled tail and undergoes other transformations in preparation for aquatic life as an adult. Newts can be significant predators on invertebrates, mole salamander larvae, and other amphibians in vernal pools[181].

In lake populations, the adults are permanently aquatic, but in populations breeding in temporary pools and in shallow permanent ponds, the adults alternate between the aquatic habitat and land. Vernal-pool-breeding populations appear to be widely distributed. In New York and the Ottawa Valley of Ontario, this species is found in small lakes and semi-permanent pools; in the latter, adults move onto land when the pools dry during drought periods[182]. In Virginia, adult newts typically overwinter on land and migrate to the pools to breed in early spring[183]. My colleagues and I have found adult and larval red-spotted newts in flooded vernal pools on Cape Cod and in central Massachusetts, and adults and juveniles on land near the drawn-down pools.

Adult newts have an extended breeding season from fall to spring, but many females only lay eggs in spring[184]. The eggs are deposited singly, attached to vegetation, sticks, or leaves, over a period of several weeks, and the larvae hatch in three to five weeks, depending on water temperatures. Newt larvae are voracious predators and feed on zooplankton, aquatic insects, and the larvae of other amphibians[185]. In mid to late summer to fall, after two to five months of aquatic life, the larvae transform into bright orange, terrestrial juveniles known as red efts. The juveniles spend several years feeding on the forest floor before transforming into mature, aquatic adults[186].

Lungless Salamanders (Plethodontidae)

Lungless salamanders include the common terrestrial redback salamander, *Plethodon cinereus*, and the four-toed salamander, *Hemidactylium scutatum*. Neither of these species is a vernal pool indicator, but both are sometimes found in association with vernal pools.

Redback Salamander[187]

The redback salamander (*Plethodon cinereus*) has a fully terrestrial life cycle, living and reproducing on land. Adults sometimes move into vernal pool basins as the pool dries, and they can often be found under woody debris or leaf litter in dry pools. It is thought that the high relative humidity, as well as invertebrate prey, such as spiders, isopods, and molluscs, attract these salamanders to the dry pool basins.

The redback salamander is small, with a length of 6 to 9 cm (2.3 to 3.5 in) and a broad, rusty-red or gray, median dorsal stripe extending down the back. All-red or erythrystic individuals that look much like the juvenile red eft stage of the eastern newt occur in some areas. This species is widely distributed in wooded habitats throughout the glaciated northeast.

Four-Toed Salamander[188]

The four-toed salamander (*Hemidactylium scutatum*) inhabits damp woods and lays its eggs in moist *Sphagnum* moss overhanging pools or woodland streams. It is sometimes found along the edges of vernal pools that have *Sphagnum* tussocks along the edges or interior of the basin.

These salamanders are small, 5 to 7.5 cm (2 to 3 in) in length, with a dark reddish-brown back and a white underside dotted with black, star-shaped spots. There is a distinct constriction at the base of the tail. This species is widely distributed in the eastern United States and Canada, but it occurs only patchily within this broad range. In the glaciated northeast, populations have been reported from all states and provinces except Prince Edward Island, Labrador, and Newfoundland, but they are considered to be stable only in Ontario, Michigan, Pennsylvania, New York, and Connecticut and are uncommon to endangered elsewhere within the region.

The four-toed salamander has an interesting life history[189]. Adults mate from late summer to late fall, and the females store the sperm over the winter within their reproductive tracts. In late spring, they lay thirty to forty eggs in nests hollowed out in *Sphagnum*. Several females may share a nest, and one of them remains and guards the eggs until they hatch. Incubation usually takes one to two months. Upon hatching, the larvae drop into the water where, like other salamanders, they prey on a variety of aquatic invertebrates. Metamorphosis is complete within six weeks. The young salamanders feed on invertebrates on land for three years before becoming sexually mature.

Co-Occurrences of Amphibians in Vernal Pools

It is common for several species of amphibians to co-occur in northeastern vernal pools[190]. As discussed in Chapter 7, when closely related species occur together, the potential for competition exists unless resources such as food or space are not limited. For example, it could be that even though all of the salamander larvae that occur in vernal pools are fierce predators, there is enough food in the water to support several species in one pool. Alternatively, species might be distributed spatially so as to minimize interactions with one another, or key events in their life cycles might be temporally offset. Multi-species communities of amphibians have been studied extensively as models for examining competitive interactions among species. Sorting out the alternatives can be quite challenging!

Wood frogs may be the only amphibians that breed consistently in the shallowest and least permanent pools, but more commonly, short-cycle, woodland vernal pools typically support breeding by two or more species, including wood frogs, chorus frogs, spring peepers, spotted salamanders, and Jefferson's or blue-spotted salamanders. Up to a point, the probability of reproductive success seems to increase for mole salamander populations in long-cycle pools. Later-breeding anurans such as gray treefrogs, American toads, and Fowler's toads also breed in vernal pools where the larvae of early spring species may already be present. Increasing hydroperiod contributes to a greater species richness, with eastern newts and green frogs becoming increasingly common in semi-permanent pools and bullfrogs appearing in well-vegetated permanent waters.

Interactions among individuals and species within amphibian communities in vernal pools are highly complex[191]. Density-dependent effects on growth and development have been identified within single- and multi-species assemblages of both anurans and salamanders. This means that the number of tadpoles or larvae present affects the growth rate, mortality, and/or size at metamorphosis. Interactions include direct competition for food and inter- and intra-specific predation. The nature of the interactions between any two species may vary among pools, depending on site-specific conditions such as the timing of oviposition and hatching and the relative sizes of tadpoles/larvae of the different species present. Studies of individual pools over time have shown dramatic year-to-year variations in the numbers of individuals of different species reaching metamorphosis, as a result of these interspecies interactions. For instance, in a pool with breeding populations of Jefferson's and spotted salamanders in Berkshire County, Massachusetts, one of the two species usually succeeds in recruiting juveniles each year, but the year's "winner" varies from one year to the next and is apparently associated with which species' eggs are deposited and hatch the earliest[192].

Some Remaining Questions about Amphibians

- What physical and/or biological features are associated with the use of particular pools by particular species of amphibians?
- How does the breeding biology of vernal pool amphibians vary with weather and climate? When do adults arrive at a pool, how long do they stay, when do they lay eggs, when do they leave, and how do these relate to snowpack, temperatures, rainfall, and other meteorological variables?
- To what extent do the amphibian communities of semi-permanent vernal pools differ from those of annual vernal pools, in terms of the species that are present, inter- and intraspecific interactions, and population biology?
- To what degree are clusters of vernal pools important to maintaining genetic diversity within populations of vernal pool amphibians?
- How do anthropogenic pools differ, if at all, in their contributions to amphibian populations? Do they differ in food composition, quantity, or quality?

Do they differ from undisturbed pools in water quality or hydrology? What features of such pools contribute to successful recruitment of juveniles into the population?

- To what extent do small, short-hydroperiod pools contribute to amphibian populations? Are these pools sinks, sources, or both, depending on the year?
- How does recruitment from a given pool vary over the long term in association with between-year changes in hydroperiod?
- What are the metapopulation genetics of vernal pool amphibian populations throughout the glaciated northeast?
- What kind of connectivity is needed between vernal pools, and between forested areas containing vernal pools, to maintain genetic exchange and population viability for vernal pool amphibians?
- Is homing back to the natal pool universal among amphibians that breed in vernal pools? What proportion of first-breeding adults stray to new pools? How does among-pool movement vary with between-year variations in flooding regime? Are there differences between males and females in breeding-site fidelity?
- What are the upland habitat needs and uses of forest amphibians breeding in vernal pools?
- How is historic land use in currently forested areas related to the distribution of terrestrial amphibians that breed in vernal pools? Do genetic analyses suggest that populations radiated from historically forested "islands" into second growth forests?
- How much protected upland is needed around vernal pools to provide adequate terrestrial habitat for vernal pool amphibians? What are the habitat characteristics that need to be provided within this area?
- What are the roles of woodland amphibians in forest ecology, as predators on soil fauna, maintainers of soil structure (through burrowing activities), and influences on seedling recruitment (indirectly through predation on seed-eating invertebrates)?
- What are the effects of forestry practices on vernal pool amphibians, and how can forestry activities be modified to avoid adverse impacts on vernal pools and their wildlife?
- How well do tunnels under roads work to allow unimpeded migration of breeding adults and post-metamorphic juveniles between upland woods and vernal pools?
- Do drawn-down vernal pools serve as summer refugia for terrestrial amphibians, including redback salamanders?

References for Chapter 11

[1] Hopey and Petranka 1994, Massachusetts Division of Fisheries and Wildlife 1988, Stone 1996. [2] Burne 2000a, Calhoun et al. 2003, Conant 1975, Palik et al. 2002, Paton and Egan 2001. [3] Bellis 1965, Herreid and Kinney 1966, Martof and Humphries 1959, Meeks and Nagel 1973, Paton 2000, Seigel 1983, Werner 1992. [4] Cook 1981. [5] Berven and Grudzien 1990. [6] Berven and Grudzien 1990. [7] Beatini 2003. [8] Johnson and Johnson

2002; Catherine Johnson, Monongahela National Forest, personal communication. [9] Hopey and Petranka 1994. [10] DeGraaf and Rudis 1984, Herreid and Kinney 1966, Meeks and Nagel 1973, Seale 1982, Seigel 1983, Wright and Wright 1949. [11] Herreid and Kinney 1967, Martof 1970, Seale 1982, Wright and Wright 1949. [12] Seale 1982, Waldman and Ryan 1983. [13] Crouch and Paton 2000. [14] Gascon and Planas 1986, Herreid and Kinney 1966, Rowe et al. 1994, Sadinski and Dunson 1992, Seigel 1983, Wilbur 1971. [15] Bellis 1965, Petranka 1998, Petranka et al. 1994, Zappalorti 1993. [16] Herreid and Kinney 1966, Seale 1982, Seigel 1983. [17] Bellis 1965, Paton 2000, Seigel 1983, Werner 1992, Zappalorti 1993. [18] Berven and Grudzien 1990. [19] Bellis 1961, 1965; Berven and Grudzien 1990. [20] Herreid and Kinney 1966. [21] Zweifel 1989. [22] Regosin et al. 2003b; Storey and Storey 1986a, 1986b. [23] Bellis 1961, 1965; DeGraaf and Rudis 1983, 1990; Heatwole 1961; Knox 1992; Marshall and Buell 1955; Stockwell and Hunter 1989; Windmiller 1996. [24] Bellis 1965, Heatwole 1961. [25] Bellis 1965, Windmiller 1996. [26] DeGraaf and Rudis 1990, Heatwole 1961, Heatwole and Getz 1960, Stockwell and Hunter 1989, Windmiller 1996. [27] Bellis 1961, 1965; Gilhen 1984; Martof and Humphries 1959; Zappalorti 1993. [28] Bachmann and Bachmann 1994, Bellis 1965, Calhoun et al. 2003, Clark 1986a, Cook 1981, Dale et al. 1985, Gascon and Planas 1986, Heatwole and Getz 1960, Jokinen and Morrison 1998, Knox 1992, Ling et al. 1986, Rowe and Dunson 1993, Sadinski and Dunson 1992, Seale 1982, Seigel 1983, Skelly 1998, Waldman and Ryan 1983. [29] DiMauro and Hunter 2002. [30] Freda and Dunson 1985a, Gascon and Planas 1986, Ling et al. 1986, Pierce and Wooten 1992, Sadinski and Dunson 1992. [31] Calhoun et al. 2003; Gascon and Planas 1986; Seale 1982; Windmiller 1990, 1996. [32] Berven and Grudzien 1990. [33] Hopey and Petranka 1994. [34] Berven and Grudzien 1990, Petranka et al. 1994. [35] Conant 1975. [36] Cook 1991, Dickerson 1969. [37] Paton 2000. [38] DeGraaf and Rudis 1990, Martof 1953. [39] Conant 1975. [40] Werner 1992. [41] Conant 1975. [42] Emlen 1968, 1976; Howard 1978. [43] Emlen 1976, Paton 2000, Seale 1980. [44] Sarah Allen, Normandeau Associates, personal communication; Munz 1920. [45] Conant 1975; Cook 1981, 1992; Miersma 2003; Northern Prairie Research Center 2002. [46] Cook 1981, Dickerson 1969, Gee and Waldick 1995. [47] Harding 1997. [48] Conant 1975, Dickerson 1969. [49] Edwards et al. 2000, Froom 1982. [50] Cook 1981, Dickerson 1969, Lykens and Forester 1987, Tyning 1990. [51] Storey and Storey 1987. [52] Cook 1981, Dickerson 1969, Paton 2000. [53] Duellman and Trueb 1986, Lykens and Forester 1987. [54] Dickerson 1969, Paton 2000. [55] DeGraaf and Rudis 1990, Marshall and Buell 1955. [56] Conant 1975, Ptacek et al. 1994, Tyning 1990. [57] Cook 1981, Paton 2000. [58] Storey and Storey 1987, Wells et al. 1995. [59] Cook 1981, Wells et al. 1995. [60] Ritke et al. 1991. [61] Kiesecker and Skelly 2000, 2001. [62] Gee and Waldick 1995. [63] Bellis 1959, Cook 1981, Holomuzki and Hemphill 1996, Petranka et al. 1994. [64] Cook 1981, Paton 2000. [65] Acker et al. 1986. [66] Gee and Waldick 1995. [67] Wilbur 1977a. [68] Paton 2000. [69] Oldham 1966. [70] Bellis 1959. [71] DeGraaf and Rudis 1990. [72] Froom 1982, cited in Storey and Storey 1987. [73] Dickerson 1969. [74] Conant 1975. [75] Freda and Dunson 1986. [76] Dickerson 1969, Tordoff 1965. [77] Freda and Dunson 1986. [78] Dickerson 1969. [79] Tordoff 1965. [80] Dickerson 1969, Tordoff 1965. [81] Conant 1975, DeGraaf and Rudis 1983, Tyning 1990. [82] Doody 1995, Loafman 1991, Stenhouse 1985. [83] Bishop 1941, Petranka 1998. [84] Bishop 1941; Brodman 1995; Aram Calhoun, University of Maine, personal communication; Paton 2000; Windmiller 1996. [85] Husting 1965, Talentino and Landre 1991. [86] Husting 1965, Shoop 1968, Whitford and Vinegar 1966. [87] Shoop 1965, 1968; Stenhouse 1985; Windmiller 1996. [88] Brodman 1995, Petranka 1998. [89] Husting 1965, Shoop 1974, Whitford and Vinegar 1966, Windmiller 1996. [90] Husting 1965, Whitford and Vinegar 1966, Woodward 1982a, Windmiller 1996. [91] Bishop 1941, Petranka 1998, Tyning 1990. [92] Bishop 1941, Brodman 1995, Calhoun et al. 2003, Kenney 1991, Portnoy 1990, Stangel 1988, Talentino and Landre 1991, Windmiller 1996, Woodward 1982. [93] Brodman 1995, Landy 1967, Portnoy 1990, Stangel 1988, Wilson 1976, Windmiller 1996. [94] Petranka 1988, Rowe et al. 1994, Ruth et al. 1993. [95] Brodman 1995, Gilbert 1944. [96] Freda 1983, Nyman 1991, Stangel 1988, Wacasey 1961, Whitford and Vinegar 1966, Windmiller 1996. [97] Brodman 1995, Voss 1993, Whitford and Vinegar 1966. [98] Ireland 1989, Shoop 1974, Talentino and Landre 1991, Werner et al. 1999, Windmiller 1996, Woodward 1982a. [99] Stangel 1988. [100] Brodman 1995, Clark 1986b, Cook 1983, Ireland 1989, Jackson and Griffin 1991, Stangel 1988, Stenhouse 1987, Windmiller 1996. [101] Albers and Prouty 1987; Clark 1986; Clark and Hall 1985; Cook 1978, 1983; Freda and Dunson 1985a, 1985b; Jackson and Griffin 1991; Ling et al. 1986; Pierce 1987; Portnoy 1990; Pough 1976; Sadinski and Dunson 1992. [102] Milam 1997, Massachusetts Audubon Society unpublished data, Rowe et al. 1994, Stout and Stout 1992. [103] Voss 1993, Walls and Altig 1986. [104] Petranka 1998, Rowe et al. 1994, Stangel 1988, Stenhouse et al. 1983. [105] Bachmann 1988, Nyman 1991, Williams 1973. [106] Freda 1983, Nyman 1991. [107] Shoop 1974, Stenhouse et al. 1983, Wilbur and Collins 1973. [108] Bishop 1941, Freda 1983, Phillips 1992, Shoop 1974, Stangel 1988, Wacasey 1961, Windmiller 1996. [109] Phillips 1992, Stangel 1988, Whitford and Vinegar 1966. [110] Horne and Dunson 1994b. [111] Shoop 1974, Wilbur 1977c, Windmiller 1996. [112] Bishop 1941, Homan et al. 2003a, Husting 1965. [113] Cook 1978; DeGraaf and Rudis 1990; Douglas and Monroe 1981; Downs 1989; Klemens 1990; Shoop 1965, 1968; Wacasey 1961; Williams 1973. [114] Homan et al. 2003b, Shoop 1965, Windmiller 1996. [115] Petranka 1998, Windmiller 1996 and personal communication. [116] Downs 1989, Kleeberger and Werner 1983, Regosin et al. 2003a, Wacasey 1961, Williams 1973, Wilson 1976, Windmiller 1996 and personal communication. [117] Kleeberger and Werner 1983, Wacasey 1961, Williams 1973, Wilson 1976, Windmiller 1996. [118] Bishop 1941, Wacasey 1961. [119] Clark 1986b, Collins and Wilbur 1979, Cook 1978, Dale et al. 1985, Kenney 1991, Ling et al. 1986, O'Donnell 1937, Portnoy 1990, Pough 1976, Rowe and Dunson 1993, Stangel 1988, Stenhouse 1985, Wacasey 1961, Williams 1973, Wilson 1976, Windmiller 1996. [120] Doody 1995, Stenhouse 1995. [121] Scott 1994, Shoop and Doty 1972. [122] Petranka and Petranka 1981. [123] Petranka and Petranka 1981, Petranka 1990. [124] Hassinger et al. 1970, Taylor and Scott 1997. [125] Stenhouse 1987. [126] Petranka and Petranka 1980. [127] Petranka and Petranka 1980, Stenhouse et al. 1983, Wilbur 1971. [128] Hassinger et al. 1970, Petranka and Petranka 1980, Seibert 1989. [129] Taylor and Scott 1997. [130] Williams 1973. [131] Bishop 1941; Klemens 1990; Minton 1954; Petranka 1989a, 1990; Shoop and Doty 1972; Williams 1973. [132] Williams 1973. [133] Stenhouse 1985, Williams 1973. [134] Jackson et al. 1989; Noble and Brady 1933; Petranka 1989a, 1990; Petranka and Petranka 1981; Seibert 1989; Shoop and Doty 1972; Stenhouse 1985; Stenhouse et al. 1983; Williams 1973. [135] Sessions 1982, Uzzell 1964. [136] Petranka 1990. [137] Anderson and Giacosie 1967; Conant 1975; Nyman et al. 1988; Petranka 1998; Uzzell 1964; Weller et al. 1978, 1979; Weller and Menzel 1979. [138] Bishop 1941, 1943; Brame 1959; Licht and Bogart 1987; Minton 1954; Stille 1954; Weller et al. 1979. [139] Brodman 1995, Nyman 1991, Piersol 1910, Talentino and Landre 1991. [140] Brodman 1995, Douglas 1979. [141] Douglas 1979, Lowcock et al. 1992. [142] Brodman 1995, Cook 1981, Talentino and Landre 1991, Wilbur 1971. [143] Kumpf and Yeaton 1932; Uzzell 1964, 1969. [144] MacGregor and Uzzell 1964; Sessions 1982; Uzzell 1964, 1969. [145] Williams 1973. [146] Uzzell 1964, Wilbur 1971.

[147]Brodman 1995, Jackson and Griffin 1991, Massachusetts Audubon Society data, Piersol 1910, citations in Uzzell 1964, Wilbur 1971, Williams 1973. [148]Cook 1983, Jackson and Griffin 1991, Sadinski and Dunson 1992. [149]Brodman 1995, Cortwright 1988, Williams 1973. [150]Licht and Bogart 1989, Nyman 1991, Talentino and Landre 1991, Uzzell 1964, Wacasey 1961. [151]Bachmann and Bachmann 1994, Nyman 1991, Wilbur 1971. [152]Wilbur 1971. [153]Licht and Bogart 1989, Uzzell 1964, Wilbur 1971. [154]Talentino and Landre 1991, Wacasey 1961, Williams 1973. [155]Bishop 1941, Lowcock et al.1992, Wilbur 1977, Williams 1973. [156]Anderson and Giacosie 1967; Bachmann and Bachmann 1994; Bishop 1941; Cook 1978; Douglas 1979; Klemens 1990; Nyman et al. 1988; Wacasey 1961; Weller et al. 1978; Williams 1973; Wilson 1976; Bryan Windmiller, Hyla Ecological Services, personal communication. [157]Horne and Dunson 1994a. [158]Bishop 1941, Douglas and Monroe 1981, Downs 1989, Wacasey 1961, Williams 1973. [159]Anderson and Giacosie 1967, Bachmann and Bachmann 1994, Bishop 1941, Brodman 1995, Freda and Dunson 1985b, Horne and Dunson 1994b, Lowcock et al. 1992, Minton 1954, Nyman 1991, Nyman et al. 1988, Piersol 1910, Talentino and Landre 1991, Wacasey 1961, Williams 1973, Wilson 1976. [160]Michael Klemens, Wildlife Conservation Society, personal communication; Piersol 1910; Weller and Menzel 1979. [161]Anderson and Giacosie 1967, Bachmann and Bachmann 1994, Brodman 1995, Cook 1978, 1983, Cortwright 1988, Minton 1954, Nyman 1991, Piersol 1910, Williams 1973, Wilson 1976. [162]Freda and Dunson 1985b, Horne and Dunson 1994, Karns 1992, Sadinski and Dunson 1992. [163]Bogart et al. 1985, Husting 1965, Petranka 1982. [164]Kraus et al. 1991, Licht and Bogart 1989. [165]Petranka 1982. [166]Bishop 1943; McWilliams and Bachmann 1989; Morse 1904, cited in Bishop 1943. [167]Conant 1975, Michael Gray, Ohio EPA, personal communication, Morse 1904, cited in Bishop 1943. [168]Wilbur 1972. [169]Petranka 1998, Zappalorti 1993. [170]Kraus et al. 1991, Morris 1985. [171]Anderson et al. 1971, Bishop 1943, Hassinger et al. 1970, New York Department of Environmental Conservation 2003, Stine 1984, Stine et al. 1954. [172]Anderson et al. 1971, Hassinger et al. 1970. [173]Anderson et al. 1971, Morin 1983. [174]Anderson et al. 1971. [175]Wilbur 1972. [176]Bishop 1943, Zappalorti 1993. [177]Wilbur and Collins 1973. [178]Bishop 1943. [179]Wilbur 1971, 1972; Petranka 1998. [180]Gates and Thompson 1982. [181]Bishop 1943, Healy 1974, Morin 1983, Petranka 1998, Tyning 1990. [182]Bishop 1941, Cook 1981. [183]Massey 1990. [184]Bishop 1941, Petranka 1998. [185]Hamilton 1940, Harris et al. 1988. [186]Bishop 1943, Petranka 1998. [187]Conant 1975, Tyning 1990, Storey and Storey 1986b. [188]Atlantic Canada Conservation Data Centre 2003, Bishop 1941, Burgason 1992, Cook 1981. [189]Cyberlizard (UK) 2004. [190]Husting 1965; Jackson and Griffin 1991; Nyman 1991; Wilbur 1972; Williams 1973; Wilson 1976; Windmiller 1990, 1996. [191]Berven 1988, 1990; Morin 1981, 1983; Petranka et al. 1994; Stenhouse et al. 1983; Wilbur 1971, 1972, 1976, 1977a, 1977b. [192]Unpublished data, Pleasant Valley Wildlife Sanctuary, Massachusetts Audubon Society, Lenox, MA.

Chapter 12

Reptiles, Birds, and Mammals

Several species of turtles are closely associated with vernal pools, and some species of snakes, birds, and mammals also take advantage of the presence of water and the abundant aquatic life that the pools support. There is little detailed information available on non-amphibian vertebrates, especially snakes, birds, and mammals, in association with vernal pools. This chapter briefly discusses what is known about the non-amphibian vertebrates of vernal pools, all of which may be considered to be non-breeding migrants who move to the pools from other habitats to feed, water, or rest.

Turtles

Spotted turtles, Blanding's turtles, painted turtles, and snapping turtles often occur in northeastern vernal pools (Plate 14). On Cape Cod, box turtles (*Terrapene carolina*) often submerge in the shallow water of vernal pools during the summer[1]. The precise role of vernal pools as sources of food, especially during the period when females are developing their eggs, and water during dry periods for turtles remains to be defined. Also unknown is the extent to which predation by turtles on amphibian eggs and hatchlings may affect reproductive success in vernal pool salamanders and anurans.

Spotted Turtle

The spotted turtle (*Clemmys guttata*) is a relatively small turtle, with an adult plastron length of about 8.0 to 12.5 cm (3 to 5 in). It is distinguished by small, round, bright yellow or orange spots dispersed widely on its black carapace, legs, head, and neck (Plate 14). The plastron is yellow with black blotches.

Distribution

In the eastern part of its range, this species is found from southern Maine and southeastern New Hampshire through the southern New England states and New Jersey, and south along the Atlantic coast to northern Florida. The range extends west through southern New York and Pennsylvania, into northern Ohio and Indiana, southern Michigan, and southeastern Ontario up to the Ottawa Valley, and there are a few records from southern Quebec[2].

Life History

Spotted turtles are largely aquatic, but they spend significant portions of their lives on land[3]. The adults overwinter in wetlands. They emerge from their hibernacula and are active in late winter and early spring in vernal pools, shallow streams, and other small aquatic habitats, and they can often be seen basking on logs and tussocks, and along the shoreline[4]. In central Massachusetts, individuals may move 20 to 550 m (65 to 1,804 ft) through uplands to vernal pools[5]. Recent research on habitat use by spotted turtles indicates that they spend significant amounts of time feeding in vernal pools in the spring, with some individuals spending three to four months in vernal pools[6]. The first time I saw a spotted turtle, it was an adult eating a spotted salamander egg mass. Adults also feed on wood frog egg masses, amphibian larvae, aquatic insects, crustaceans, worms, fingernail clams, pulmonate snails, and algae[7].

Adult spotted turtles become reproductively mature at a plastron length of around 8.1 to 8.3 cm (3.25 in), or around seven years of age[8]. Most mature females leave pools and wetlands in June and move to upland meadows and fields, where they excavate nests and lay their eggs. On average, nest sites in central Massachusetts are 182 m (597 ft) from wetland edges[9]. Some females may nest in emergent wetlands, depositing their eggs in decaying sedge tussocks[10]. The eggs hatch in August or September, and the young may overwinter in the nest.

During dry periods, adults of both sexes may aestivate. Both wetlands and uplands are used as aestivation sites[11]. In two central Massachusetts populations, some individuals aestivate in emergent and shrub-scrub wetlands, but most aestivate in upland habitats, either in forest or along forest-field edges more than 60 m (195 ft) from wetlands[12]. Of twenty-six spotted turtles studied, twenty-four moved more than 60 m (195 ft) from wetlands to aestivation sites. The range of movement for all twenty-six individuals was 13 to 412 m (43 to 1,352 ft), with a mean of 178 m (584 ft). Individual animals aestivated continuously for two to ninety-three days. The duration of aestivation appears to be related to rainfall, with individuals aestivating for significantly longer periods in dry years than in wet ones.

Spotted turtles spend the winter in a state of dormancy, with adults moving to hibernacula in wetlands by mid fall[13]. Emergent, shrub-scrub, and forested wetlands are all used as overwintering sites by this species. Adults overwinter under logs or tree roots in mud and water, with an air space between the water surface and the overlying log or roots. Several spotted turtles may use the same hibernaculum, and individuals may return to the same overwintering sites in successive years. If air temperatures remain warm, individuals may move out of hibernacula into flooded vernal pools to forage in October and November[14].

Habitats and Home Ranges

Spotted turtles use a variety of wetland habitats, and they also spend substantial amounts of time in uplands[15]. Home range estimates vary, averaging 2 to 4 ha (3 to 10 ac) in Ontario and Massachusetts, and 0.5 to 0.8 ha (1.2 to 2 ac) in Illinois, Massachusetts, and Pennsylvania (see Table 17). In central Massachusetts, the smallest home range was 0.2 ha (0.5 ac), and the largest 53.1 ha (131 ac). Adults traveled an average maximum distance of 265 m (869 ft; median 226 m, 741 ft) from hibernacula. Between-year differences in the home ranges of individual spotted turtles appear to be associated with precipitation, with reduced movement and increased time spent in aestivation in dry years.

Spotted turtles seem to spend disproportionately more time in vernal pools than would be expected based on the availability of vernal pools as opposed to other habitats in the landscape. In contrast, the amount of time spent in shrub-scrub upland forests is significantly less than expected based on habitat availability. However, uplands are used for significant periods of each year, for aestivation and nesting. Based on this information, it has been estimated that buffers extending 412 m (1,352 ft) from wetlands are necessary to provide habitat for spotted turtle nesting, aestivating, overwintering, and foraging in vernal pools[16].

Blanding's Turtle

Blanding's turtle (*Emydoidea blandingii*) is a moderately large turtle, with a shell length of 13 to 27 cm (5 to 10.5 in). It has a bright yellow chin and throat, and tan-to-yellowish flecks and streaks marking its dark carapace. The plastron is yellow with a large, dark blotch on each shield, and hinged, somewhat like that of a box turtle.

Distribution

This species is widely distributed through much of the glaciated northeast, occurring in southeastern Minnesota, most of Wisconsin and Michigan, southern Ontario, southwestern Quebec, eastern Iowa, northern Illinois, Indiana, Ohio, the northwestern corner of Pennsylvania, eastern Massachusetts, southeastern New Hampshire, and southern Maine. It is abundant in the central part of its range and uncommon in New England. Occasional individuals are found in southern New York, southern Connecticut, and Nova Scotia[17].

Life History

Blanding's turtle, like the spotted turtle, is an aquatic turtle that spends significant amounts of time foraging in vernal pools. Adults overwinter in wetlands, where they may be dormant in mud and debris or active under the ice[18]. Studies of populations in Maine, Massachusetts, and Wisconsin have shown that, like spotted turtles, Blanding's turtles travel to vernal pools to feed in spring, foraging on amphibian egg masses and larvae, crustaceans, aquatic insects, molluscs, annelids,

Table 17. Home Range and Migratory Distances for Spotted Turtles

Location	Home Range	Migratory Distance	Reference
Massachusetts			
	mean: 0.7 ha (1.7 ac) (3 individuals)		Graham 1995
Massachusetts			
	mean: 3.5 ha (8.6 ac) median: 2.0 ha (4.9 ac) (2 populations)	mean: 265 m (869 ft) median: 226 m (741 ft)	Milam 1997, Milam and Melvin 2001
Ontario			
	mean: 3.2, 3.8 ha (7.9, 9.3 ac)		Litzgus and Brooks 1995
Pennsylvania			
	mean: 0.5 ha (1.2 ac) (large population, nesting areas not included in area)		Ernst 1970
Illinois			
	mean: 0.8 ha (2 ac) (8 individuals, April to June)		McGee et al. 1989

and plants[19]. Adults reach sexual maturity at twelve to fifteen years of age[20]. Although courtship has been observed throughout the year, mating occurs at vernal pools in late spring and females move to upland nesting sites in June and July[21]. Nests are in loamy, sandy, and gravelly soils of upland fields, and females produce from six to seventeen eggs per clutch[22]. Eggs hatch in late summer, and the young may overwinter in the nest.

Habitats and Home Ranges

This species is typical of marshes, ponds, and bogs, although it is also found in well-vegetated coves of large lakes and in backwaters of rivers. It prefers habitats with extensive submerged and/or emergent vegetation. Uplands, particularly open areas without woody vegetation and canopy cover, are important for nesting areas, and wooded uplands also provide migration corridors between wetlands and from wetlands to nesting sites. Individuals may move a kilometer (just over three-fifths of a mile) between wetlands, or between wetlands and nesting sites[23]. As is the case with spotted turtles, Blanding's turtles use vernal pools significantly more often than would be predicted based on habitat availability[24]. It seems likely that foraging in vernal pools provides these turtles with an important source of protein and energy.

Painted Turtle

Painted turtles (*Chrysemys picta*) (Plate 14) are among the most familiar aquatic turtles. Achieving a length of up to 25 cm (10 in), these turtles are characterized by an olive or black carapace with each plate outlined in yellow and with yellow and red markings along the sides. The head, neck, face, and legs are patterned with bright yellow and red stripes and spots. The plastron is yellow.

Distribution

Painted turtles are abundant and are widely distributed throughout the glaciated northeast. Three of the four recognized subspecies of painted turtles in North America occur in our area. The midland painted turtle (*Chrysemys picta marginata*) is found from the Upper Peninsula of Michigan, eastern Wisconsin, and Illinois through the midwestern United States, southern Ontario, and southern Quebec, through Pennsylvania, New York, and New England to southern Maine. The western painted turtle (*Chrysemys picta belli*) is found in the Upper Peninsula of Michigan, Wisconsin, Illinois, Minnesota, and southwestern Ontario. The eastern painted turtle (*Chrysemys picta picta*) is found in eastern Quebec, eastern New York and Pennsylvania, New England, Nova Scotia, and southern New Brunswick[25]. Intergrades are common in areas where subspecies overlap.

Life History

Adult painted turtles overwinter in permanent waters in mud, and they are sometimes active under ice. They emerge in early spring and can be observed basking on logs and rocks in and around ponds, marshes, lakes, and vernal pools[26]. Adults are omnivorous, feeding on a variety of living and decomposing animal and plant material. Females move to uplands in June and July to lay their eggs. The eggs hatch in late summer, and the young usually overwinter in the nest.

Habitats and Home Ranges

Painted turtles have wide habitat preferences, occurring in marshes, shallow coves of lakes, ponds, slow-flowing streams, and bogs. They are often seen in vernal pools, particularly those that are long-cycle or semi-permanent or near other water bodies. While highly aquatic, adult painted turtles may move from one water body to another. In a Michigan population, adults were observed to move up to 144 m (472 ft) from overwintering ponds to other ponds and wetlands in the spring[27]. In my studies of vernal pools in Massachusetts, I have found painted turtles regularly from early spring until pool drying in mid summer.

Snapping Turtle

Snapping turtles (*Chelydra serpentina*) are large, up to 47 cm (18.5 in) in length. They are dark gray or black, with large heads, long saw-toothed tails, small plastrons, and coarse scutes on their shells. These turtles are found in permanent ponds, lakes, streams, river backwaters, marshes, and semi-permanent and temporary ponds throughout the glaciated northeast. They are omnivorous feeders and, like other turtles, may take advantage of the abundant amphibian and invertebrate food in vernal pools. Snapping turtles have been observed to prey on the ringed salamander *(Ambystoma annulatum,* a relative of northeastern mole salamanders from the Ozarks), in Missouri ponds[28].

Adult snappers overwinter in mud on the bottom of permanent waters and become active in early spring. Breeding occurs from April to November, with egg-laying usually occurring in mid June. Upon hatching, the young snapping turtles migrate to permanent water[29].

Snakes

Among the species of snakes that have been reported from vernal pools are garter snakes (*Thamnophis sirtalis*), ribbon snakes (*Thamnophis sauritus*), and water snakes (*Nerodia sipedon*). Garter snakes and ribbon snakes prey on newly metamorphosed anurans, and water snakes feed extensively on tadpoles, adult frogs, and salamanders[30].

Birds

Waterfowl, wading birds, and woodland birds feed and/or water at vernal pools. Yellowlegs (*Tringa melanoleuca*, *T. flavipes*), little blue herons (*Florida caerulea*), and green herons (*Butorides striatus*) feed on anuran tadpoles in vernal pools on the Coastal Plain in New Jersey[31], and great blue herons (*Ardea herodias*) and American bitterns (*Botaurus lentiginosus*) prey on *Ambystoma annulatum* in Missouri[32]. I have casually observed yellowlegs, great blue herons, wood ducks (*Aix sponsa*), and mallards (*Anas platyrhynchos*) in New England vernal pools. I am unaware of any systematic surveys of avian uses of vernal pool habitats.

Mammals

Mass mortality in breeding American toads has been reported as a result of predation by raccoons (*Procyon lotor*) in Pennsylvania[33] and striped skunks (*Mephitis mephitis*) in Maryland[34]. Nine species of bats, including the federally endangered Indiana bat, feed on insects emerging from vernal pools in Kentucky[35]. No mammals have been identified as typical or indicative of vernal pools. A survey of shrews and other small mammals typical of northeastern forests was conducted next to three pools and in an adjacent upland forest in central Massachusetts.

There were no differences in the overall abundance of small mammals in the two areas, although smoky shrews (*Sorex fumeus*) were three times more common in the pond-side areas than in the upland[36]. However, this was a small study and took place during a dry year when all study pools dried too early for juvenile amphibians or most invertebrates to emerge successfully. To date, there have been no large-scale surveys or long-term studies to quantify the use of vernal pool habitat by mammals.

Some Remaining Questions About Non-Amphibian Vertebrates

- What is the role of vernal pools in the life history and population ecology of reptiles such as Blanding's and spotted turtles?
- To what extent do turtles, snakes, wading birds, waterfowl, and predatory mammals feed on amphibian eggs, larvae, juveniles, and adults in vernal pools; on pool-breeding amphibians during the terrestrial phases of their lives; and on pool invertebrates?
- Are insects that emerge from vernal pools an important food for insectivorous birds or bats?
- Are there relationships between the distribution of small mammals and/or reptiles and the distribution of vernal pools in the landscape?

References for Chapter 12

[1] Colburn unpublished data; John Portnoy, US National Park Service, Cape Cod National Seashore, personal communication. [2] Conant 1975, Cook 1981, DeGraaf and Rudis 1983, Glowa 1992, Litzgus and Brooks 1998, Milam 1997, Tyning 1990. [3] Glowa 1992, Graham 1995, Joyal et al. 2001, Milam 1997, Milam and Melvin 2001. [4] DeGraaf and Rudis 1983, Tyning 1990. [5] Milam 1997, Milam and Melvin 2001. [6] Glowa 1992; Graham 1995; Joyal 1996; Joyal et al. 2001; Milam 1997; Milam and Melvin 2001; Donald Schall, ENSR, personal communication. [7] Babcock 1938, DeGraaf and Rudis 1983, Glowa 1992, Milam 1997, Milam and Melvin 2001. [8] Ernst 1970, 1976; Glowa 1992. [9] Milam 1997, Milam and Melvin 2001. [10] DeGraaf and Rudis 1983, Milam 1997, Milam and Melvin 2001. [11] Babcock 1938, DeGraaf and Rudis 1983, Ernst 1970, Graham 1995, Joyal 1996, Joyal et al. 2001, Milam 1997, Milam and Melvin 2001. [12] Milam 1997, Milam and Melvin 2001. [13] Joyal 1996, Litzgus et al. 1999, Milam 1997, Milam and Melvin 2001. [14] Milam 1997, Milam and Melvin 2001. [15] Ernst 1970, Graham 1995, Litzgus and Brooks 1998, McGee et al. 1989, Milam 1997, Milam and Melvin 2001. [16] Milam 1997, Milam and Melvin 2001. [17] Babcock 1938, Conant 1975, DeGraaf and Rudis 1983, Graham 1992, Joyal 1996, Tyning 1990. [18] DeGraaf and Rudis 1983. [19] Babcock 1938, Casper 2002, Graham 1992, Joyal 1996, Joyal et al. 2001, Lagler 1943, Rowe 1992. [20] Congdon et al. 1983; Graham and Doyle 1979, cited in Graham 1992. [21] DeGraaf and Rudis 1983, Graham 1992. [22] Carr 1952, Graham 1992. [23] Casper 2002; Haskins unpublished data, cited in Graham 1992. [24] Graham 1992, Joyal 1996, Joyal et al. 2001. [25] Conant 1975, Cook 1981, DeGraaf and Rudis 1983, Tyning 1990. [26] Cook 1981, DeGraaf and Rudis 1983, Etchmerger 1992, Tyning 1990. [27] Sexton 1959. [28] Peterson et al. 1991. [29] DeGraaf and Rudis 1983,

Tyning 1990. [30] Arnold and Wassersug 1978; Carpenter 1952; DeGraaf and Rudis 1983, 1986; Thomas Tyning, Berkshire Community College, personal communication. [31] Hassinger et al. 1970. [32] Peterson et al. 1991. [33] Seale 1982. [34] Groves 1980. [35] Biebighauser 2003. [36] Brooks and Doyle 2001.

Chapter 13

Energy Flow, Seasonal Cycles, and Variations in Community Composition

The interactions among different components of the biological community in a vernal pool are as diverse as the individual organisms. Pools are dynamic systems that undergo dramatic fluxes of energy and materials and experience seasonal and between-year changes in their physical environment and biota. Solar energy is fixed by algae and plants and cycled via various pathways through the pool fauna. Periodically, there are massive inputs of proteins and carbohydrates in the form of amphibian egg masses, and these supplement the plants and detritus as a food source for some pool animals. There are similar large-scale exports of energy and living materials as amphibian larvae and immature aquatic insects complete their development and leave the pools. Accompanying and driving these fluxes of energy and materials are the seasonal and year-to-year patterns of fluctuating water levels, varying water chemistry and temperatures, and appearances and disappearances of different species of algae, plants, microorganisms, and macroscopic animals.

This chapter synthesizes information on the sources of the energy that sustains vernal pools and on how energy flows within the biological community of the pools. It also examines patterns in the physical and biological changes that occur in pools over the course of a year.

Energy Flow

As in other ecosystems, the growth of microbial, plant, and animal life in vernal pools is powered ultimately by the sun (Figure 39). Primary producers, or autotrophs, including photosynthetic protozoans, algae, and plants, use solar energy to combine water, carbon dioxide, and nutrients into the organic carbon compounds that are the building blocks of cells and the basis for life. Primary

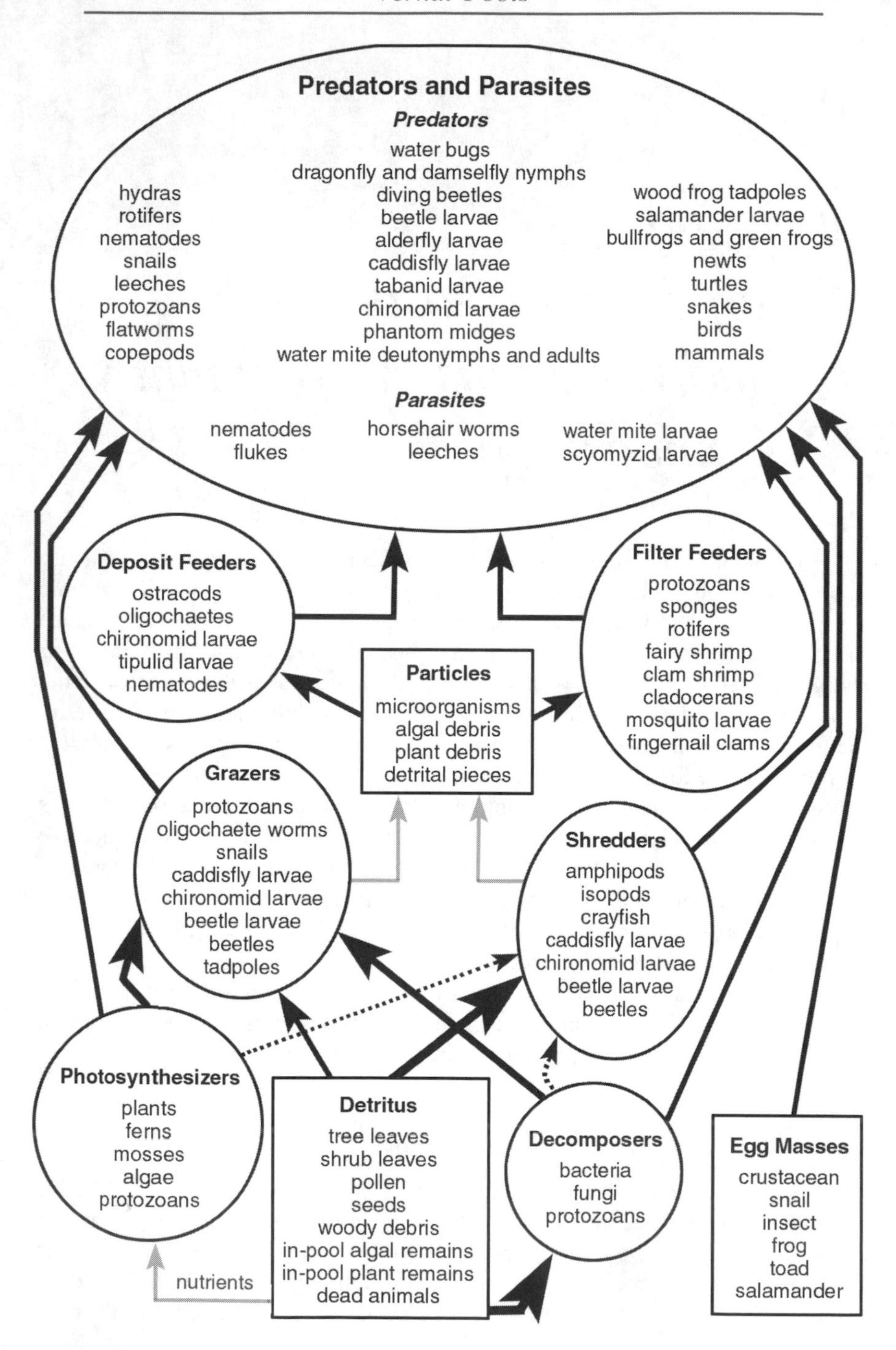
Predators and Parasites
Predators
water bugs
dragonfly and damselfly nymphs
diving beetles
beetle larvae
alderfly larvae
caddisfly larvae
tabanid larvae
chironomid larvae
phantom midges
water mite deutonymphs and adults
hydras
rotifers
nematodes
snails
leeches
protozoans
flatworms
copepods
wood frog tadpoles
salamander larvae
bullfrogs and green frogs
newts
turtles
snakes
birds
mammals
Parasites
nematodes
flukes
horsehair worms
leeches
water mite larvae
scyomyzid larvae
Deposit Feeders
ostracods
oligochaetes
chironomid larvae
tipulid larvae
nematodes
Filter Feeders
protozoans
sponges
rotifers
fairy shrimp
clam shrimp
cladocerans
mosquito larvae
fingernail clams
Particles
microorganisms
algal debris
plant debris
detrital pieces
Grazers
protozoans
oligochaete worms
snails
caddisfly larvae
chironomid larvae
beetle larvae
beetles
tadpoles
Shredders
amphipods
isopods
crayfish
caddisfly larvae
chironomid larvae
beetle larvae
beetles
Photosynthesizers
plants
ferns
mosses
algae
protozoans
Detritus
tree leaves
shrub leaves
pollen
seeds
woody debris
in-pool algal remains
in-pool plant remains
dead animals
nutrients
Decomposers
bacteria
fungi
protozoans
Egg Masses
crustacean
snail
insect
frog
toad
salamander

consumers including herbivores and detritivores feed on the plant-derived organic carbon, releasing some of the stored energy and recycling the molecules as they build the cells of their own bodies. A given volume of primary consumers represents a more concentrated source of energy and nutrients than the same volume of plant-derived material, and predators take advantage of this more concentrated food, recycling the carbon yet again. Studies of how energy flows from the sun into producers, and from producers into consumers, provide insights into some of the factors that determine which plants and animals occur where, and what conditions contribute to good habitat for different species.

There has been relatively little research on the sources and flows of energy in vernal pools and on the overall trophic structure of the pool community. The two studies that have been published indicate that detritus is the base of the food web[1] (Figure 39). Most of the primary consumers in pools feed heavily on dead and decaying plant material, rather than on living plants, and fungi and bacteria play a critical intermediary role in decomposing the detritus so that it can be digested by animals. This finding is consistent with studies of forest ecosystems overall, as well as with studies of other shallow aquatic systems within the glaciated northeast[2]. In temperate forests, photosynthesis by trees and other plants provides energy and nutrients for herbivores that feed on leaves, seeds, and fruits. The food web of the forest floor is detritally based, fueled by the annual deposits of fallen leaves and woody debris. It has been widely recognized since the 1960s that these annual inputs of leaf litter are also key to the energetics of temperate stream systems. Detritus derived largely from *in situ* photosynthesis by algae, phytoplankton, and/or macrophytes is the base of the food web in north temperate lake systems, freshwater wetlands, and snowmelt pools in the far north. Much of our current understanding of the energetics of vernal pool ecosystems is based on inferences from the community composition of the fauna and on analogies with other aquatic systems in northeastern forests, especially temperate woodland streams.

Figure 39 (at left). The food web in vernal pools includes many pathways through which energy from living plants, algae, and detritus is cycled through microorganisms, invertebrates, and vertebrates. In this schematic food web, the solid black arrows represent transfers of energy from food sources (for example, detritus) to consumers (for example, shredders). The width of each arrow provides a rough guide to the relative importance of the food source to the consumer. Dotted arrows indicate minor pathways for energy transfer. Light gray arrows represent indirect effects of food consumption on food sources; for instance, the decomposition of detritus releases nutrients that contribute to the growth of primary producers such as algae and vascular plants.

Energy Sources

Organic carbon enters vernal pools through photosynthesis by primary producers in the pool (autochthonous or self-generated carbon), and by transport from the surrounding forest (allochthonous or externally generated carbon). The proportions of autochthonous and allochthonous carbon vary from pool to pool. Just as vernal pools occur along a broad gradient of flooding frequency and duration and span a wide range of sizes, so too do they vary widely in their primary food sources. At one extreme are pools that are fed by inputs of leaves from the surrounding forest each autumn and have little or no in-pool primary production; at the other extreme are systems that depend heavily on *in situ* photosynthesis by algae and aquatic plants. This continuum depends on the extent of canopy closure above the pool and the amount of sunlight that reaches the pool (Figure 40). Small pools with closed canopies have little algal growth and in-pool plant photosynthesis. Vernal pools in more open settings, those with larger surface areas and less canopy cover, and those that receive higher nutrient inputs from naturally fertile soils or as a result of watershed disturbance by development or forest clearing, have more algae and greater within-basin plant growth. Often they have dense macrophyte beds and/or extensive mats of filamentous algae[3].

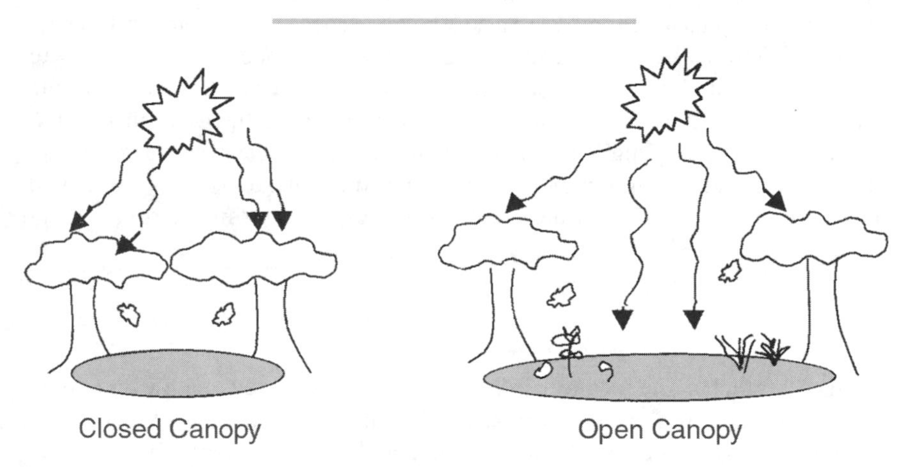

Figure 40. Pool size and the extent of canopy closure help determine how much of a pool's energy comes from in-pool photosynthesis and how much depends on leaves and other detritus that originated outside of the pool. A small pool under a closed canopy (left) receives proportionately more detritus from outside sources than from in-pool photosynthesis (such as from algae) and detritus, whereas a larger pool under an open canopy (right) receives proportionately less detritus from outside sources than from in-pool photosynthesis and detritus.

Algae

Algal production in vernal pools varies with canopy closure, and the amount of algae affects the composition of the aquatic community. For example, snails tend to be more common in pools with extensive algal growth, and higher snail densities are associated in turn with a greater abundance of trematode worms (see Chapter 7).

The tadpoles of many anuran species feed on algae, and their ability to complete development in vernal pools may be affected by algal production. In a group of pools in a research forest in Michigan, where amphibian populations have been studied for decades, the amounts of algae have decreased and spring peeper populations have become locally extinct as the forest has grown up and the canopy has closed over some pools. The peepers grow more slowly in closed canopy pools than in open ones where algae are abundant. Controlled feeding experiments show that the peeper tadpoles do not grow as well on food from closed canopy pools as they do when fed food from open canopy pools. Wood frogs also grow more slowly in closed canopy pools than in open ones, but they do well in pools with closed canopies and food webs that rely on inputs of leaves from forest trees rather than on algae for their organic carbon source, and their breeding populations persist in pools from which peepers have disappeared[4].

The gradient of canopy cover and algal production in vernal pools may be analogous to that in forest streams. In small forest streams, where the dense canopy limits the penetration of sunlight as it does in small vernal pools, there is very little algal growth, and for many years algae were believed to be insignificant in the overall energy budget. Research in the 1980s in Bear Brook, a small headwater stream with a fully closed canopy in the Hubbard Brook Experimental Forest in New Hampshire, showed that algae can be important even in small, highly shaded streams. Algal photosynthesis in spring, before the trees leaf out, contributes to 75% of the growth and development of caddisflies in the stream, and almost one-tenth of the species in the stream eat algae[5]. Thus, although it produces only a small amount of the total carbon in the stream, algal photosynthesis is a very important contributor to the biodiversity of the stream's fauna. Similarly, although herbivores make up a small proportion of the fauna of many woodland vernal pools, it appears quite possible that algae growing on the surfaces of leaves and other substrates in early spring, when the water is clear and the canopy has not yet closed, contribute to the development of some species of vernal pool wildlife even in pools with closed canopies.

One approach to looking at energy flows in ecosystems involves measuring primary production and monitoring the oxygen consumption, or respiration, of different components of the community. Oxygen consumption is a measure of energy use, because the cells of aerobic plants and animals use oxygen in breaking down organic carbon compounds and synthesizing new cellular components. By measuring the energy use by plants, different groups of animals, bottom sediments, and so on, researchers can determine how much of the carbon that is fixed by plants and algae is used by different segments of the biological community.

One such study showed that in some pools, algal photosynthesis contributes significant amounts of detritus to the food web. The study examined the annual metabolism of a moderately large (3,500 m^2, 37,600 ft^2), long-cycle, fall-filling pool in Amherst, Massachusetts, during an unusually wet year when the pool did not dry[6]. The pool has an open canopy and is bordered by a 2 m (6.5 ft) fringe of buttonbush (*Cephalanthus occidentalis*). *Chara*, or stonewort, a large, plant-like alga, covers the pool bottom, and its photosynthesis is the main source of primary production in the pool. Filamentous algae and epiphytic diatoms are also present, but they do not contribute significantly to the gross primary production. Most of the oxygen consumption in the pool is associated with the benthic, or bottom, sediments and is presumed to be largely microbial. Because microbes are largely involved in the decomposition of organic matter, this result indicates that the primary production by the stonewort is deposited into the pool sediments where it is broken down by decomposers. The pool sediments have a high organic content, which further shows that significant amounts of detritus are deposited in the pool each year. The organic-rich sediments and high benthic respiration rates suggest that in this pool, nutrient cycling and energy flow occur primarily through the decomposition of detrital algae.

In-Pool Plant Growth

As we saw in Chapter 5, the extent of plant cover among vernal pools varies greatly, depending on hydroperiod and canopy closure. Some pools have no plants on the bottom and others have the bottom fully covered by vegetation including ferns, mosses, herbaceous plants, and woody species. In a small (835 m^2, 9,000 ft^2), short-cycle, spring-filling vernal pool in an Ontario woodland, 94% of the basin supports some vegetation. Herbivores may eat some of the living plants, but most of the growth dies back and becomes detritus in the pool basin. The dominant plants contribute 119 g/m^2 (0.4 oz/ft^2) ash-free dry weight of detritus annually. If the area of the pool that does not have plants is included in the calculations, in-pool vegetation contributes 94 g/m^2 (0.3 oz/ft^2) over the entire pool. This represents 41% of the total detritus available and is equivalent to 71% of the leaf detritus entering the pool[7].

Leaf-Litter Inputs

Annual leaf fall is a significant component of ecosystem dynamics in northeastern forests. It contributes to nutrient cycling and energy flow on the forest floor, in headwater streams, and in vernal pools. In a second-growth New Hampshire forest, leaf litter contributes up to 600 g/m^2 (2 oz/ft^2) of ash-free dry weight to the forest floor annually[8]. Autumnal leaf deposition into the channel of Bear Brook is similar, approximately 300 g/m^2 (1 oz/ft^2), and accounts for more than

44% of the energy inputs to the stream[9]. The annual leaf-litter input to a short-cycle, spring-filling, woodland vernal pool that is surrounded by trees and has some open canopy in Ontario, is of the same order of magnitude, approximately 133 g/m^2 (0.5 oz/ft^2), and accounts for 59% of the detritus in the pool[10].

Amphibian Egg Masses[11]

The seasonal breeding migrations of amphibians essentially represent a transfer of energy from the land into the water in the form of egg masses. During the terrestrial phase of their life cycles, amphibians prey on soil invertebrates that feed on forest leaf litter, and females convert much of the energy derived from this food into eggs that they subsequently lay in vernal pools. Hundreds of thousands of eggs are deposited into pools over the course of a few weeks in the spring (see Chapter 11). The communal rafts of wood frog egg masses often contain as many as 100,000 eggs and can easily cover an area of 1 m^2 (10 ft^2) to a thickness of 10 cm (4 in) or more, or a volume of 0.1 m^3 (3.6 ft^3). Comparable numbers of eggs are produced by breeding populations of spring peepers and toads, although the peeper eggs are deposited individually and those of toads are in long strings, rather than masses. Breeding by mole salamanders can result in an aggregate of 10,000 or more eggs in woodland vernal pools.

The mature eggs of frogs and salamanders are commonly 2 to 3 mm (approximately 0.1 in) in diameter. After the eggs have been laid and the surrounding jelly layers have absorbed water, each egg plus its gelatinous coat has a diameter of 5 to 20 mm (0.2 to 0.8 in), depending on the species. The central yolk consists of proteins, phospholipids, and fats, and the jelly is composed of glycoproteins plus water. Amphibian egg masses thus represent a very substantial additional source of energy for a variety of pool animals including caddisfly larvae, leeches, and turtles.

Decomposition of Detritus

Detritus is composed of dead algal and plant material, as well as the remains of dead animals. Most plant detritus is composed of cellulose, lignins, and other complex compounds that are not readily broken down by animals. Decomposition by fungi, bacteria, and protozoans plays a crucial role in making the carbon compounds in detritus available to the animals in vernal pools.

Decomposition includes the leaching of nutrients from fallen leaves and other dead plant materials, the colonization of detritus by bacteria and fungi, the gradual breakdown of the detritus by the microbial decomposers, and a corresponding increase in microbial biomass. Invertebrates contribute to decomposition by feeding on the decaying materials and their associated microflora, breaking them down, changing their nutrient content, and increasing their surface area so that they are more readily attacked by the microbial decomposers.

Leaf Litter

The decomposition of leaf litter in vernal pools combines characteristics of terrestrial and aquatic decay processes[12].

In forests, terrestrial leaf detritus, when conditioned and decomposed by fungi and bacteria, provides an abundant and nutrient-rich source of food for forest organisms. In late summer and early autumn, trees resorb from their leaves much of the nitrogen, phosphorus, sugars, and proteins. Once the leaves fall to the ground, they undergo a rapid leaching of their remaining weight and nutrients, with a loss of up to 10% of their initial dry weight during the first month. The amounts of nitrogen, phosphorus, and protein then start to increase, reflecting the growth of microorganisms as bacteria and fungi colonize the leaves and start to break them down, and the leaves become progressively more nutrient-rich — and microbe-rich — over a period of eight to twelve months. The rate of leaf decomposition and of nutrient accumulation within the leaves varies among different species of trees. Soil invertebrates feed upon the conditioned leaves, further contributing to their decomposition.

A parallel process is seen in leaves that fall into water. These leaves also experience rapid leaching of nutrients and organic materials and colonization by bacteria and fungi. The initial weight loss by leaching is more rapid in water than in air, with leaf weights decreasing up to 25% in the first twenty-four hours. Nitrogen within the leaves is rapidly immobilized, becoming sequestered within microbial biomass. Smaller amounts of protein and nutrients accumulate in the decomposing leaves in water than on the forest floor, indicating a lower level of microbial growth.

As on the forest floor, the rate of leaf decomposition in water varies with tree species[13]. Conifer needles and the leaves of deciduous trees undergo similar leaching and colonization by microorganisms, but the rates of decomposition for conifer needles tend to be much slower. Temperature affects microbial processes and can affect decomposition, although in streams, decomposition takes place rapidly even at freezing temperatures. Decomposition rates are also affected by the nutrient content of the water, pH, and water movement[14]. Depending on the specific conditions, the rates of leaf decomposition in water may vary by several orders of magnitude.

Physical fragmentation by macroinvertebrate shredders contributes to decomposition in water by increasing the surface area available for microbial colonization. Large densities of aquatic invertebrates increase the rates of decomposition and affect the extent to which nitrogen is released into the water[15].

***In Situ* Detritus**

As with the tree leaves, detritus derived from in-pool plant growth is colonized by bacteria and fungi, experiences initial leaching of mass and nutrients, undergoes an increase in protein, nitrogen and phosphorus, decays at rates that vary with plant species, and is fed upon by macroinvertebrates[16].

Differences Between Annual Vernal Pools and Semi-Permanent Pools

In fresh waters, as on the forest floor, leaf decomposition is mediated largely by fungi, although the numbers of bacteria increase over time as decomposition progresses. The fungi that colonize leaves and other detritus in woodland streams differ from those on the forest floor and are largely members of a group known as aquatic hyphomycetes. The species of fungi that colonize leaves in semi-permanent pools and in annual vernal pools once they become flooded are primarily aero-aquatic fungi, a group that is characteristic of conditions where oxygen is limited[17].

The nutritional value of leaves that have been decomposed by terrestrial fungi differs from that of leaves that have been decomposed exclusively by aquatic fungi. Leaves that decompose in spring-filling vernal pools experience several months of conditioning by terrestrial fungi before inundation. Such leaves develop higher protein levels and are more palatable to invertebrates such as caddisfly larvae than leaves that decompose in semi-permanent pools and are not subject to a period of terrestrial conditioning. It appears probable that the ready availability of high-quality foods contributes to the high productivity of spring-filling vernal pools and may help mitigate the life history limitations imposed by the short hydroperiod in these pools[18].

Detritus in Food Webs

Detritus contributes to the growth of primary producers in vernal pools and is the primary food for many kinds of consumers (Figure 39). The nutrients that are released into the water from detritus when a vernal pool floods are taken up by epiphytic algae on the surfaces of plants and decaying leaves, and by filamentous algae within the water column. Fungi and bacteria colonize the detritus rapidly, and their activity increases thc availability of protein and other nutrients and makes the detritus more palatable to invertebrate shredders and grazers. As discussed in Chapter 4, seventeen species of bacteria were identified in a small snowmelt pool in Quebec soon after the pool thawed[19]. The bacteria fed on the organic compounds leached into the water and on the decaying detritus itself. The rapid proliferation of the bacteria provided rich food for 194 species of protists. The protists, in turn, supported one species of rotifer, five species of crustaceans, four species of mosquito larvae, thirty-one other insect species, and wood frogs.

Many aquatic invertebrates are specialized for feeding on microbially conditioned detritus. In the process they break the leaves and other materials into smaller pieces. These species are characterized functionally as "shredders," because of their role in physically fragmenting detritus[20]. They include common vernal pool species such as log-cabin caddisflies, haliplid beetles, adults of some hydrophilid beetles, and some chironomid midge larvae. Members of another functional feeding group, known as "scrapers," graze on the film of fungi and bacteria (and algae, if present) on the decaying leaves. Snails and many anuran tadpoles including spring peepers and toads fall in this group. Other species, characterized as "collectors" or "gatherers," feed preferentially on small particles broken off of the decaying leaf materials or released by the shredders and scrapers. These include deposit feeders such as oligochaete worms, ostracods, cranefly larvae, and some chironomid midge larvae, which ingest sediment and extract nutrients from it, and filter-feeders such as rotifers, fairy and clam shrimp, cladocerans, mosquito larvae, and fingernail clams, which feed on decomposer bacteria, other microorganisms, and small particles of detritus in the water.

The importance of decaying leaves and other detritus to the vernal pool food web is reflected in the dominance of shredders and collectors among permanent residents, early migrants, and hatch-on-flooding species in early spring[21]. The functional-feeding-group structure of vernal pools in spring is similar to that in small headwater streams, whose faunas include high proportions of shredders, scrapers, and collector-gatherers[22].

Predator-Prey Relationships

Vernal pools support a wide variety of predators, including carnivorous protozoans that feed on microscopic herbivores, large salamander larvae and dragonfly nymphs that prey on all manner of smaller animals in the pools, bats and tree swallows that feed on insects as they emerge, and racoons and water birds that prey on invertebrates and breeding amphibians (Figure 39). Over time during the season, as shredders and collector-gatherers transform decomposing detritus into increasing amounts of invertebrate and tadpole biomass, the numbers and diversity of predators in vernal pools increases. The sweep of a dip net in a vernal pool in late spring or early summer is likely to yield a diverse array of predators including dragonfly and damselfly nymphs, tabanid larvae, phantom midge larvae, larval diving and water scavenger beetles, adult diving beetles, giant water bugs, backswimmers, water striders, wood frog tadpoles, mole salamander larvae, and newts. The variety of predators and their sheer numbers are evidence of the vast numbers of individuals of filter-feeders and deposit-feeding collectors in pools (Plate 8).

There are multiple levels of predators, and the food web consists of many food chains with intersecting links. In the following example of a hypothetical food chain, prey are listed first, with the arrow pointing to a predator: herbivorous protozoan ➡ carnivorous protozoan ➡ rotifer ➡ copepod ➡ damselfly nymph ➡ dytiscid beetle larva ➡ dragonfly nymph ➡ salamander larva ➡ garter snake ➡ red-shouldered hawk. Many other chains of predator-prey relationships can be found in pools. By the time a snake, water bird, or raccoon feeds on a metamorphosing amphibian, or a tree swallow captures a dragonfly on the wing and carries it to its nestlings, the carbon fixed by a pool-side tree will have been recycled many times through the bodies of many different kinds of microbes and macroscopic animals.

Different species come and go as the season progresses. Predatory protozoans and rotifers feed on bacteria and on other microscopic organisms soon after the pool floods. Flatworms are common in winter and early spring and feed widely on a variety of other animals. Diving beetle larvae and adults, water scavenger beetle larvae, and phantom midge larvae are present early in the spring, and as the water warms they are joined or replaced by dragonfly and damselfly nymphs, backswimmers, and salamander larvae. Predators can play significant roles in structuring the aquatic community and in determining whether other species reproduce successfully in a given year.

Annual Cycles in Northeastern Vernal Pools

The successional patterns, life history adaptations, and community composition in vernal pools have long interested ecologists. The following discussion considers three important studies that evaluate patterns in one or more vernal pools throughout the year and presents a synthesis that may be applied to vernal pools of different hydrologic classes, landscape positions, and origins throughout the glaciated northeast.

Comparison of Three Michigan Pools

Roman Kenk's study of the biological communities of four ponds in Michigan remains a classic source of information about vernal pools[23]. His study sites include two annual vernal pools, a vernal pool that is continuously flooded but lacks fish (possibly a semi-permanent pool as defined herein), and a permanent pond that does not support vernal pool fauna. He sampled the pools over the course of one season, from fall through the drying of the annual pools early the next summer, and he revisited the pools in the spring of the following year. Both of the annual vernal pools are short-cycle pools, drying in late June. They typically start to fill in late autumn/early winter, reaching maximum size in March. During a dry year, both remained dry during the winter and filled in spring.

Kenk identifies three distinct phases during the yearly cycle of the annual vernal pools.

(1) Winter or cold-water stage, from late November/early December to the end of March. This stage is characterized by frozen conditions and a depauperate fauna including protozoans, rotifers, oligochaetes, ostracods, copepods, ceratopogonids, gastropods, sphaeriids, small numbers of isopods and amphipods, and the flatworm *Phagocata velata*.

(2) Spring stage, from early April following ice melting to drying in late June/early July. Taxa characteristic of this phase include fairy shrimp (early spring), clam shrimp (late spring), leeches, cladocerans, beetles, odonates, hemipterans, water mites, caddisflies, and amphibians. The fauna increases in total numbers and number

of taxa soon after ice-out. Proportionally high numbers of predators such as larval and adult beetles, water bugs, and odonates are present as water levels shrink.

(3) Dry stage, from late June to late autumn/early winter. The fauna consists of terrestrial species of isopods, collembolans, molluscs, oligochaetes, beetles, and hemipterans, and aestivating aquatic snails, sphaeriids, and ceratopogonids.

Seasonal Cycles in an Ontario Pool

Another classic study, this one carried out by D. Dudley Williams in a spring-filling, short-cycle vernal pool in southern Ontario, provides a detailed analysis of changes in the biological community over the course of one year[24]. Some members of the fauna are present continuously within the pool; others appear soon after flooding and are present for only a few weeks; still others appear in mid season; and some are present only in the drying phase, two to three weeks before the water disappears entirely. A final group consists of species that are present only during the dry phase.

As in Kenk's Michigan pools, the relative abundance of organisms and the number of taxa increase rapidly within the first two weeks after flooding, remain relatively constant until mid June, and then decrease steadily as water levels decline. Different species are found in the pool at different times within the annual cycle, and species replace one another in a predictable manner.

Detritivore-herbivores (largely shredders and collector-gatherers) are the predominant functional feeding group when the pool floods, and they remain important as long as water is present, representing from 25 to 45% of the organisms present. Predators increase in both overall abundance and number of species as the season progresses from initial flooding to late spring. Predators then decrease in relative abundance as the pool dries and are present in low numbers during the dry phase. During the dry phase, most of the community consists of detritivores and herbivores that feed in the dry basin. Omnivorous scavengers also increase in relative abundance during the dry phase.

Seasonal Cycles in Relation to Life History, With Detailed Studies of Four Ontario Pools

As discussed in Chapter 6, Glenn Wiggins and his colleagues consider the fauna of vernal pools from the perspective of life history strategies. They provide a comprehensive review of the literature, faunal data for a number of vernal pools in Ontario, and detailed results of field sampling over three seasons in four pools in southeastern Ontario[25].

The seasonal patterns observed in these Ontario pools are similar to those observed by Kenk and Williams. The numbers of organisms increase rapidly after flooding, remain high through May, and decline as water levels drop. The increase through the early spring occurs through a combination of the emergence of aestivating larvae and adults from the sediments, the hatching of eggs and drought-resistant cysts, and the immigration of adult insects and parasites from distant overwintering sites. Species richness is lower in spring-filling, short-cycle pools than in fall-filling and semi-permanent pools, reflecting the limited number of species that can survive extended drying in pools that are flooded for as few as three months of the year.

An Overview of Annual Patterns

Certain successional and developmental patterns appear in vernal pools throughout the glaciated northeast. It is possible to identify the following phases over the course of a year: dry phase; newly flooded or coldwater phase; early spring phase; mid spring phase; late spring/early summer phase; and drying phase.

Dry Phase

The dry phase is probably the most difficult to characterize. In some pools the substrate becomes completely dry and terrestrial plants cover the bottom during summer and/or early fall. In many pools, particularly long-cycle pools, the substrate may remain saturated but without enough standing water to support a characteristic aquatic fauna. These pools often support wetland plants. In partially drying pools, most of the basin dries, but a small proportion remains flooded.

In fall and early winter, leaves and dead plants cover the substrate, and terrestrial bacteria and fungi initiate the decomposition process. Caddisflies, marbled salamanders, and other species that breed in the dry pool basin, mate and deposit their eggs in anticipation of the reflooding of the pool. Many aquatic species lie dormant in the sediment as cysts, eggs, or aestivating juveniles or adults. Terrestrial species including millipedes and centipedes, sowbugs, redback salamanders, spiders, pulmonate snails, and slugs often can be found in the pool depression, under bark, logs, or other debris.

Newly Flooded Phase

The key events of the newly flooded phase are (1) the inundation of detritus, (2) the emergence of animals that have been lying dormant in the sediment, and (3) the hatching of eggs and cysts of hatch-on-flooding species.

The newly flooded or coldwater phase extends from flooding to ice melting. It may start any time from late fall to early spring, depending on the flooding regime of the pool. This phase is comparable to Kenk's winter or coldwater stage. The water is clear and cold, decreasing from fall temperatures in the teens and single numbers Celsius (forties and thirties Fahrenheit) to around freezing in winter. Detritus is colonized by aquatic fungi, bacteria, and other microbial decomposers, representing a shift from the terrestrial decomposers that were active before flooding, and nutrients are released from the conditioned detritus. There is some algal photosynthesis on the surfaces of leaves and other substrate materials, depending on the availability of nutrients, the amount of canopy cover, and, if the pool is flooded during winter, on the depth of snow and ice overlying the pool. Upon flooding there is an immediate emergence of both permanent residents that have spent the dry period burrowed into the sediment (some beetles, pulmonate snails, fingernail clams, and oligochaete worms), and hatch-on-flooding species (fairy shrimp, ostracods, some copepods, cladocerans, flatworms, mosquito larvae, caddisflies, some other insects, and marbled salamander larvae). Detritus starts to be shredded and grazed by invertebrate detritivores. Overall, growth rates and activity are low for most species during this phase.

Pool animals that emerge during the newly flooded phase are active in the cold water and can be seen under ice. The planarian *Phagocata velata* completes its life cycle during this phase and disappears from the fauna soon after ice-out. It is often possible to find molluscs and water beetles along the thawed edges of frozen pools in the middle of winter, and limnephilid caddisflies are readily observed crawling along the pool bottom. In semi-permanent pools, overwintering spotted salamander larvae may be present, continuing to grow and preparing to leave the pool at a large size after the pools thaw.

The newly flooded phase varies in duration and spatial extent, depending on a pool's hydrologic cycle. In spring-filling pools this phase only briefly precedes or is concurrent with the early spring phase. In fall-filling pools it persists for as long as six months, from the appearance of water to ice-out. If a small proportion of the basin fills in fall or winter and retains enough water under ice to support aquatic life until the entire pool fills in spring, the newly flooded phase begins upon first flooding and is repeated in other sections of the basin as water levels rise. At the beginning of the early spring phase in such pools, detritus may be at different stages of decomposition in different parts of the basin and, depending on the elevation at which eggs are deposited or aestivating stages are found, several life stages of the same species may be present together, reflecting successive increases in water levels and emergences of animals from dormancy.

Early Spring Phase

The key events of the early spring phase are (1) the rapid decomposition of detritus, (2) algal photosynthesis, (3) the rapid growth of pool animals, and (4) the appearance of early spring migrants, especially vernal-pool-dependent amphibians.

The early spring phase extends from ice melting to approximately two to three weeks later. It typically begins between mid February and mid May, depending on the location of the pool. Water temperatures arc still quite cold although they increase somewhat by the end of the phase. The water remains relatively clear, and algal photosynthesis on the surfaces of leaves and other substrates in the pool may be extensive. In some pools, thick growths of filamentous algae appear. Detritus breakdown, grazing, and shredding continue at a greatly increased rate. Permanent residents and hatch-on-flooding species become more active, and many reach maturity during this phase. At different times during this period we see the appearance of thermally cued hatchers and early spring migrants (e.g., beetles, *Culex* mosquitoes, wood frogs, chorus frogs, spring peepers, and ambystomatid salamanders — the amphibians are present as adults for a short period, then as eggs). Turtles may move to pools to feed on amphibian eggs and invertebrates as air temperatures warm.

Mid Spring Phase

The mid spring phase starts about two to three weeks after ice melting and lasts as long as two months, until the trees start to leaf out. It typically occurs from mid April to early June but may begin earlier or later depending on seasonal weather conditions and geographic location. This phase is characterized by a rapid increase in water temperature and an increase in faunal richness. Algal photosynthesis peaks and may then decline in pools with canopy cover as trees leaf out. Usually there is a noticeable increase in water color. Permanent residents continue to grow, and they reproduce when mature. Hatch-on-flooding species continue to grow rapidly, and many of these species mature, reproduce, and enter a dormant stage; for example, fairy shrimp usually reach peak numbers but disappear by the end of this phase. There is continued emergence of thermally cued hatchers as threshold temperatures are reached, and these also develop rapidly. The eggs of fast-growing early spring migrants hatch (*Culex* mosquitoes, wood frog, spring peeper, and dytiscid beetles), and many of these species are ready to transform and leave the pools by the end of this phase. Salamanders are still present as egg masses, but they may hatch toward the end of this phase. As air and water temperatures increase, later migrants move into the pools. These include more beetles, hemipterans, toads, and treefrogs — depending on the species, the amphibian adults are present for periods of varying lengths, and their eggs are present until hatching. Turtles forage in pools throughout this period.

Late Spring/Summer Phase

The late spring/summer phase extends from leaf-out, approximately two months after ice-out, to pool drawdown. In short-cycle pools this phase is relatively short, and in long-cycle pools it lasts as long as six to seven months. Water levels decline and the pool's surface area decreases during this period, but there is ample aquatic habitat available. Depending on the hydrologic class of the pool, this phase may experience a rapid turnover of species with maturation and dispersal of the next generation, or there may be continued recruitment of new species and several generations of existing taxa. During this phase, the eggs of species requiring warm water temperatures hatch. Amphibian larvae grow rapidly. First wood frogs, then peepers and toads transform, leaving the pool. Salamander larvae also grow rapidly, but they do not leave the pool until water levels decline substantially during the drying phase. The fauna is dominated by salamander larvae; permanent residents such as clam shrimp, oligochaete worms, lestid damselflies, and midge larvae; beetles; and hemipterans such as backswimmers and water boatmen. If enough light penetrates through the canopy, aquatic and emergent plants leaf out and bloom along the shoreline and in the basin.

Drying Phase

The drying phase encompasses the period when the pool is decreasing rapidly in surface area and depth. A relatively small area of the pool basin may retain water during this phase, and that part of the pool will remain in the late spring/summer phase of development with a more diverse fauna. The same is true of semi-permanent pools. In the drying portion of the pool, the fauna consists mostly of permanent residents and drying phase species. Depending on the length of time that has passed since pool flooding and egg deposition, most amphibians will either have completed metamorphosis and left the pool or be stranded and die. Some salamander larvae may remain in the pool until it is almost fully dry, even if the pool retains water into late summer or fall. As the water level recedes, the damp shoreline becomes vegetated in successive zones if the pool receives sufficient sunlight.

Variations in Community Composition Over Space and Time

It is evident from the discussion of seasonal cycles that the aquatic community in vernal pools changes continuously through the season, reflecting the life cycles of individual species and the movements of migrants. Both the species and the number of individuals present vary widely. There is much year-to-year variation in the fauna of many vernal pools, as well. This variation reflects the responses of resident species to changes in hydrology, water quality, and predators/competitors and differences in colonization by migrants from one year to the next[26].

Several authors have explored the idea that the theory of island biogeography may apply to temporary waters in the glaciated northeast and elsewhere. This theory holds that the number of species present on an island are associated with the size of the island and its distance from the mainland or other sources of potential colonizers[27]. In applying the theory to vernal pools, the idea is that the pools are aquatic islands that differ in size and occur at different distances from other pools.

Evidence supporting the application of island-biogeography theory to vernal pools is mixed. In some studies, the numbers of individuals and species of predatory insects and amphibians that colonize vernal pools appear to be positively correlated with pool sizc[28]. In tcmporary pools in southcrn California, thc numbcrs of vascular plant species are correlated with pool size, and the diversity of rotifers and microcrustaceans is correlated with pool hydroperiod[29]. For some amphibian species, including wood frogs in northeastern vernal pools, leopard frogs in grassland landscapes, and pool frogs in Scandinavia, the presence and persistence of breeding populations within pools is highly correlated with the presence of other pools nearby[30].

Marshall Laird, whose studies of microorganisms were discussed in Chapter 4, argues, in contrast, that extremely small, temporary habitats support remarkably rich assemblages of bacteria, diatoms, and protozoans. He suggests that, at least for small taxa, the theory of island biogeography may not be applicable, and he cites experimental studies with small microcosms which suggest that species richness of microorganisms may vary inversely with pool size[31]. The absence of a relationship between pool size and the numbers of breeding amphibians or invertebrate species in a number of studies also suggests that island-biogeography theory may not be readily applied to vernal pools[32].

It is clear that isolation and size are not the only factors influencing taxonomic richness and species composition in northeastern vernal pools. As we have seen, hydroperiod is a primary factor implicated in the distribution of many invertebrate and amphibian taxa. Female amphibians select oviposition sites based on a variety of factors including the presence or absence of other species. Land uses, the presence and density of roadways, the extent of habitat fragmentation, the proximity to other breeding populations, and the availability of non-breeding habitat required by organisms during part of their life cycles affect taxonomic richness in birds, mammals, turtles, amphibians, and invertebrates[33]. Land use within the watershed is thus another important factor affecting the taxonomic richness and densities within vernal pools.

Remaining Questions About Energy Flow and Seasonal Cycles

Energy Flow

- How do inputs of allochthonous and autochthonous detritus differ in pools that occur in different positions within the landscape, that have different surface areas, and that are surrounded by different plant communities?
- How important is algal photosynthesis in the energy budgets of vernal pools?
- Do the species of animals in vernal pools and their trophic relationships vary quantitatively along a continuum of pool size and canopy cover, and hence of relative importance of allochthonous leaf litter inputs vs. *in situ* production by plants and algae?
- How do nutrient processing and nutrient cycling vary in vernal pools along a continuum of flooding duration?
- How do the microbial communities that decompose detritus differ in species composition and biomass in pools with different hydroperiods, and on detritus from different plant species? How do such differences affect the nutritional value of the decomposed detritus for pool animals?
- How do variations in weather affect the duration of different pool stages and the energy flow within the pool?

- Does prior-year hydrology affect the kinds and amounts of organic carbon that are available in a vernal pool?
- What are the pathways and relative distributions of carbon flow within the different trophic levels, and different taxa?
- How high is the productivity of vernal pools, and how much do vernal pools contribute to forest carbon and nutrient dynamics?
- How important are amphibian egg masses as food for pool wildlife?

Seasonal Cycles

- How consistent are the seasonal patterns of appearances and disappearances of species in individual pools from one year to the next?
- How are patterns of species occurrence, and of succession over time, associated with physical and chemical variables such as water level, pH, temperature, nutrients, and color?
- To what extent is the timing of the life cycles of individual species tied to physical habitat, and to what extent to inter-specific interactions with potential predators and competitors?
- Do the hydrologic conditions of prior years influence the species that occur in pools in later years?

References for Chapter 13

[1] Bärlocher et al. 1978, Cole and Fisher 1978. [2] Benfield 1996; Bormann and Likens 1979; Brinson et al. 1981; Gosz et al. 1972, 1973, 1976; Hynes 1970; Laird 1988; Likens et al. 1977; Minshall 1967; Rich and Wetzel 1978; Vannotte et al. 1980; Webster and Benfield 1986. [3] Bachmann and Bachmann 1994, Cole and Fisher 1978, Gates and Thompson 1982, Williams 1983. [4] Skelly 1998; Skelly and Golon 2003; Skelly et al. 1999, 2002. [5] Burton et al. 1988, Fisher and Likens 1973, Mayer and Likens 1987. [6] Cole and Fisher 1978. [7] Bärlocher et al. 1978. [8] Gosz et al. 1976. [9] Fisher and Likens 1973. [10] Bärlocher et al. 1978. [11] Torrey 1967. [12] Bärlocher et al. 1978, Ostrofsky 1997, Rowe et al. 1996, Sedell et al. 1975, Webster and Benfield 1986. [13] See Webster and Benfield 1986. [14] Aerts et al. 1995, Hill and Periotte 1998, Meyer and Johnson 1983, Webster and Benfield 1986. [15] McDiffett and Jordan 1978, Stewart 1992. [16] Bärlocher et al. 1978. [17] Bärlocher et al. 1978, Bärlocher and Kendrick 1974. [18] Bärlocher et al. 1978. [19] Laird 1988. [20] Cummins and Klug 1979; Merritt and Cummins 1996a, 1996b. [21] Wiggins et al. 1980, Williams 1983. [22] Benfield 1996, Hynes 1970, Merritt and Cummins 1996b, Minshall 1996, Webster and Benfield 1986. [23] Kenk 1949. [24] Williams 1983. [25] Wiggins et al. 1980. [26] Daborn 1974, Donald 1983, Schneider and Frost 1996. [27] Brooks et al. 1998, Ebert and Balko 1980, MacArthur and Wilson 1967. [28] Brooks 2000, Burne 2000, Roth and Jackson 1987, Seale 1982. [29] Ebert and Balko 1980. [30] Berven and Grudzien 1990, Burne 2000, Gulve 1994, Pope et al. 2000. [31] Dickerson and Robinson 1985, Laird 1988. [32] Brooks 2000, Calhoun et al. 2003, Gascon and Planas 1986. [33] Burke and Gibbons 1995, Dunning et al. 1995, Ehrlich and Murphy 1987, Fahrig et al. 1995, Findlay and Houlahan 1997, Henderson et al. 1985, Pope et al. 2000, Semlitsch 1998, Taylor et al. 1993.

Chapter 14

Protecting Vernal Pools

In recent years, much of the interest in vernal pools has been directed toward conserving these habitats and their wildlife. The simplest reason for protecting vernal pools is based on the assumption that the diversity of life should be maintained. Given this assumption, vernal pools and other temporary waters deserve to be conserved because they support many species that do not live in other habitats. Many of the species that occur in vernal pools do not reproduce successfully in permanently flooded waters that contain fish. If vernal pools are destroyed or degraded, animals that occur nowhere else will disappear from them, and if large numbers of pools are lost, these species will become extinct.

Most of the focus to date has been on protecting amphibian populations that depend on vernal pools. Over the past few decades, there has been an increase in public awareness of frogs and salamanders as indicators of environmental quality and a heightened sense that they are in need of protection. The calling of spring peepers is a widely recognized sign of spring, and many people are familiar with wood frogs and mole salamanders and find them both attractive and interesting. The efforts of naturalists and herpetologists to make people more aware of the links between these amphibians and small, temporary pools have contributed to an interest in vernal pool protection at the local level in many communities. News reports about worldwide declines in amphibian populations and discoveries of deformed frogs in a number of locations in the glaciated northeast have contributed to a sense of urgency and have stimulated support for conservation.

Biologists have long known about the relationships between habitat loss and the disappearance of species, but the broad focus on protecting vernal pools and the species that depend on them is relatively new. It has been more than a century since the first Audubon societies advocated for the protection of migratory birds, since the first national, provincial, and state parks were established in Canada and the United States, and since conservation was accepted as a valid and important goal of governmental management of public lands. It has been thirty years since the Endangered Species Act was passed by the US Congress. However, public support for protecting invertebrates and amphibians has lagged behind the interest in protecting birds and mammals. The tools and resources available to protect vernal pools and their wildlife are limited.

Because vernal pools are small, intermittently flooded, and broadly distributed across the landscape, in many cases lying in the midst of well-drained uplands, and because the species that depend on them are not generally perceived as "charismatic" or economically valuable, protecting them is not a simple task. The challenges involved are both practical and political. On the practical side, the physical basins of the pools need to remain intact, and so do their hydrologic regimes, water quality, food sources, and biological connections with other pools, upland forests, and permanent waters. Also, because amphibians that breed in vernal pools are terrestrial during most of their lives, conservation of these species requires the protection of forests as well as pools. It is no trivial task to identify objectively the areas that are critical as habitat, and to devise effective strategies to maintain the important habitat characteristics of pools and forests in the face of a changing landscape. On the political side, there is by no means universal support for the regulation by government agencies of activity on private lands, or for lower tax assessments on property that is set aside for wildlife habitat, or for the commitment of public funds to the conservation of habitat for amphibians and aquatic invertebrates, or even for the commitment of private conservation dollars. Not everyone believes that the protection of vernal pools and their wildlife is an important public purpose. Even where there is agreement about the value of conservation, there are often disagreements about the best way to accomplish the goal.

Despite these challenges, there are many efforts underway to protect vernal pools throughout the glaciated northeast, from New Jersey to the Maritime Provinces and from Ohio to Minnesota and Ontario. These range from education programs designed to increase public awareness and encourage voluntary action to protect pools, to regulatory reviews carried out under state and local wetlands protection laws and ordinances, to the listing of rare species of invetebrates and salamanders under endangered-species laws, to land acquisition focused on providing habitat for a wide range of species including turtles and vernal-pool-dependent amphibians.

The problems facing vernal pools in the glaciated northeast have parallels elsewhere. The details differ from one geographic area to another, but the same fundamental conservation issues pertain to temporary waters in southern pine forests, prairies, montane meadows, bottomland floodplains, desert playas or canyons, or northeastern deciduous woodlands, whether they are in Australia, Europe, Africa, or North America. Indeed, the conservation of amphibians, reptiles, and macroinvertebrates that need a habitat mosaic of small, isolated breeding pools plus large areas of upland is of global interest. Temporary waters everywhere are threatened by habitat alteration or destruction, and by the introduction of species that compete with or prey on pool-dependent wildlife. Conservation professionals worldwide are working to identify effective ways to protect temporary pools.

This chapter discusses vernal pool conservation from several perspectives. It recaps the current understanding of relationships between vernal pools and the rest of the landscape, reviews threats to vernal pools and their wildlife, considers a variety of ways to approach conservation, and discusses policy issues that need to be addressed as part of formulating conservation strategies. This discussion does not provide detailed, step-by-step recommendations for conservation action. Rather, it identifies some issues that I believe need to be addressed explicitly if we are to develop effective approaches for the long-term protection of the animals that depend on vernal pools and adjacent forests. I hope it will stimulate discussions and new thinking among regulators, conservationists, naturalists, lawyers, environmental educators, foresters, natural resource managers, legislators, planners, scientists, and others who are interested in protecting vernal pool wildlife, and that it will contribute meaningfully to successful conservation programs.

Threats to Vernal Pools and Vernal Pool Wildlife

Vernal pools and their wildlife are threatened directly by filling, draining, dredging, shoreline clearing, mosquito control, pollution, and the introduction of invasive plants and animals. Pools are affected indirectly, and some pool-dependent species directly, by development, forestry, and clearing for agriculture and recreation in their watersheds.

Physical Destruction of Pools and Surrounding Habitat

The historic and ongoing loss of vernal pools to filling, drainage, and other destructive activities is difficult to quantify. The vernal pools found in the glaciated northeast today are only a subset of the pools that were present several hundred years ago, and most, if not all, have been physically altered by human activity over the past four centuries, if not earlier (Plate 15). We do not know whether pools in northeastern forests were used extensively by Native Americans, but the large-scale changes made in the forest landscape by European settlers from the seventeenth century onward had many dramatic effects. The values of vernal pools as wildlife habitat for amphibians and reptiles, large branchiopods, and other invertebrates were not appreciated by pioneers trying to wrest a living from the land. Seasonal flooding rendered vernal pools unsuitable for growing crops, but annual drawdown prevented their use by beavers, muskrats, and waterfowl as habitat or by settlers for irrigation, stock watering, or fish production. Therefore, as first trappers and hunters, and later farmers, moved inland from the coast, many vernal pools were destroyed by ditching and plowing, or by dredging.

By the mid 1800s, most of southern and central New England and much of the Maritime Provinces had been deforested, with up to 85% of the landscape cleared and converted to fields and pastures, and with most of the rest used as woodlots[1] (Plate 6). This wholesale clearing removed the forests that supported woodland amphibians, regulated temperatures and hydrology, and supplied the base of the food web in vernal pools. Deforestation moved with the pioneers in the nineteenth century into western New York, Pennsylvania, Ohio, and other lower midwestern states, western Quebec, and southern Ontario. Throughout this period, vernal pools and forest-dependent animals presumably persisted in the woodlots and uncleared forest patches that remained scattered across the landscape.

In many formerly cleared areas, farm abandonment and the regrowth of forests in the last century (Plate 7) have been accompanied by the reestablishment of amphibian populations and other forest-dependent components of the fauna. Today, over much of the glaciated northeast these second-growth forests are disappearing, as a new wave of clearing opens the way for new residential and commercial development.

The destruction of vernal pools represents a significant threat for fairy shrimp, clam shrimp, and other permanent residents with limited dispersal abilities. The late Denton Belk, who spent his professional life studying large branchiopods, noted that in Europe, the former USSR, and the North American Great Plains, the loss of habitat has resulted in substantial reduction in the distribution of fairy shrimps, and many species are now restricted to areas that are distant from human activities. This theme of the loss of ephemeral pools recurs consistently in literature from around the world[2].

From their extensive studies of fairy shrimp (*Eubranchipus*) in Ohio and Illinois, Ralph Dexter of Kent State University and his colleagues[3] presented descriptive information on 110 vernal pools. In these surveys, conducted more than half a century ago, more than 52% of the pools studied had been altered by human activity. Five of the pools were destroyed during the course of the studies. Comments on the loss of amphibian breeding habitat also appeared in the literature as early as the 1940s[4]. Roman Kenk noted that vernal pools were often located along roadsides and in fields and pastures in southern Michigan and commented, "within a single generation...many of the old ponds and woods pools have disappeared." In New York, the felling of trees and tilling of the soil led to an often radical alteration of the character of the land surrounding vernal pools, and in Indiana, the small, shallow forest ponds that provided suitable breeding habitat for mole salamanders had been extensively filled, drained, or converted into permanent waters unsuitable for salamanders by the end of the first half of the twentieth century.

Such destruction continues today. It is difficult to estimate the numbers of pools that have been lost, or the rates at which they are disappearing, because few data are available on their overall numbers and distributions. However, the following observations, coupled with broader data on rates of forest fragmentation and development, suggest that losses may be heavy. A re-survey of many of the sites studied by Dexter in the 1940s found that pools continue to be destroyed, and the disappearance of seasonally flooded pools in Ohio and Indiana may have contributed to declines in the distributions of some species of clam shrimp and fairy shrimp[5]. Surveys of vernal pools in Maine found that in a Penobscot County study area dominated by industrial forestry, most vernal pools occur in depressions associated with forest harvesting activities, and anthropogenic pools outnumber natural pools by a factor of 3.7 in heavily managed forests[6]. In the suburban town of Reading, Massachusetts, no vernal pools are found in the highly developed downtown area. The densities of pools are lower, and pools are significantly larger, in densely developed residential sections of the town than in less densely developed areas, suggesting that smaller pools that formerly occurred in these areas have been filled or otherwise lost[7].

Hydrologic Alterations

Changes in water sources, depth, volume, timing of filling or drying, and degree of fluctuation in response to precipitation events can alter a pool's suitability for the survival of individual species and can greatly affect the composition of the biological community. Many such changes can be avoided, especially if people become more aware of vernal pools and their wildlife. For example, vernal pools are often either drained or deepened because of a misplaced assumption that they are "waste places" or that "real ponds" need to contain fish.

Increased Water Depth, Volume, and Duration of Flooding

Land-use changes such as clearing or increases in impervious area can significantly increase surface runoff in the watershed of a vernal pool. As a result, the pool may receive more water, its depth fluctuations may become more pronounced, and the timing and duration of flooding may be lengthened. As low spots in the landscape, vernal pools are the logical place for diversion of runoff and they often receive direct discharges, as well as the effects of overall increases in watershed runoff. Such direct inputs of storm flows may produce sudden changes in water levels over a short period of time. As discussed earlier, altered flooding regimes can change the timing of the life cycles of pool fauna and allow the survival of some drought-intolerant species in formerly inhospitable pools, with resulting shifts in predator-prey dynamics and community composition.

Private and public property owners often dredge and/or impound vernal pools to increase depth and volume and decrease the frequency of annual drawdown, and in the process they often convert temporary pools into permanent ponds. Often fish are stocked into such altered ponds, making the habitat unsuitable for successful reproduction of vernal-pool-dependent species. Even without the presence of fish, the composition of the biological community changes, with some species that are characteristic of annual vernal pools excluded by the presence of predators that cannot withstand seasonal drying[8].

Water Withdrawal

Changes in hydrology associated with the development of water supply wells, removal of water for irrigation, or re-direction of runoff flows away from vernal pools can affect populations of amphibians and invertebrates by decreasing the hydroperiod below the threshold required for successful completion of development. A pool that normally remains flooded most of the year, and in which mole salamanders and toads reproduce successfully, may become suitable only for species with rapid growth and short development times, such as spring peepers and wood frogs. A pool that naturally is flooded for only two to three months each year may be drawn down so early as to become inhospitable to even wood frogs and fairy shrimp.

Isolation

Although by definition they are physically isolated from other surface water bodies, vernal pools do not occur in hydrologic or ecological isolation from other waters or from uplands[9]. Activities that destroy the connections among vernal pools, and between pools, upland woods, and permanent waters, threaten the long-term survival of pool-dependent wildlife populations (Figure 41).

Relationships Among Vernal Pools

Vernal pools that occur in clusters may be connected hydrologically by seasonal streams, periodic high-water events that merge adjacent pools, and common groundwater sources. Regardless of their hydrologic relationships, individual vernal pools are interconnected genetically and ecologically with other pools through the dispersal of insects, amphibians, and other wildlife.

We have seen that vernal-pool-dependent amphibians show high fidelity to their breeding pools, returning consistently to the same locations year after year. Studies on many different species show that there is little or no straying by individuals between pools, even when an area of forest contains many pools that are used by amphibians. At the same time, the number of species breeding in a given vernal pool depends on the presence of other pools nearby, suggesting a relationship among populations in different pools. In fact, many amphibians that breed in temporary waters constitute metapopulations. For each isolated breeding population, some dispersal to and recruitment from other pools contributes to long-term population stability[10]. It is estimated that other pools within a radius of 1,000 m (3,300 ft) are needed to maintain gene flow and inhibit genetic differentiation in wood frogs[11]. The literature suggests that comparable distances also apply for other vernal pool vertebrates, such as ambystomatid salamanders. Thus, if some pools become isolated so that they cannot exchange individuals with other pools, it may spell local extinction for their amphibian populations.

There is ample evidence that many species of water beetles, water bugs, caddisflies, and other insects readily colonize newly flooded habitats. Depending

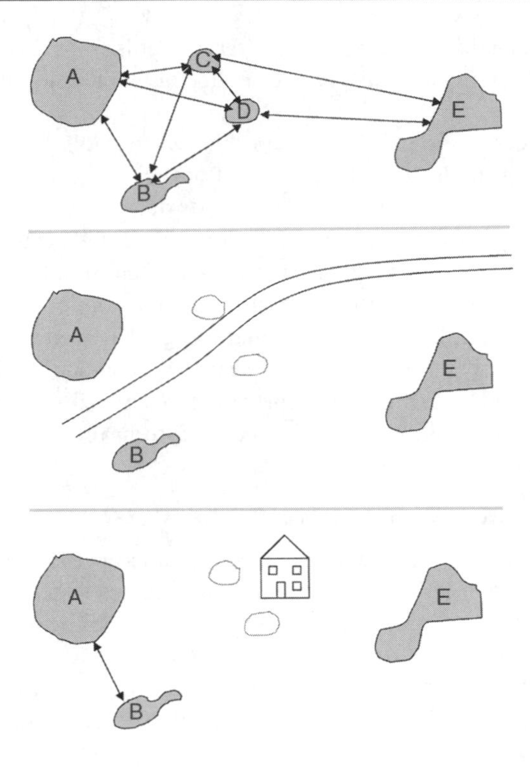

Figure 41. The isolation of individual vernal pools can cut off genetic exchange among populations of amphibian and other species that occur as metapopulations in multiple pools. Over time, this can reduce the genetic variability and viability of those populations. In this hypothetical example of an undisturbed cluster of five pools (A, B, C, D, and E) that support wood frogs (top), all local populations of the frog theoretically can exchange genes as a result of the dispersal of individuals, as indicated by the arrows, either directly between pools or through intermediary pools. For example, the breeding populations in pools A and B may exchange genes with that in pool E through the populations in pools C and D. The construction of a highway through the cluster of pools (center) destroys pools C and D and isolates the remaining pools so that the frogs cannot migrate between them, thereby increasing the probability that the local populations in these pools will become extinct. The construction of houses or other structures (bottom) could also destroy pools C and D and isolate the populations in pools A and B from the one in pool E. In this example, the connections between pools A and B are maintained, but the genetic exchange in all three of the remaining pools is significantly lower than before construction.

on the species, aquatic insects move into vernal pools from early spring through the dry period. Some insects disperse from drying pools into others with longer hydroperiods.

Mobile species of animals also move between pools from one year to the next depending on hydrologic conditions. Pools that are normally short-cycle support species that require longer hydroperiods during unusually wet years, and drought-tolerant species colonize normally long-cycle and semi-permanent pools during extended dry periods. Landscapes that contain numbers of vernal pools with different hydroperiods thus provide not only opportunities for genetic exchanges between pools, but also potential refugia between which animals may move during climatic extremes of drought or extended flooding. To date, the numbers of pools that would constitute an appropriate reserve for the maintenance of metapopulations of amphibians, or for the long-term protection of vernal pool invertebrates, have not been determined.

Relationships Between Vernal Pools and Other Water Bodies

The movements of animals also connect vernal pools to other waters and wetlands. Amphibians and turtles move from permanent waters to vernal pools to feed and/or breed. Birds and mammals forage in the pools or use them as resting areas during migration. Invertebrates such as water beetles, water bugs, and water mites overwinter in permanent waters and breed in vernal pools.

The seasonal use of vernal pools by wildlife species that have traditionally been associated with permanent wetlands has only recently begun to be appreciated. We now realize that Blanding's and spotted turtles regularly move to vernal pools to feed in spring, returning to permanent wetlands to overwinter. Several species of water beetles and water bugs, including water boatmen and backswimmers, along with their associated parasitic water mites, similarly overwinter in permanent lakes, ponds, and streams and fly to vernal pools to breed and feed on the abundant aquatic life there as water temperatures warm. As more is learned about the species of animals using northeastern vernal pools, and about their natural history and ecology, these relationships between vernal pools and other types of aquatic habitats may come to be better understood.

Many vernal pools are also hydrologically connected to other surface waters by groundwater or by intermittent streamflows during high-water periods. The periodic inundation of floodplains provides a direct connection between some pools and nearby rivers.

Relationships Between Vernal Pools and Uplands

Connectivity between vernal pools and upland habitats needs to be factored into conservation planning for pool-dependent vertebrates. Vernal-pool-breeding amphibians require upland woodland habitat for most of their lives. Many features of vernal pools including hydrology, water quality, the relative importance of terrestrial leaf litter and within-pool plant production, and the animal species that are present, depend in part on watershed characteristics such as forest cover and composition, land uses, habitat fragmentation, and the density and proximity of roads. Studies on leopard frogs, for example, have shown that the population distributions and densities of this species can only be understood in relation to the availability of both the small ponds in which it breeds and the upland fields and meadows that provide habitat during most of its life cycle[12]. Parallel relationships between breeding pools and upland forests may be expected for the woodland amphibians that breed in vernal pools. Similarly, turtles that feed in vernal pools before moving into uplands to lay eggs need a combination of wetlands, pools, open sandy nesting sites, and aestivation areas.

Watershed Alteration

Watershed alteration by residential and commercial development, forestry, and land clearing for agriculture or recreation changes the hydrology, water quality, and energetics of vernal pools. Watershed alteration may also affect vernal pool fauna by reducing the amount of upland habitat available for vernal-pool-breeding amphibians, by increasing access to interior areas for predators, and by contributing to road-associated mortality of amphibians and turtles.

Development

The replacement of forests by housing, commercial and industrial development, schools, hospitals, airports, roads, and other construction directly destroys vernal pools and upland habitat for vernal pool amphibians and nesting or aestivating turtles. It also eliminates forested connections between vernal pools, thereby reducing the ability of animals to move between remaining pools, and isolates pools from other habitats.

Development also affects pools by altering hydrology. Land clearing, road construction, and increases in impervious surfaces may increase surface runoff, potentially increasing pool hydroperiods. Changes in surface contours and the redirection of runoff flows to prevent increases in downstream flooding may divert water flows either into or away from vernal pools, respectively increasing or decreasing the hydroperiod. In either case, changes in the hydroperiod alter the fauna of vernal pools and affect the ability of amphibians and other species to reproduce successfully in the pools.

Water quality also changes with development. Vernal pools with significantly disturbed watersheds have a higher pH, more mineral substrate, and more algae, on average, than do pools whose watersheds are not developed. The fauna of pools may change dramatically in response to development, depending on the individual species' tolerances of increased turbidity, higher nutrient levels, and dissolved contaminants[13].

Roads affect vernal pools through changes in hydrology and water quality and through the traffic-asssociated mortality of animals that are migrating between pools and uplands. Roadway crossings have been shown to affect the macroinvertebrate fauna of streams in forested wetlands[14]. Vehicles on roads kill large numbers of turtles, migrating adult amphibians, and dispersing juvenile amphibians[15]. This high mortality of adults has important implications for the populations, because in nature the mortality in these species is concentrated in the young, and the long lives of the adults help ensure that each individual will have a good probability of reproducing itself at least once. In Rhode Island, declines in wood frog populations are associated with increases in road density[16]. Decisions about where to locate roads in relation to vernal pools and uplands deserve a high level of attention in conservation planning.

Forestry

Forest cutting may alter the inputs of autumn leaves, change the nutrient content of inflowing surface water and groundwater, increase the penetration of sunlight, contribute to increased algal production, raise water temperatures, increase sedimentation into vernal pools, and remove upland habitat used by amphibians (Plate 16). Ruts created by skidders and other heavy equipment can serve as barriers to amphibian migration and dispersal, contribute to erosion and sedimentation, and sometimes channel runoff into or away from vernal pools. Industrial forestry practices often change the species of trees, the age structure, and the availability and sizes of woody debris in forest stands. These changes are of interest, because the distributions of amphibians and reptiles differ within different forest types[17]. The abundance and species richness of salamanders are usually greater in deciduous than in coniferous forest stands. Various studies have shown correlations between the densities and activity of salamanders, including mole salamanders, and the depths of leaf litter and presence of large woody debris[18].

The abundance of redback salamanders and juvenile red-spotted newts is lower in disturbed forests than in undisturbed stands[19]. Clearcutting has been associated with significant short- and long-term declines in salamander populations in Florida pine forests, a variety of forest types in the southern Appalachians, and in the northwestern United States, although there is disagreement about the significance of these findings and uncertainty about their applicability to forest management practices in the glaciated northeast[20]. The effects of partial cutting have been less well studied. The elimination of some or all canopy cover, reduction in leaf litter, and loss of large woody debris on the forest floor as a result of forest harvesting all contribute to a decrease in habitat quality for vernal-pool-dependent amphibians[21].

Forest harvesting may also create small depressions that fill with water in spring but do not remain flooded long enough for the successful reproduction of vernal-pool-dependent amphibians. Such depressions can serve as population "sinks" where straying individuals expend the effort to breed without a chance of success. For example, in comparison with uncut forests, managed forests in Maine have a relatively high proportion of small, artificially produced depressions that serve as sinks for dispersing amphibians[22].

In many forests that are managed for commercial harvesting, insecticides are routinely sprayed by aircraft to control forest pests. The effects, if any, of such spraying on vernal pool amphibians and invertebrates have not been evaluated.

Clearing for Agriculture and Recreation

Throughout most of the glaciated northeast, the overall trend over the past fifty to one hundred years has been toward the net loss of agricultural land, not of new clearing, but some old farms are being reclaimed for agricultural uses. On existing farms, the tendency has been toward expanding the sizes of fields and the elimination of hedgerows and windbreaks. The construction of golf courses and of recreational fields for soccer and softball has skyrocketed during the same time period. Like development and forestry, the conversion of forest land into pastures, croplands, or orchards, or to recreational uses such as playing fields and golf courses, alters hydrology, produces erosion and sedimentation, and increases water temperatures, nutrient concentrations, and, often, pesticide levels. It also directly removes upland habitat used by amphibians.

Mosquito Control

Mosquitoes that are dangerous to public health or domestic animals rarely breed in vernal pools. Nonetheless, vernal pools have long received the attention of mosquito control authorities, and many pools have a history of physical modification and of chemical treatments. Mosquito controls can severely disrupt ecosystem function in vernal pools, and the effects on wildlife are disproportionately large in comparison to the limited nuisance represented by the mosquitoes emerging from most pools. Unless otherwise noted, the following discussion of mosquito control practices and their impacts is based on reviews and summaries of the literature from Massachusetts and New Jersey[23].

Water Control Methods

Formal mosquito-control programs in the glaciated northeast started in the early years of the twentieth century. Early efforts focused on eliminating larval habitat by cleaning ditches, removing streamside vegetation, filling small depressions, and draining wetlands and small woodland pools. In southern New England, I have often found vernal pools with the remains of drainage ditches that were dug many years ago. The elimination of habitat for mosquitoes also eliminates habitat for fairy shrimp, wood frogs, salamanders, and other vernal pool wildlife.

Surface Films

During World War II, petroleum oils including diesel, fuel oils, and kerosene were used to control mosquito larvae in efforts to combat malaria, and they were applied widely in the northeast for mosquito control after the war. The oils disrupt the surface tension and prevent the larvae and pupae from obtaining oxygen from the atmosphere, effectively suffocating them. A variety of commercial oil films are currently used to control mosquitoes in this way. They include Flit-MLO, a straight-chain or branched alkane oil; Agnique MMF, a monomolecular surface film, Poly(oxy-1,2-ethanediyl),α-isooctadecyl-ω-hydroxyl; and GB-1111 or Golden Bear Oil, a cycloalkane oil that is used to kill mosquito pupae at sites where pesticides were not applied early enough to kill larvae.

Surface films designed to suffocate mosquito larvae can have substantial effects on other members of the aquatic community of vernal pools. Flit-MLO, commonly used by mosquito control districts, kills air-breathing aquatic insects such as dytiscid beetle larvae and adults, hydrophilid beetle adults, and adult backswimmers and water boatmen. It also affects surface-dwelling insects such as water striders and whirligig beetles. Surficial films have also been found to be toxic to clam shrimp, but they do not appear to harm microcrustaceans, snails, or non-air breathing insects[24]. The effects of these films on amphibian eggs or larvae have not been assessed.

Chemical Larvicides

Chemical larvicides have been used extensively since the 1950s. Different states and provinces have had different regulations governing pesticide use, and individual mosquito control districts have had their own preferences. Synthetic organic pesticides became the chemicals of choice for mosquito control throughout North America in the 1950s. Over time, there has been a continuing trend for highly toxic and persistent pesticides to be withdrawn from use and replaced with more selective and less toxic chemicals or with biologically based toxins.

The effects of chemical larvicides on non-target organisms in vernal pools have not been well studied and must be inferred from toxicity testing on taxa generally not found in vernal pools. Of the pesticides commonly used today, methoxychlor persists in soil, is toxic to crustaceans and to a number of non-target insect species, and bioaccumulates in molluscs, decapod crustaceans, and mosquito larvae. Pyrethroids are also toxic to non-target aquatic crustaceans, including *Daphnia*, and a wide range of aquatic insects. Studies in the Sahel of the effects of insecticides on invertebrates in temporary ponds have found fairy shrimp and backswimmers to be highly sensitive to pyrethroids, and cladocerans to organophosphates[25]. In general, both organochlorines and organophosphates are toxic to non-target aquatic insects, crustaceans, and molluscs.

Few data are available on the effect of mosquito control pesticides on amphibians. No toxicological or ecological data are available on the impact of treatment of vernal pools with chemical larvicides on wood frog and *Ambystoma* adults, eggs, or larvae. Vertebrate toxicity data are available for some species of fish, as well as for birds and small mammals.

Organochlorine pesticides are generally more toxic to fish than are organophosphates; methoxychlor has a relatively low toxicity to fish and other vertebrates tested compared with other organochlorines. The organophosphate chlorpyrifos is acutely toxic to fish and to birds. Fenthion is less toxic to fish, but highly toxic to birds; its use in Massachusetts was largely suspended after it resulted in a massive bird kill in 1969. Malathion, the most commonly used organophosphate larvicide, has lower toxicity to fish but has a more-than-additive effect in combination with other pesticides. It does not appear to bioaccumulate in, or be toxic to, tadpoles. Pyrethroids are acutely toxic to fish, but generally harmless to birds and mammals.

This review has found no studies evaluating the effects of mosquito control pesticides on food webs in vernal pools.

Biological Controls

Since the mid 1970s there has been increased interest in using more selective controls, including insect growth regulators such as diflubenzuron and methoprene (Altosid). These products do not kill the larvae, but they prevent them from transforming into adults, thereby blocking the mosquito's life cycle. They also affect other, non-target insects.

Most recently, mosquito control projects have been using the biological *Bacillus thuringiensis* var. *israelensis* (*B.t.i.*), a microbial control specific to dipteran larvae. *B.t.i.* affects mosquitoes (Culicidae), phantom midges (Chaoboridae), non-biting midges (Chironomidae), and biting midges (Ceratopogonidae and Dixidae), but it does not appear to have any direct effect on most non-dipteran insects, molluscs, crustaceans, or treefrog or toad tadpoles. Direct effects on non-target species in streams where *B.t.i.* is applied for black fly control appear to be minimal. Specialist predators that feed on black flies in such streams appear to decline if there is insufficient alternative prey available, and species that scavenge upon dead organisms increase; there is thus some potential for the restructuring of the aquatic community[26]. The effects of *B.t.i.* in relation to food webs in vernal pools have not been evaluated.

Pollution

Many aquatic animals are highly susceptible to degraded water quality. This is a result of their need to maintain permeable membranes for the exchange of oxygen with the water. Such membranes allow the ready movement of toxic materials into the body. In recent years, reports of deformed frogs with missing limbs and eyes, or with extra legs, often in odd locations on the body, have sparked concern about environmental quality. Most of these reports have involved leopard frogs, but wood frogs and other species have also been affected. Deformities in amphibians have been reported since the 1700s, and there is some disagreement as to whether the recent observations of deformities represent an actual increase in the phenomenon. Most current evidence implicates infection by trematodes or other parasites, but some studies find no linkages between trematode parasitism and developmental abnormalities. There may be geographic differences in the causes of the deformities. Some data suggest that exposure to pesticides such as atrazine may contribute to deformities, especially in conjunction with trematode infection[27].

Acidic water can contribute to increased levels of metals such as aluminum in vernal pools. At high concentrations, metals contribute to embryonic deformities and to lowered hatching success in amphibian eggs. For many years, the deposition of acids in rain and snowfall has been of great concern throughout parts of the glaciated northeast where soils are poorly buffered[28]. Some amphibian populations appear to be adversely affected by acidic runoff into breeding pools. The atmospheric transport of other pollutants, including pesticides and heavy metals including mercury, has dispersed toxic materials all over the globe, but the effects of such atmospheric pollutants have not been evaluated in vernal pools.

Biological Introductions and Removals

In western North America, introductions of fish into fishless lakes, and of bullfrogs into ponds, have had well documented adverse effects on native species of amphibians. Such effects have not been studied extensively in the glaciated northeast. When fish, bullfrogs, or other drought-intolerant animals that prey on vernal-pool-dependent species are introduced into semi-permanent and permanent pools, the structure of the biological community changes, as does the ability of resident species to reproduce successfully. Introductions of aquatic plants that spread rapidly and take over habitat formerly occupied by open water or by native plants also have the potential to affect the habitat value of vernal pools. Fungal and viral diseases that attack and kill amphibians can be introduced inadvertently by naturalists and researchers on boots and sampling equipment[29]. The long-term effects of such introductions are unknown.

Harvesting of breeding amphibians, especially salamanders, from vernal pools for sale to biological supply houses and the pet trade has been anecdotally related to declines in local populations. Researchers and school groups studying vernal pools also need to consider how their activities may affect populations of pool wildlife.

Approaches to Conservation

Until recently, because vernal pools are small, shallow, and transitory and their inhabitants are relatively inconspicuous, they have been largely ignored in programs for wildlife conservation. This has changed in the past fifteen years. Traditional approaches to the conservation of wildlife have involved state, provincial, or national limits on hunting, fishing, and collecting; the development of voluntary measures that can be implemented by farmers, foresters, and other land managers; and the setting aside of wildlife reserves, refuges, and sanctuaries. All of these approaches have been considered for vernal pools, and some of them hold great promise for certain situations, but none of them has been applied widely.

In some parts of the glaciated northeast, efforts to protect vernal pools have emphasized a regulatory approach. This involves governmental review of projects that affect wetlands, water quality, or rare species. Each project is approved or denied depending on whether it is possible to avoid, minimize, or mitigate any adverse effects.

Vernal pools are aquatic ecosystems that provide wildlife habitat. The protection of wildlife habitat is one of the goals of the US Clean Water Act and of state and municipal laws protecting wetlands and water quality. Therefore, the regulations that were developed under these laws have been used increasingly to protect pools.

Proposed activities that may affect plants or animals whose populations have been identified as at risk of extirpation are subject to special regulatory review under endangered-species laws. In a few instances in which vernal pools provide habitat for species that have been listed as endangered, these laws have been used to protect the pools.

Existing environmental laws and regulations tend to focus narrowly on a single medium such as water, air, or land, or on a single species. Biological systems do not recognize artificial boundaries between terrestrial and aquatic habitats, and the challenge in protecting vernal pools and their wildlife lies in developing a new approach to conservation that recognizes and maintains connections between forests and pools over large landscapes. Efforts to conserve vernal pools will fail unless they include the associated uplands, nearby annual and semi-permanent vernal pools, other freshwater wetlands, and larger, permanent water bodies. Further, such efforts must be based on a solid understanding of the biology of the species of conservation interest, as well as of the specific habitat characteristics of individual vernal pools. A clear understanding of the nature of threats to pools and the most effective ways to mitigate those threats is also needed.

Regulatory approaches that have been applied to vernal pool protection are reactive, rather than proactive, and they typically deal with problems one project at a time, without regard to the long-term, cumulative effects of multiple projects. While the review of individual projects will continue to play a role, effective conservation programs will need to incorporate many other approaches that have not been used extensively to protect wildlife habitat. Landscape planning, based on the distribution of vernal pools, forests, soils, surface waters, and other habitats in relation to land ownership and existing human uses of the land, can be coupled with the proactive use of land-use tools such as zoning, tax incentives for owners to place conservation restrictions on their property, and land acquisition. Best management practices for forest cutting, road construction, and other activities that alter the land surface can also be implemented to minimize adverse effects on vernal pools and upland habitats. All of these approaches can play a role in assuring the long-term protection of vernal pools and their wildlife.

Whether we succeed in developing new approaches to conservation, especially approaches that look at the whole landscape and consider the cumulative effects of multiple disturbances, will be influenced by legal precedents, existing ways of thinking about environmental protection, and economic pressures in society. Discussions of the history of the conservation movement, the major legislative and social changes that have occurred over the past fifty years in relation to environmental protection, the ongoing policy debates about private property rights versus the public interests in clean air and water and in wildlife, and the disconnect between the market-driven assessment of the "highest and best use" of land and the needs of wildlife are beyond the scope of this book. It is important to recognize, though, that history, the law, public opionion, and individual economic choices will all affect the ultimate success of efforts to achieve the long-term protection of animals that require vernal pools and intact forests as habitat.

Existing Protection of Vernal Pools — Wetlands Regulatory Programs

Thanks in part to public awareness of vernal pools as breeding sites for wood frogs and mole salamanders, conservation efforts have focused on amphibians, but few effective approaches to ensure the sustainability of amphibian populations have been implemented. To date, most of the attempts to protect amphibian breeding pools have involved programs that regulate activities in or near wetlands. The regulatory status of vernal pools — whether or not an individual pool is defined as a wetland and is subject to regulatory review — varies geographically according to definitions established in different federal, state, provincial, and municipal laws. Most vernal pools are small, and they are often isolated from direct contact with other surface waters. They may or may not be vegetated with wetland plants. The underlying soils usually have features such as a gray color and mottling, reflecting long periods of saturation, and are usually classified as hydric soils. All of these factors influence whether activities affecting a pool can be regulated.

Historically, many vernal pools, by virtue of their small size and seasonal nature, have not been protected explicitly under state and federal wetlands regulations. In recent decades, some jurisdictions have increased their regulation of activities affecting vernal pools. In the United States, recent restrictions on federal regulatory authority over small, isolated wetlands have raised concerns that wetlands regulations may not be effective tools for accomplishing the conservation of temporary pools and the amphibians, large branchiopods, and other species that depend on them. Comparable concerns are felt in Canada.

United States Federal Wetlands Regulatory Program

The federal wetlands protection program is administered by the US Army Corps of Engineers, with review input from the US Environmental Protection Agency and the US Fish and Wildlife Service, under Section 404 of the federal Clean Water Act. Federal wetlands are defined as areas with plant communities that are dominated by wetland plants, hydric soils, and wetland hydrology. Federal jurisdiction is predicated on a value for interstate commerce. Such a value is presumed to be present for wetlands that are adjacent to surface water bodies. For many years, isolated wetlands have also received protection based on their importance for migratory birds and other wildlife whose ranges extend beyond state lines.

The level of protection provided under the federal program varies both regionally and, within a region, by state. For example, in EPA Region 1, which covers New England, the US Army Corps of Engineers has adopted a Programmatic General Permit for each state. Unless they meet special criteria of size or impact, most projects receive what is essentially an automatic permit under the General Permit, without an extensive review. Each state can establish the criteria that determine when an individual review is required. In Massachusetts, vernal pools can be granted special protection that takes them out of the General Permit. The state established a program for identifying and certifying vernal pools and classified all certified vernal pools, regardless of their size, as Outstanding Resource Waters under the state's Surface Water Quality Standards. No discharges are permitted into such waters, and any proposed project that will affect a certified vernal pool that is classified as a federal wetland is theoretically subject to individual permit review by the Corps. There are limitations to such protection: a pool must first be certified, a process that depends on volunteers, and it must be classified as a federal wetland. Projects that will affect non-certified vernal pools or pools that do not qualify as federal wetlands can go forward under the General Permit.

Protection for isolated wetlands is uncertain under the Clean Water Act. On January 9, 2001, the United States Supreme Court ruled in "Solid Waste Authority of Northern Cook County vs. US Army Corps of Engineers" (SWANCC) that federal regulatory authority does not extend to small, isolated, seasonal wetlands that provide habitat for migratory birds and other wildlife. The decision has been interpreted as removing the majority of upland vernal pools, as well as prairie potholes and other wetlands isolated from major surface waters, from regulatory jurisdiction under the Clean Water Act. Thus, many vernal pools are not federal wetlands.

In 2003, further restrictions on federal protection of small, isolated wetlands were proposed. As of this writing, the proposed changes have been withdrawn. However, it cannot be denied that small wetlands represent problems for developers, governments, and small land-owners focused on human uses of the land. Political pressure against the protection of vernal pools, and especially against the protection of pools plus associated uplands, is unlikely to diminish in coming years.

State and Provincial Programs

In addition to the federal law, a few states have their own wetlands protection programs. Most of these do not directly address the regulation of activities affecting vernal pools. In New England, most states have adopted regulatory language in their state wetlands or water quality protection programs explicitly to allow for the regulation of activities affecting vernal pools, and many other states and Canadian provinces have initiated major efforts to identify vernal pools and increase public awareness about them. Wisconsin responded to the SWANCC Supreme Court decision by passing a law that, as of May 8, 2001, took state authority over isolated wetlands and requires issuance of a water quality certification, or permit, for proposed discharges into vernal pools. New Jersey has initiated a comprehensive program of vernal pool identification, certification, and regulation. In states where municipalities have home-rule authority, including the New England states and New York, cities and towns may adopt local ordinances protecting vernal pools[30].

Limitations of Protection Under Wetlands Regulatory Programs

At best, wetlands regulatory programs as they exist today are of limited value for protecting vernal pools, both because vernal pools are such small habitats and because significant areas of upland need to be protected in addition to the aquatic habitat itself. Most wetlands statutes and regulations establish size thresholds below which impacts are presumed to be insignificant. Further, the legislative authority for protecting water resources generally does not extend regulatory jurisdiction to non-wetlands areas.

Most vernal pools do not meet size thresholds under wetlands protection programs. In Maine, impacts affecting an area smaller than 4,300 ft^2 (400 m^2) of wetland are presumed to be insignificant and are exempt from review, and activities affecting pools between 4,300 ft^2 and 15,000 ft^2 (400 m^2 and 1,370 m^2) receive only minimal review. Thus, many small vernal pools are still unprotected. Of 304 surveyed pools, 86% were too small to be subject to any regulatory protection under the state law[31]. In Massachusetts, permanent ponds are regulated if they have a surface area of at least 10,000 ft^2 (929 m^2). Isolated, intermittently flooded water bodies must annually attain a volume of at least 0.25 ac-ft (1,234 m^3) with a minimum average depth of 6 in (15 cm) to be regulated. More than 25% of 156 amphibian breeding pools examined in eastern Massachusetts and 93% of 67 vernal pools that provide breeding habitat in west-central Massachusetts fail to meet these minimum criteria[32]. It is evident that if vernal pools are to be protected through wetlands regulations, specific size thresholds considerably smaller than those used for most wetlands need to be established.

Wetlands regulatory programs also do not provide for protection of the upland forest near the pools. As we have seen, vernal-pool-dependent amphibians are forest animals and require upland habitat as well as breeding pools, and the uplands also regulate pool hydrology, light, temperature, and food. Interconnections among vernal pools, and connectivity between pools, other wetlands and water bodies, and upland forests, are also integral to the long-term survival of amphibian populations and other pool-dependent species including turtles, water beetles, water mites, and others.

In general, wetlands protection laws provide only limited jurisdiction over activities outside of wetlands. They are thus powerless to protect the upland forest habitat required by vernal pool amphibians. In the United States, the constitutional protection of private property rights limits governmental regulation of land-use activities to situations where there is a clear and overriding public interest. For a project to be subject to federal regulation, there must be a connection to interstate commerce or other evidence that the activity has effects across state lines. The states are authorized by their constitutions to regulate activities within their borders. Most cases that have set legal precedents have focused on public interests associated with surface water bodies, particularly with navigable waters that cross state boundaries. Public interests recognized by the courts as justifying programs that regulate activities affecting wetlands and surface waters include public health (e.g., as affected by water quality), public safety (in relation to flood damage, for example), and fisheries and wildlife (through state constitutions and, at the federal level, through the constitutional provision relative to interstate commerce).

Vernal pools serve important wildlife habitat functions, and they are increasingly valuable as educational and recreational resources. However, it is unclear whether the expansion of federal wetlands regulatory authority over significant areas of upland, in the name of the protection of vernal pool amphibians, would be able to withstand court challenges. In practice, even where the regulation of upland activities affecting vernal pools is explicitly permitted, as under Connecticut's Inland Wetlands and Water Courses Act[33], or under local bylaws, municipal boards responsible for issuing permits are generally unwilling to significantly restrict activities not immediately adjacent to regulated wetlands resources.

Protection of Vernal Pools — Alternative Mechanisms

In addition to traditional wetlands regulatory programs, there are other approaches to protecting habitat for vernal pool wildlife. Many have the potential to be more successful in providing long-term protection to all of the habitat components, both upland and aquatic, that are critical for the long-term survival of vernal pools and vernal pool wildlife. Such approaches include zoning bylaws, local regulatory review of development projects, transfers of development rights, implementation of best management practices for forestry activities, changes in mosquito control practices to reduce damage to vernal pool wildlife habitat, conservation restrictions or conservation easements through which land owners limit current and future uses of their land, land acquisition, and establishment of ecological management plans for protected lands.

Zoning Tools

In many states and provinces, local governments have broad authority to adopt local zoning ordinances or bylaws. These local laws identify the permissible uses of land in different parts of the municipality and regulate lot sizes, setbacks, and other development parameters. Several zoning options can be useful for conservation. Overlay districts restrict approved activities based on the need to protect identified natural, historic, or other resources. Cluster zoning reduces the minimum lot size to provide for higher than usual densities of development in some parts of a site in exchange for the set-aside of undeveloped areas. Incentive zoning may be similar to cluster zoning but provides added incentives, such as increased overall densities of development, for developing some areas and not others. All of these tools can be applied to the protection of vernal pools[34].

Development Review

Depending on the location and the type of activity involved, local municipalities, county supervisors, and/or state or provincial agencies are responsible for reviewing development proposals to ensure that the public health, safety, and welfare will be protected. Natural resources, including vernal pools and their associated uplands, and the criteria governing their protection, can be incorporated into the regulatory guidelines for such reviews. Criteria might include rules regulating the discharge of runoff, providing for the maintenance of adequate buffers, and ensuring connections between pools and uplands.

Forestry Best Management Practices

With increased understanding of the importance of upland forests to vernal pool amphibians, efforts have been underway to develop best management practices for forestry activities in the vicinity of vernal pools. Such practices are intended to minimize direct alteration of the pools themselves, to reduce changes in the hydrologic regime, and to maintain some suitable upland habitat for amphibians.

In Massachusetts, for example, forestry best management practices are required for work near certified vernal pools and may be voluntarily implemented near pools that have not been certified[35]. The regulatory requirements are: (1) no more than 50% of the basal area may be cut within 50 ft (15 m) of a certified vernal pool, (2) slash may not be deposited in the pool basin, and (3) foresters are encouraged to avoid creating ruts or other disturbances of the land surface that may interfere with water flow or the migration of amphibians.

In Maine and Michigan, best management practices for forestry activities near vernal pools are voluntary. Recommendations for forest harvesting within 500 ft (150 m) of vernal pools in Maine include: (1) avoiding conversions of one forest type into another, (2) operating heavy machinery when soils are frozen, (3) limiting skidder traffic to a few defined trails, (4) avoiding whole-tree harvest practices, (5) leaving dead wood on the ground, (6) leaving dead, standing trees, (7) conducting only light partial cuts within 100 ft (30 m) of pools, and (8) maintaining a shelterwood including hardwoods, shrubs, and herbaceous plants in part of the remaining upland[36]. These practices are designed to help minimize disturbance of the forest's organic layer, maintain a variety of sizes of downed logs and branches in varying states of decay, and maintain shade in the upland habitat used by vernal pool amphibians.

Conservation Restrictions and Conservation Easements

In general, a conservation restriction or conservation easement is a formal document that restricts or limits certain uses of land in order to protect specified conservation objectives, while reserving other uses for the landowner. Which term is preferred varies in different legal jurisdictions; in Canada and some states, conservation easement is the more common term. By placing a conservation restriction on a piece of property, the landowner can ensure that the land will be protected in perpetuity while continuing to own it and maintaining the rights to carry out identified activities on it. The conservation restriction takes the form of a legal interest that is conveyed by the landowner to a qualified conservation agency or organization; however, the landowner continues to own the land and can make use of the land consistent with the terms of the restriction. The landowner may realize income, property, and estate tax benefits from the donation of a conservation restriction.

The conservation restriction is recorded at the registry of deeds or other repository of land records and "runs with the land," meaning that it binds not only the present landowner but also future owners of the property. In addition to defining uses which are limited or restricted, the terms of the conservation restriction typically identify those uses of the property which are retained by the owner, as well as the rights of the holder of the restriction to access the property for monitoring compliance with the provisions of the restriction, for ecological inventories and management, or for other purposes as may be specified in the document.

Conservation restrictions can be attractive conservation tools because they can assure permanent protection of the land while maintaining private ownership and providing potential tax benefits to the owner and his or her heirs. Beyond that, conservation restrictions offer the landowner and conservation organization or agency an opportunity to draft a document specifically tailored to the particular circumstances of the parties and the land. That includes the ability to create a document that is designed to protect a subtle, yet extremely important, ecological resource such as a vernal pool.

Conservation restrictions designed to protect vernal pools might include language specifying the locations of pools and associated woodlands, identifying restrictions on forest cutting or other landscape changes, and listing activities that have the potential to alter pool hydrology, water quality, shading, or leaf-litter inputs, or upland conditions including soil characteristics, plant community integrity, and woody debris and other forest-floor features that contribute to amphibian habitat[37].

Land Acquisition

The most effective way to ensure that vernal pools and associated woodlands will be permanently protected is to purchase a tract of land containing a complex of vernal pools and high-quality upland habitat. Most conservation programs, whether through federal, provincial, state, or local government agencies, or through non-governmental organizations, have not included vernal pools in their planning criteria for land acquisition. Two projects in the greater Chicago area of Illinois have demonstrated that land protection projects that include uplands and ephemeral wetlands can be valuable for public education and awareness, as well as for conservation of amphibian populations and aquatic biodiversity[38]. Acquisition of land for other primary purposes, including recreation and watershed protection, sometimes can also protect important habitat for vernal pool wildlife.

Ecological Management Planning and Implementation

It is unsafe to assume that vernal pools and associated wildlife on land that is held by public or private conservation organizations or agencies are automatically protected. Both public and private conservation lands are subject to proposals for changes in use and/or management, including trails, campsites, nature centers, cellular communications towers, forest clearing or thinning, and water supply development. By documenting important wildlife resources, including vernal pools and upland habitat, identifying management actions and/or restrictions on activities that are necessary to protect those resources, and implementing an ongoing program of ecological monitoring and management, holders of conservation lands can help ensure that vernal pools and their wildlife will be protected for the future.

Pool Restoration or Creation

In some situations, the restoration of pools that have been filled can contribute to perpetuation of populations of amphibians inhabiting adjacent uplands, and it can also create habitat for invertebrates that may disperse from other vernal pools or nearby perennial waters. In Illinois, the restoration of a pool abutting a forest that provides habitat for the state's only known population of one of the Jefferson's/blue-spotted salamander hybrids resulted in successful breeding in the first year and increasing biodiversity over time[39]. Also, creation of new vernal pools may enhance habitat in areas where pools have been lost or where local conditions have not contributed to pool formation.

There is little literature about vernal pool creation or restoration in the glaciated northeast. In general, forested wetlands are the most difficult to establish, and overall the success of wetland creation in this part of the world depends on the extent to which the appropriate hydrology is established and maintained[40]. In California, where the temporary waters known as vernal pools are best known for their unique plant species, there have been many projects that have created or restored pools, and there is an extensive literature discussing specific projects and techniques[41]. The US Forest Service has had experience with creating vernal pools in Kentucky, Ohio, and Minnesota for many years and has developed an excellent manual for use by people interested in constructing pools[42].

Key Policy Questions

Vernal pool conservation is still in a very early state of development. Before effective strategies to protect pools can be developed and implemented, many issues need to be resolved. These include conservation goals, approaches for locating and identifying vernal pools, criteria for defining whether a given pool should be given protection, details of how to protect pools, mechanisms for protecting associated upland areas, and criteria for evaluating whether conservation goals have been achieved. The following discussion identifies a series of issues, presents a brief background summary of each, and identifies choices that need to be made. It then suggests questions that need to be considered and discussed in the context of developing conservation policy, depending on the specific choices that are selected. This list is not intended to be definitive, but rather is presented in the hope that researchers, regulators, educators, legislators, conservation advocates, and the public will consider these questions and identify others.

The detailed technical information in the rest of the book provides answers for some of these questions. Other questions involve political and societal choices, rather than scientific understanding. Still other questions can only be answered through further study of vernal pools.

Why Protect Vernal Pools?

If the importance of vernal pools as wildlife habitats is the primary impetus for conservation efforts, the first move is to decide which wildlife species are of particular interest. Is the overall purpose of vernal pool conservation the maintenance of biological diversity — of both vertebrates and invertebrates — and of ecological functions within pools? Are vertebrates, particularly amphibians, the primary focus? Are rare species the main concern? Is the long-term protection and enhancement of populations within a local area, or region, a goal? The way we choose to answer these questions will chart the direction of conservation.

Biodiversity

If the conservation of biodiversity is a goal — and I hope my bias in this direction is evident! — much more needs to be learned about the biological communities of vernal pools. We need to know the species that pools support if we are to evaluate the success of conservation efforts. Relatively few studies have provided the level of taxonomic resolution necessary to distinguish readily how annual vernal pools differ from semi-permanent and permanent ponds, how pools with intermittent inlets and outlets differ from floodplain pools and from completely isolated basins, how vernal pools differ from other intermittently flooded habitats such as floodplain forests, wooded swamps, and shrub swamps, or how the biological communities of pools vary geographically. Those studies that do examine community composition find tremendous taxonomic richness in vernal pools, as well as evidence that regional biodiversity increases with increasing numbers of vernal pools. Relationships have been identified between community composition and habitat characteristics, including hydrology, but no clear patterns in terms of characteristic types of vernal pools supporting distinctive associations of organisms have yet emerged. A better understanding of such patterns would contribute substantially to planning for conservation.

Amphibians

Most of the conservation efforts in the glaciated northeast emphasize amphibians that depend heavily on vernal pools for breeding. Vertebrates have the greatest public appeal, and a conservation policy that is based on these animals is apt to be more readily accepted and, hopefully, understood. The importance of vernal pools as breeding habitats for wood frogs and at least some species of mole salamanders is well established, and our more complete knowledge of the habitat requirements of amphibians means that it should be relatively straightforward to define the steps that are necessary to protect habitats for these species.

However, even for amphibians there are many unanswered questions. For example, to what extent are clusters of vernal pools and contiguous woodland needed for the long-term maintenance of genetic diversity and viable amphibian populations? Further, while the literature suggests that the proximity of other pools is an important factor influencing the suitability of a vernal pool as amphibian habitat, the relative importance of vernal pools located at different points along the hydrologic gradient of flooding has not been resolved.

How Can Vernal Pools Be Found?

Because vernal pools are small, are flooded seasonally, and occur in woods, often under tree canopy cover, the mere problem of locating them presents a challenge. Decisions about the methods that will be used to locate pools will influence the kinds and numbers of pools that will potentially be protected, because each available method has strengths and limitations.

Older, 1:24,000-series USGS topographic maps work well for identifying depressions that extend below the groundwater table in well-stratified glacial deposits, and this technique has been used effectively on outer Cape Cod[43]. On updated maps at 1:25,000 scale, many of the smaller pools are no longer mapped.

This method also misses many perched vernal pools and those within larger forested or shrub-dominated wetlands.

Vernal pools are often identified in the course of herpetological inventories and through a wide variety of ground-based surveys. When such surveys are carried out during periods of anuran calling, pools may sometimes be located by the presence of vocalizing spring peepers, wood frogs, or toads. Ground-based surveys are highly labor intensive and expensive, but they do provide a high probability of finding vernal pools if they are carried out thoroughly.

Remote sensing has great potential as a tool for identifying vernal pools. In Massachusetts, where the pre-identification of vernal pools through field documentation of physical features and the presence of indicator species has been part of the state's wetlands regulatory structure since 1987, the number of pools formally identified through the year 2000 ("certified" under state regulations) was relatively small (about 2,000). Most of these pools were identified on the ground, rather than through remote sensing. Subsequent pre-identification of potential vernal pools through aerial photos provides a first step toward a comprehensive inventory[44]. Approximately 29,000 potential pools were identified through this study. The survey missed an as yet unknown number of small pools, and it also failed to identify pools under heavy conifer cover and in portions of the state where the available photos were of inadequate quality. Ground-level surveys are still needed to verify the accuracy of the photo identification and to determine how many small pools were missed.

Aerial photography has been used to identify vernal pools in north-central Minnesota with an estimated error of commission (incorrectly identifying wetlands as pools) of 20%, and an estimated error of omission (missing pools) at less than 10%[45]. New Jersey is in the midst of mapping its pools using aerial photography. Field checking of mapped pools against criteria including basin characteristics, the duration of flooding, the presence of indicator wildlife species (using a diversity of species, including "obligate" and "facultative" taxa), and the absence of fish, suggests the mapping has a high level of accuracy, with an estimated error of commission thus far of about 15%[46].

In forested landscapes in Maine and Massachusetts, color infra-red aerial photos allow the accurate identification of many vernal pools, but they miss pools under coniferous canopy, and anthropogenic pools associated with roadway construction and forestry activities are also often missed[47]. Depending on landscape characteristics and local vegetation, the accuracy of identification of vernal pools from aerial photos thus varies with geographic location. However, the results of such studies to date are promising, and it seems likely that at least a proportion of potential vernal pools can be pre-identified through remote sensing with a high degree of confidence.

In black-and-white, low-altitude aerial photos taken in early spring, vernal pools often stand out as white, snow- and/or ice-covered patches on the darker forest floor[48]. Pool identification using such photos is straightforward and requires less technical sophistication than the interpretation of color infra-red photos. However, low-altitude photographs taken at the appropriate time of year are not available for many locations. Also, as with infra-red photos, pools under coniferous trees are apt to be missed.

What Physical Features of Pools Are Important?

As we have seen, vernal pools are highly variable, spanning a wide range of hydrologic conditions, sizes, and physiographic types. Different habitat characteristics support different aggregations of species. Once biological conservation goals have been specified, and potential vernal pools have been identified, the next step is to define the physical criteria that will be used to choose those pools that are critical to achieving the goals.

Physical criteria used to define vernal pools have typically included a combination of basin characteristics, size and depth specifications, and hydrological conditions. Because pools are so highly variable, relying on these criteria to determine whether a particular water body is a vernal pool is fraught with difficulties. Most efforts to document whether individual shallow depressions are vernal pools have focused on short-term, single-season observations. Many problems associated with accurately identifying the habitat status of a depression at any one time have been identified, and many questions need to be addressed in the process of identifying appropriate criteria for accepting or rejecting individual basins as vernal pools.

Should every shallow, isolated, seasonally flooded woodland water body qualify as a vernal pool? Are there specific physical or basin characteristics that must be present, such as complete isolation from other surface waters or wetlands, or minimum water depths, regardless of the composition of the biological community? Are there physical or basin characteristics that should *a priori* disqualify a water body? Alternately, are there physical characteristics that indicate *a priori* that a basin is a vernal pool and that can be used to identify pools without the need to assess the biological community? Is there a minimum frequency of flooding that must be present? Should semi-permanent pools, which remain continuously flooded for several years and sometimes support a biological community similar to permanent ponds, be given the same kind of protection as annual vernal pools, which normally dry each year? Should pools that have intermittent inlets or outlets but are seasonally isolated be considered differently than pools that are completely isolated in uplands? What about pools that occur in depressions within large wetland systems, and those that are located in floodplains and are periodically flooded by larger water bodies? Should the proximity of other vernal pools be a factor in identifying the long-term viability of a vernal pool? Should the availability of contiguous upland forest be a criterion?

Pool Size: Are Large Vernal Pools More Valuable Than Small Ones?

The overall volume of vernal pools is small (remember, the median surface area is less than 1,000 m^2, or about 0.25 ac, and maximum depths average 100 cm, or just over 3 ft). Regulatory protection of wetlands and water bodies is often based on minimum size thresholds, and most vernal pools fail to meet these thresholds.

Small wetlands are often critical to the maintenance of animal populations[49]. The extent to which vernal pools, and particularly small vernal pools, contribute to the sustainability of amphibian populations and the preservation of biodiversity has not been determined in the glaciated northeast, but analysis of data from isolated wetlands on the southeastern Atlantic Coastal Plain suggests that elimination of small breeding pools poses serious threats to biodiversity[50].

It has been suggested by some that conservation should focus on the identification and protection of large vernal pools. This suggestion derives from the observation that in some localities, the number of amphibian egg masses and the species richness of breeding amphibians appear to be positively correlated with pool size, and pool size, in turn, is positively correlated with hydroperiod. Large pools represent a small proportion of vernal pools, and achieving protection of a significant number of such pools may be easier to accomplish politically than protecting the full range of pool types. It has also been argued that small vernal pools commonly dry before amphibians can complete larval development and metamorphosis, and that such pools may act as sinks that disperse amphibian genetic potential[51].

On the other hand, the extent to which a vernal pool is isolated from other pools and ponds is negatively correlated with amphibian breeding activity and species richness[52]. The elimination of small pools, which constitute most of the water bodies within the landscape, would effectively increase the isolation of the relatively few, larger pools that were targeted for protection. Further, in some geographic areas there is no apparent relationship between pool size and the number of breeding amphibians or egg masses, or the taxonomic richness in other groups of animals or plants. Also, small vernal pools with short hydroperiods contribute substantially to invertebrate biodiversity by providing habitats for animals that do not occur in pools with long hydroperiods. Before a decision is made to eliminate small vernal pools from consideration for conservation purposes, the following questions need to be considered.

Metapopulation dynamics and genetic diversity. First, what is the contribution of small vernal pools to the metapopulation dynamics of amphibians and other species? Do small pools contribute to amphibian recruitment over time? Small, isolated wetlands on the southeastern Coastal Plain are important sources of recruitment[53]. Vernal-pool-dependent amphibians tend to be long-lived species with a reproductive strategy that involves multiple breeding attempts and high pre-adult mortality. The literature indicates a high degree of plasticity in breeding, with both males and females of several species skipping years, and with years of high breeding effort interspersed among years with little evidence of reproductive activity. Adequate population studies have not been carried out to provide information on the role of pools of different sizes in population maintenance over time.

Similarly unexplored is the closely related question of the role of clusters of pools with different hydroperiods in maintaining gene flow and inhibiting genetic differentiation. Are small pools with short hydroperiods in fact genetic sinks, or do they play a role in maintaining genetic diversity within metapopulations of amphibians or other animals? Long-term data on populations of wood frogs and ambystomatid salamanders using clusters of pools with different hydrologic characteristics are needed to assess these questions.

Many species of aquatic invertebrates in vernal pools, including large branchiopods and some water beetles, may also function as metapopulations. The questions posed for amphibians also apply to these species.

Predator distributions. Second, do larger pools tend to support predators that depress amphibian recruitment? Large pools are often semi-permanent or permanently flooded. In addition to obligate vernal pool inhabitants, such pools support a high proportion of vertebrate and invertebrate species that require long hydroperiods, and often more than one year, to complete development. Such species include some aeshnid and libellulid dragonflies, some phryganeid caddisflies, and facultative amphibians such as green frogs and bullfrogs, all of which may compete with or prey upon vernal pool species. Some species characteristically found in annual vernal pools in Wisconsin were excluded from long-hydroperiod pools by the presence of predators[54]. Large dragonfly nymphs have been shown to play an important role in controlling populations of bullfrogs, chorus frogs, and other amphibians in a variety of habitats[55]. In eastern Massachusetts, a large, semi-permanent pool with tens of thousands of breeding wood frogs, spring peepers, green frogs, spotted salamanders, and other amphibian species produced few metamorphosing juveniles for several years, until the season following a summer in which the pool dried; large libellulid dragonflies, present the previous springs, were absent after drying and in their absence high recruitment of amphibians was observed[56]. The inference is that dragonfly predation may be implicated in the low recruitment of amphibians during the previous years.

Pool size and biodiversity. Third, how do pools of different sizes contribute to regional biodiversity? Potentially, the conservation of vernal pools and their wildlife is fundamental to overall preservation of biodiversity in northeastern forests, and not just to the conservation of a few selected amphibian species. Large annual vernal pools and long-hydroperiod semi-permanent pools have greater α diversity than some smaller pools, but β diversity throughout a given geographic area depends on the entire suite of taxa occurring in vernal pools. The elimination of small pools could threaten those species that are specialized for small, short-cycle pools and are not found in large, semi-permanent sites.

Pool Hydroperiod: Are Annual Vernal Pools as Valuable as Semi-Permanent Pools?

The question of the relative importance of annual and semi-permanent vernal pools is intimately connected to both the problem of how vernal pools are defined and the question of whether large pools are more valuable than small ones. Semi-permanent pools and some fishless, permanent ponds often support breeding by some vernal pool amphibians and invertebrates. Such pools have been investigated extensively as part of regional surveys and population studies of vernal pool amphibians, and much of our understanding of the biology of mole salamanders is derived from literature on these habitats. These pools also include a variety of animals that appear to require the absence of fish predation but lack mechanisms to avoid or withstand pool drying.

The fauna of semi-permanent pools includes species typical of permanent waters, and the community composition and community dynamics may be very different from those in annual vernal pools. Plans for the conservation of vernal pools need to address the differences between annual and semi-permanent pools. The questions already identified about metapopulations, predator distributions, and biodiversity in relation to pool size need to be asked as well in relation to pool hydroperiod. In addition, it is worth considering whether semi-permanent pools at some point along the hydrologic continuum become so similar, biologically, to permanent ponds that they should be excluded from discussions pertaining specifically to the conservation of vernal pools.

What Biological Criteria Define Important Vernal Pool Habitat?

Conservation organizations and agencies concerned with education and regulation in the glaciated northeast have tended to use the presence of "indicator" wildlife species, such as breeding mole salamanders (*Ambystoma*) or wood frogs (*Rana sylvatica*), or of fairy shrimp (*Eubranchipus*) or clam shrimp (*Lynceus brachyurus*), rather than the biological community as a whole, as the primary criterion for defining a particular site as a vernal pool[57] (Plate 1).

Should breeding activity by indicator species be used as the primary criterion for determining whether a water body is to be protected as a vernal pool? Does the absence of indicator species mean that a pool is not worthy of protection? Is the presence of a single vernal-pool-indicator species sufficient? If indicator species are used, should there be threshold numbers of individuals or of egg masses of a particular indicator species, or a minimum number of indicator taxa? Should there be a minimum number of years within a given time period during which indicator taxa are present? Are there other biological criteria, relating either to the community as a whole or to evidence of successful reproduction by indicator taxa, that should apply before a pool is determined to be a vernal pool? Are there other biological criteria, such as the presence of established fish populations or of other organisms characteristic of perennial ponds, that should argue against a water body being recognized as a vernal pool?

There are several limitations to the "indicator species" approach. The presence of an indicator taxon alone does not conclusively identify a site as a vernal pool. Most species of vernal-pool-dependent amphibians are known to breed, at least in small numbers, in a variety of other aquatic habitats (although the success of such breeding has not been demonstrated). In their first year of breeding, some individual amphibians disperse to new potential breeding sites instead of returning to their pool of origin, and some of these may breed in ephemeral water bodies, permanent waters, or recently created depressions in which reproduction is unlikely to be successful.

Conversely, the absence of a particular vernal-pool-indicator taxon does not conclusively eliminate a site as a vernal pool. Vernal-pool-dependent amphibians and/or fairy shrimp do not necessarily appear in a given vernal pool every year, even though the pool may be important breeding habitat, may be important for the long-term maintenance of species populations, and may support a variety of other organisms that depend on the unique habitat provided by vernal pools.

There are many species of aquatic insects, molluscs, and other invertebrates that are typically found in annual vernal pools, are adapted to seasonal drawdown, and appear to require breeding pools that lack predatory fish (Table 1). These species can also be considered to be vernal-pool-indicator species, although for practical reasons associated with difficulties in identifying invertebrates to species, they have not been used extensively for the purpose of identifying vernal pools.

How Important Are Pool Origins and Alterations?

The landscape of the glaciated northeast has been subject to manipulation by industrial humankind for several hundred years. It is likely that most vernal pools have experienced some alteration over time, either as a result of watershed and atmospheric changes or from direct influences on the pools themselves. From the perspective of conservation, should existing or potential function as wildlife habitat be the primary criterion for determining that a depression serves as a vernal pool, regardless of the site's origin or history of alteration? Should artificial pools or altered depressions be disqualified from protection? If the history of alteration is a consideration in deciding whether a particular water body qualifies for protection, are there gradations of alteration, either in terms of degree, type, or elapsed time since the alteration took place that should be considered?

Should Conservation Efforts Focus on Individual Pools, Pool Clusters, or Landscapes?

Another important policy issue relates to the scale of conservation efforts. Attempts to protect vernal pools through wetlands and endangered-species regulations have dealt with one pool at a time. In general, a regulatory approach will necessarily involve responding to threats to individual pools as they arise. An alternative approach involves identifying areas that contain a mixture of aquatic habitats, including vernal pools, as well as associated upland forest habitat, and working to protect the broader habitat matrix. Such an approach would mean that some vernal pools, and populations of organisms dependent on those pools, would be denied protection in favor of larger complexes of pools and uplands for which a long-term protection strategy can reasonably be developed. Should conservation efforts emphasize large-scale strategies to the exclusion of regulatory approaches? Is there a balance between single-pool and broad-scale strategies that will provide appropriate and effective long-term protection for vernal-pool-dependent wildlife?

How Can Uplands Integral to Vernal Pool Conservation be Protected?

The most difficult policy issues involve the critical relationship between vernal pools and associated woodlands. Recurring repeatedly throughout this book is the theme that effective protection of an individual vernal pool and its wildlife requires protection of the upland habitat, as well as the pool. Ultimately, protection of vernal-pool-dependent amphibians requires not only the conservation of individual, local populations, but also the maintenance of genetic exchange among the various local populations that make up metapopulations. Thus, upland protection involves both maintaining protected areas around individual breeding pools, and ensuring that the connections between pools persist over time. Effective conservation at the population level also requires protection of the larger forest ecosystem, including the interconnections among pools. As noted earlier, traditional regulatory programs that focus only on aquatic habitat cannot effectively conserve the large areas of upland that woodland amphibians and nesting turtles need as habitats, nor can they maintain the hydrologic conditions and flow of terrestrial carbon that are critical to the long-term stability of the vernal pool ecosystem.

How Much Upland?

Raymond Semlitsch is a herpetologist at the University of Missouri who has conducted substantial research on amphibian conservation in the central and southeastern United States. Using information from the literature on average distances traveled by amphibians from breeding pools, he has made estimates of the minimum area necessary to protect 95% of amphibians that depend on isolated seasonal wetlands on the southeastern Coastal Plain. He suggests that a minimum radius of 164 m (538 ft), or approximately 8.4 ha (20 ac) of upland "life zone" should be protected from alteration around each pool. He further suggests that an additional "buffer zone" of up to 50 m (164 ft) may be necessary to avoid averse impacts from land uses outside the critical habitat areas represented by the upland life zone and the breeding pool[58]. The combined life zone and buffer zone radius of 214 m (700 ft), covering an area of approximately 14 ha (35 ac), may be adequate to support a significant proportion of the local population of at least some species breeding in a given vernal pool.

In two populations of spotted salamanders in Massachusetts, in which 14 to 40% of individuals occur at distances greater than 200 m (660 ft) from their breeding pool, and in one of which 20% occur farther than 300 m (1,000 ft), a larger life zone than that recommended by Semlitsch would be required to protect an area supporting 95% of the population[59].

As discussed previously, in addition to protecting life zones around vernal pools, corridors of forested land need to be maintained between pools to allow for genetic exchange. Ideally, areas of continuous forest containing numbers of pools should be protected, to minimize adverse effects from predators and other factors associated with forest edges. Based on densities of vernal pools reported from the literature and the genetic distance of 1,000 m (3,300 ft) calculated for wood frogs, a minimum effective size for such reserves might be 500 ha (1,235 ac), with an estimated minimum diameter of at least 2,500 m (1.5 mi).

In practice, it is difficult to find on-the-ground support for setting aside the large areas estimated to be needed to protect vernal pool amphibians. Often, however, compromise solutions can be proposed. For example, recommendations for voluntary management of forestry habitat for vernal pool amphibians in Maine urge the limited harvest of trees within 30 m (100 ft) of pools, covering an area of 3.1 ha (1.4 ac), and only partial harvest in a life zone with a radius of 122 m (400 ft), an area of 4.7 ha (13 ac)[60]. Zoning regulations can include requirements for the set-aside of habitat in exchange for higher densities of development on the remaining portions of a site.

What Kind of Upland Habitat?

The kind of upland habitat that is protected will influence the outcomes of conservation efforts. Differences in habitat quality affect the overall sizes of amphibian breeding populations as well as the densities of individuals within upland habitats surrounding vernal pools. Upland habitat characteristics identified as important for amphibians include soil type (especially in terms of drainage and suitability for burrows), the availability of leaf litter and large woody debris, and aspect. The length of the growing season in a particular area, and localized variations in degree-days and precipitation, are also likely to affect the suitability of an area as upland habitat for amphibians.

The identification of appropriate areas of upland habitat has practical importance. If a development site has some areas of well-drained soils suitable for septic systems and other areas of bedrock and poorly drained till, setting aside huge areas of the latter will not help salamanders that need well-drained soils for burrows.

What Mechanisms Will Allow Protection of Upland Habitat?

Existing laws generally do not provide regulatory authority over activities in uplands. To protect uplands associated with vernal pools through regulatory review of activities, new legislation will be needed. Other options include the implementation of tax incentives for the set-aside of upland habitat, the inclusion of habitat-protection requirements in local zoning and subdivision bylaws, the voluntary protection of property through conservation restrictions, and the acquisition of areas that include vernal pools and uplands.

Next Steps

As should be evident from the preceding discussion, much remains to be done if the rich wildlife legacy provided by northeastern vernal pools is to be conserved for future generations. Direct action is crucial to prevent irremediable losses of large branchiopods, amphibians, and uncommon invertebrates that cannot tolerate extended watershed disturbance. Research is needed to address unanswered questions and to inform conservation efforts. Public education is also critical for establishing and maintaining political and economic support for conservation action.

Conservation Action

Small, seasonally flooded wetlands and the wildlife species that depend on them are disappearing from the landscape. Vernal pools are being filled, dredged, and drained. Their watersheds are being altered by residential, commercial, agricultural, and recreational development, fragmented by transportation projects, and dramatically modified by industrial forestry activities. Rates of conversion of woodlands to other uses have skyrocketed across the continent. The spread of mosquito-borne diseases, such as West Nile Virus, has increased public support for mosquito control activities, many of which target vernal pools.

Multiple stressors are affecting amphibian populations worldwide. These include climate change, with altered precipitation patterns and temperature regimes; increases in ultraviolet radiation; exposure to atmospheric deposition of mercury and other metals; predation from domestic pets as well as native species such as skunks and raccoons that thrive in proximity to human settlements; and highway mortality of migrating adults and dispersing juveniles moving between vernal pools and uplands. Combined with the alteration and loss of habitat, these stresses add up to threats that are profound and require immediate action.

To date, most approaches to vernal pool conservation have focused narrowly on providing a level of regulatory protection to selected, individual, amphibian breeding pools. Inadequate attention has been paid to the need for untouched areas around pools to maintain canopy cover, leaf inputs, and shelter for animals. Virtually no effective mechanisms have been devised for protecting the necessary upland forest habitat — the "life zone" defined by Semlitsch — and for maintaining additional buffers between these critical upland habitats and developed areas. We have been blind to the metapopulation dynamics of amphibians that use complexes of pools and uplands, and as a result conservation efforts have failed to maintain connectivity among pools with different hydroperiods. Further, we have failed to address the role of vernal pools in maintaining some turtle populations, and to consider the habitat requirements of the great diversity of invertebrates that depend on vernal pools.

The situation is not all bleak, however. There is a solid, and growing, critical mass of people who care about amphibians, invertebrates, and small woodland pools. The term "vernal pool" has begun to enter into the popular consciousness, and it is seen in news articles and popular magazines and on TV shows. Local schools, wildlife and nature clubs, and conservation committees have the ability to organize coalitions of interested people in their communities. Many states and provinces have conservation organizations, schools, and agencies that are actively creating awareness of vernal pools through public education and outreach programs, efforts to locate and identify pools, and projects to protect them. And national government agencies at many levels include people who are knowledgeable about wildlife, care about small wetlands, and are committed to their protection. All of these are strengths that can be built upon.

There are many different approaches to effective long-term conservation of vernal pools and of species that depend on them and their associated woodlands. Some specific suggestions include the following.

- The *a priori* identification of conservation goals for vernal pool protection within different legal jurisdictions, for example, at the level of the individual landowner, local municipality, state or provincial agency, or national government. Lead agencies or groups can serve as the catalyst for such efforts, as well as playing important roles in getting information out to the public.
- The establishment of programs designed to pre-identify vernal pools on the landscape through a combination of remote sensing, map analysis, and ground surveys. Many such programs are already underway at the state or municipal level, either through government agencies or by efforts of conservation organizations.
- The development of criteria by which pools that serve important habitat functions can be identified.
- The identification of a variety of means by which pools can be protected from destruction and alteration, and the implementation of those that fit local conditions.

• The identification and preservation of contiguous areas of woodland containing clusters of vernal pools, other wetlands, and surface waters. This can be accomplished through a combination of direct acquisition, conservation easements and restrictions, and regulatory restrictions. Such a program would have conservation benefits that would extend well beyond those favoring vernal-pool-dependent species. Many agencies and organizations currently engage in conservation planning. In regions where pool mapping is already underway, it should be relatively straight forward to seek out and include information about the occurrence of vernal pools and intact forest units in such planning. In other locations, pressure from the conservation community to develop mapping programs, or the use of other means of identifying potential pool sites as discussed above, can provide a first approximation that can drive additional, site-based surveys.

• The adoption by local zoning authorities of bylaws and regulations that encourage (and in some pre-selected areas mandate) cluster development, and that ensure that conservation set-aside land has value for wildlife and includes upland as well as wetland areas. Local legislation always needs strong advocates behind it, but with adequate public education the support for vernal pool protection can be mustered.

• The development of workshops, press releases, radio programs, and publications to educate the public and decision makers about the diversity of kinds of vernal pools; the uniqueness of the wildlife species that depend on them; the importance of contiguous, unfragmented upland habitat and of maintaining connections to other vernal pools within the landscape; the adaptations of vernal pool species to (and their dependence on) variable hydrologic conditions; and the importance of the exclusion of fish.

• The education of resource managers, decision makers, and the public about the effects of watershed changes, forestry practices, runoff discharges, pool dredging and drainage, and mosquito control programs on vernal pool amphibians and other wildlife.

• The incorporation of education about vernal pools into the curricula of local schools, with the inclusion of pools as field-study areas where students can learn about biodiversity, life histories, hydrology, weather, soils, and ecology; develop skills in mapping, drawing, and writing; and participate in local discussions about land use, government, and conservation.

• The implementation of effective best management practices for forest cutting to maintain adequate canopy cover, provide adequate on-ground leaf litter and large woody debris, and avoid damage to breeding pools from skidders and staging operations.

• The development of conservation tax incentives for landowners who maintain areas for wildlife rather than primarily for commodity production or active recreation.

• The adoption of changes in mosquito control programs so as to avoid alteration of vernal pools while targeting species of mosquitoes that represent important threats to public health.

Learning More

There are many opportunities to learn more about vernal pools, both by reading and through hands-on experience in the field.

Field guides, summary volumes, monographs, and original research papers contain detailed information, some of which is presented in this book. The citations in individual chapters and the bibliography at the end of this book can serve as a useful starting point for those who wish to read more about specific topics. For information on what is going on in your local area, and for updates on the literature, the World-Wide Web can be a useful source. Do a search on your favorite search engine using the words "vernal pool" and the name of your state, province, region, or community. Searches on the names of individual species also yield a wealth of sites. Because websites and their addresses change over time, and because new sites are appearing regularly, a comprehensive listing of such sites is not provided here. The following two sites and one list-serve are available as of early 2004 and are good places to start a search.

A helpful website for general information on pools in New England, with a focus on Massachusetts, is *http://www.vernalpool.org/.* This site has been created by the Vernal Pool Association, an environmental club at Reading Memorial High School, located in the northeastern part of the state. Much of the information on the site has been generated by students. This website provides an extensive list of links to other websites dealing with vernal pools throughout the northeast, as well as to sites providing information on a wide variety of wildlife species that depend on pools. The group also sponsors poster contests and other activities for schools.

As of November, 2003, there is also a vernal pool list-serve. You can subscribe by sending an email to *vernalpool-subscribe@yahoogroups.com*. Do not put anything in the subject or message spaces. You will receive a confirmation message back. Once you have responded to the confirmation, you can send in questions to *vernalpool@yahoogroups.com,* and learn from the collective knowledge of others who are interested in vernal pools.

For information on vernal pools in California and other parts of the continent, a good resource is *http://www.vernalpools.org/*. (Note the similarity of the web address to the site discussed above.) This website provides up-to-date information on current issues, literature on a variety of topics, links to other sites focusing on California vernal pools, and links to sites providing information on pools in Washington, Virginia, New England, and other parts of North America.

Nothing compares to experiencing the life of vernal pools in the field. Most states and provinces in the glaciated northeast have one or more organizations or agencies involved in vernal pool education, research, or conservation. For people unfamiliar with natural history, field visits to vernal pools with a knowledgeable naturalist can serve as a great introduction to local pools and the amphibians and other animals in them. Some key questions meriting further research have been identified in the preceding chapters. Those who are already comfortable making field observations can carry out their own studies of local pools, adding to the body of knowledge about them in the process. The long-term collection of data by amateur naturalists, classrooms, and scientific researchers can make important contributions to our understanding of vernal pools. Clearly, many questions remain to be answered!

References for Chapter 14

[1] Foster et al. 1998. [2] Belk 1998, Mura 1993, Vekhoff and Vekhova 1995. [3] Dexter et al. 1943 ff. [4] Bishop 1941, Kenk 1949, Minton 1954. [5] Weeks 2002b, Weeks and Marcus 1997. [6] Calhoun et al. 2003, Dimauro and Hunter 2002. [7] Kenney 1991. [8] Schneider and Frost 1996. [9] For a comprehensive discussion of isolation in relation to vernal pools, see Leibowitz 2003 and Zedler 2003. [10] Andren et al. 1989, Burne 2000a, King et al. 2000, Stone 1992. [11] Berven and Grudzien 1990. [12] Pope et al. 2000. [13] Kenney 1991, Thompson and Gates 1982. [14] King et al. 2000. [15] Fahrig et al. 1995, Trombulak and Frissell 2000, Tyning 1990. [16] Paton 2002. [17] Bennett et al. 1980, DeGraaf and Rudis 1990, Heatwole and Getz 1960. [18] deMaynadier and Hunter 1995, 1998; Enge and Marion 1986; Petranka et al. 1993; Pough et al. 1987. [19] Pough et al. 1987. [20] Ash 1988, Ash and Bruce 1994, Bury 1983, McLeod and Gates 1998, Means et al. 1996, Petranka et al. 1993, Welsh 1990. [21] deMaynadier 2000; deMaynadier and Hunter 1995, 1998. [22] DiMauro and Hunter 2002. [23] Edman and Clark 1988, New Jersey Agricultural Experiment Station 1960, Nisbet 1972, Spencer 1967, Stickel 1967. [24] Nayar and Ali 2003, Takahashi et al. 1984. [25] Lahr et al. 2000. [26] Wipfli and Merritt 1994a, 1994b. [27] Gillilland and Muzzall 2002; Johnson and Sutherland 2003; Johnson et al. 1999; Kiesecker 2002; Sessions 1998; Sessions and Ruth 1990; Sessions et al. 1999a, 1999b. [28] Andren et al. 1989; Clark 1986a, 1986b; Cook 1978, 1983; Dale et al. 1985; Dunson et al. 1992; Freda and Dunson 1985a, 1985b, 1985c, 1986; Gosner and Black 1957; Grant and Licht 1993; Ling et al. 1986; Pierce 1985, 1987; Pierce et al. 1984, 1987; Pierce and Harvey 1987; Pierce and Wooten 1992. [29] Berger et al. 1998, Bollinger et al. 1999, Collins 2003, Daszak et al. 1999, Milius 2000, Speare and Berger 2000, Speare et al. 2000. [30] Calhoun 1997, Calhoun et al. 2003, Christie 2002, Colburn 1996, Fellman 1998, Flannagan 2003, Gernes and Helgen 1999, Hausmann 2002, Helgen et al. 2002, Hoskins 1998, Huckery 1998, Ingwerson and Johnson 2002, Jackson 1998, Kenney 1994, Massachusetts Division of Fisheries and Wildlife 1988, New Jersey Division of Fish and Wildlife 2003, Nolan 1998, Ohio Environmental Council 2003, Tesauro 2001, Vermont Department of Environmental Conservation 2003. [31] Calhoun et al. 2003. [32] Stone 1992, Windmiller 1990. [33] Fellman 1998, Hoskins 1998, Nolan 1998. [34] Nolan 1998. [35] Massachusetts Department of Environmental Management, 1995. [36] Calhoun and deMayadier 2003; deMaynadier 1996, 2000. [37] Some language relating to possible allowed and limited activities can be found in a draft conservation restriction developed by the Massachusetts Audubon Society, 208 South Great Road, Lincoln, MA 02159, USA, 781-259-9506, *http://www.massaudubon.org*. [38] Mauger 2002. [39] Mui and Szafoni 2002.

[40] Hammer 1992, Kusler and Kentula 1990, Mitsch and Gosselink 2000. [41] A good bibliography is available at the California Vernal Pools website: *http://www.vernalpools.org*. [42] Biebighauser 2003. [43] Portnoy 1987. [44] Burne 2001. [45] Palik et al. 2003. [46] John Heilferty, New Jersey Department of Environmental Protection, personal communication. [47] Brooks et al. 1998; Burne 2000a, 2000b; Calhoun and deMaynadier 2003; Calhoun et al. 2003; Stone 1992; Tesauro 2001. [48] Steven Roble, Virginia Division of Natural Heritage, formerly Massachusetts Natural Heritage and Endangered Species Program, personal communication. [49] Gibbs 1993. [50] Semlitsch and Bodie 1998. [51] Ling et al. 1986. [52] Burne 2000, Stone 1992, Whitlock et al. 1994. [53] Semlitsch and Bodie 1998, Semlitsch et al. 1996. [54] Schneider 1990, 1997. [55] Smith 1983, Werner and McPeek 1994. [56] Bryan Windmiller, Hyla Ecological Services, personal communication. [57] Brooks et al. 1998, Calhoun et al. 2003, Massachusetts Division of Fisheries and Wildlife 1988, Stone 1992. [58] Semlitsch 1998. [59] Windmiller 1996, 2000. [60] Calhoun and deMaynadier 2003.

Afterword

Vernal pools are out there in the woods near you! I hope that you will get out and explore them. No matter what time of year it is, there are plants and animals in the pools, evidence of water sources, and signs of maximum water depths. There is tremendous variation between pools, and from year to year in a single pool. We have barely begun to scratch the surface of understanding about how vernal pools function and their broader roles in the landscape. Woodland habitats for amphibians, wetlands and upland nesting areas for turtles, nearby lakes and streams for overwintering water beetles, and other vernal pools for genetic exchanges at all taxonomic levels are there to be discovered, in addition to the unique features of each individual pool itself. Anything you learn can add to our knowledge of these fascinating and important wildlife habitats. And for you and those you share your explorations with, vernal pools can bring new richness to life and an increased appreciation of the magic and mystery of the natural world we have inherited. The more we know, and the more we care, the more likely it is that vernal pools and their wildlife will continue to delight generations to come.

Glossary

Aero-aquatic fungi: A group of fungi common in streams and ponds. These fungi have complex life cycles and use flooded and non-flooded habitats for different life stages.

Aeshnid: A dragonfly in the family Aeshnidae, the darners.

Aestivate: To spend a hot or dry period in a state of dormancy.

Aestivation: The process of withstanding summer heat and/or drought through dormancy; the state of being dormant during summer.

Allochthonous: Originating somewhere else; for example, leaves from trees growing outside vernal pools, as opposed to plants growing within the pools, serve as important sources of energy for pool animals.

Allopatric: Not occurring together, as in the distributions of two species that do not occur in the same locations.

Alpha (α) diversity: Species richness, or the total number of species found in a particular site or individual habitat.

Ambystomatid: A mole salamander, belonging to the family Ambystomatidae.

Amplexus: The mating position of frogs and toads, in which the male clasps the female from the rear in a sort of hug. Both the eggs and sperm are discharged into the water, and fertilization takes place externally. A mating pair is said to be in amplexus or amplexed.

Anhydrobiosis: A dormant state occurring in response to stresses such as heat, extreme cold, and drying, in which most or all of the body water is lost, and in which most or all metabolic activity ceases. Once the organism is rehydrated, metabolism recovers and normal activities can be resumed. It can occur at any stage in the life cycle of some rotifers, tardigrades, and nematodes.

Annual vernal pool: A shallow, isolated, non-flowing woodland water body that attains its maximum depth and volume in spring, remains flooded for a minimum of two months, and annually loses all or most of its water volume and surface area; and in which the biological community lacks fish and includes species requiring the absence of fish predation and adapted to seasonal drying.

Antennae: Paired sensory appendages, usually somewhat long and flexible, located on the heads of insects and crustaceans. (Antenna is the singular of antennae.)

Anthropogenic: Created by humans, or produced as a result of human activity.

Anurans: Amphibians that do not have tails as adults: frogs, toads, treefrogs.

Apterous: Wingless.

Ash-free dry weight: A measure of the organic (non-mineral) biomass of living organisms. Ash-free dry weight is determined by first drying plant or animal material used in studies of energy fixation and carbon flow. It is determined in three steps. First, the sample of interest is dried at 50° C until all the water is removed, and weighed. The sample is then burned at 550° C to remove all the organic material, and the ash is weighed. The organic fraction is determined by subtracting the weight of the mineral ash from the original dry weight.

Astatic: Changing, variable; literally, not static.

Autochthonous: Originating locally; for example, in some vernal pools, autochthonous detritus, derived from plants growing within the basin, contributes to the food web of the pool.

Autotroph: Literally, self-feeding; an organism such as an alga, plant, or photosynthetic protozoan that can produce its own food using solar energy.

Benthic: Occurring on the bottom of a water body.

Benthos: Living organisms, including bacteria, fungi, plants, invertebrates, and vertebrates on the bottom of lakes, ponds, streams, rivers, wetlands, and the sea.

Beta (β) diversity: The variation in species composition, as opposed to numbers of species, across a number of habitats, associated with differences in physical habitat conditions. An example might be variations in species composition along a hydrologic gradient from spring-filling, short-cycle to fall-filling, long-cycle vernal pools.

Biota: Living organisms, including bacteria, algae, protozoans, mosses, ferns, higher plants, and animals. Often used to refer to all of the species found in a particular location or habitat.

Bivoltine: Having two generations per year.

Blastoderm: An early stage in embryonic development.

Brachypterous: With reduced wings.

Carapace: In crustaceans and some other arthropods, an expanded, shield-like structure that covers many of the segments on the front part of the body. In turtles, the upper, usually domed, part of the shell.

Carnivore: An organism that feeds on flesh; a meat-eater.

Carnivorous: Describing or pertaining to a carnivore.

Carrying capacity: The number of organisms for which a habitat can sustainably provide food and other resources.

Cercaria: The free-swimming larvae of parasitic trematode worms which seek out the second intermediate host and infect it.

Cerci: Paired appendages or structures, usually pointed, at the end of the abdomen of some insects and crustaceans.

Chironomid: A non-biting midge belonging to the family Chironomidae.

Cloaca: Derived from the Latin word for "sewer," a structure in birds, reptiles, and amphibians into which the reproductive, excretory, and digestive tracts all discharge, and which has a single opening out of the body through which wastes and reproductive products pass.

Cluster development: A type of development specified in a zoning bylaw, in which the density of buildings or living units is higher than the usual zoning density, in exchange for having part of the property set aside in an undeveloped state for conservation purposes.

Clutch: A batch of eggs produced at a given time by a breeding female.

Cosmopolitan: With a worldwide distribution.

Cryoprotectant: A substance that acts as an antifreeze to protect an animal's body fluids from freezing upon exposure to cold temperatures.

Cyst: A structure consisting of an organism surrounded by a non-living, usually hard, coating that protects the organism from drying and injury and allows survival during adverse conditions.

Cytogenetic techniques: Methods for evaluating the genetic identity of organisms by looking at chromosomes in cells. Modern techniques can include comparisons of DNA from different animals to determine how closely related they are.

Desiccation: Drying up, as in the loss of water from cells or body fluids.

Detritus: Non-living material derived from plants and animals, including leaves, conifer needles, wood, fruits, flowers, pollen, frass, and dead animals.

Detritivore: An organism that feeds on decaying leaves and other dead plant material.

Detritivorous: Describing or pertaining to a detritivore.

Deutonymph: The second instar of juvenile water mites, and the only free-living nymph stage in the mite life cycle. (All other stages of nymphal development are parasitic.)

Diapause: A period in which development is halted (e.g., a pause in the development of eggs into larvae, the growth of larvae, the maturation of gonads), usually starting and ending in response to environmental cues and allowing an animal to avoid unfavorable conditions.

Diploid: With two pairs of chromosomes, one from each parent; the usual genetic make-up of animals. (Some hybrids, such as certain salamanders, may have one or more extra sets of chromosomes — see definitions of triploid and tetraploid.)

Dormancy: A physiological state in which activity and metabolism are reduced, usually allowing an organism to survive a period of adverse environmental conditions such as high summer temperatures, pond drying, or winter cold. Both diapause and aestivation are forms of dormancy.

Dorsolateral: Along the sides on the upper (back or top) portion of the body.

Drawdown: A lowering of water level due to natural drying, drainage, or, in the case of dammed water bodies, change in outflow through the dam.

Dytiscid: A diving beetle, family Dytiscidae.

Ectoparasite: An animal that is parasitic on the external surfaces of another animal.

Elytra: The hardened outer pair of wings in beetles.

Encystment: The process by which an organism produces hard structures or encases itself in a protective coating that will allow it to survive adverse conditions.

Endemic: Restricted in distribution to a particular area or location.

Energy budget: An accounting of all energy entering, moving within, and leaving a system.

Ephemeral: Short-lived; lasting for only a short period of time.

Ephippium: A resting egg produced by cladoceran crustaceans, allowing the species to remain dormant in unsuitable environmental conditions. (Ephippia is the plural of ephippium.)

Epiphyte: A plant or plant-like organism that grows on the surface of another plant, as in an alga growing on the stem or leaf of an aquatic plant.

Epiphytic: Describing or referring to an epiphyte.

Erythristic: Red colored.

Evapotranspiration: The return of water to the atmosphere through evaporation from the land surface and through transpiration, the release of water from plant leaves in the process of plant growth and metabolism.

Facultative: Having the ability to do something, such as breeding in temporary water or undergoing a period of summer dormancy, but not requiring it as part of the life cycle.

Filter-feed: To obtain food by straining small particles from the water using fringed legs, mucus nets, silken webs, ciliated gills, or other sieve-like structures.

Flashiness: In hydrology, a rapid response by a water body to a storm event, with an increase in water volume soon after the storm followed by a rapid decrease in that volume.

Flashy: Describing or pertaining to flashiness in a water body.

Food web: A conceptual understanding, often presented as a diagram, of the feeding relationships of plants and animals in a community — essentially, a model of which organisms produce what products and consume what resources.

Frass: Debris and droppings produced by insects.

Functional feeding group: A way of classifying animals based on the way they obtain their food. For example, shredders tear food into smaller pieces, collectors filter particles from water or ingest sediment and extract nutrients, scrapers/grazers scrape or chew food off of surfaces, piercers suck fluids from plants or animals, and engulfers swallow their prey whole.

Gamma (γ) diversity: Regional diversity; or, the total number of species occurring in all habitat types in an area or region.

Gastrula: An early stage in embryonic development, during which cell layers start to differentiate.

Genome: Genetic material contributed by one parent.

Gynogenetic: Developing from only the maternal genetic material, without contribution from the male.

Herbivore: An animal that feeds on plants.

Herbivorous: Describing or pertaining to an herbivore.

Hermaphroditic: With both male and female reproductive structures present in the same organism.

Heterotroph: An organism that does not produce its own food but instead feeds on other organisms, such as protozoans, algae, plants, or animals.

Hibernaculum: An overwintering spot. (Hibernacula is the plural of hibernaculum.)

Homing: A behavior that involves migrating or traveling back to a place of origin, often to breed. Animals that home tend not to accept alternative destinations and will go to great lengths to reach the end-point to which they were heading.

Hydrogeomorphic (HGM) method: A method of classifying wetlands based on their position in the landscape and their hydrology.

Hydroperiod: Duration of flooding.

Hydrophytes: Plants that preferentially occur in water or wet soils.

Hydrology: The physical study of water, especially the presence and movement of water, including distribution (for example, in precipitation, groundwater, streamflow, lakes, and ponds) and timing of occurrence.

Hyphomycetes: Fungi in the class Hyphomycetes, or molds. This is a large group of fungi, some of which are important as decomposers of leaves on land, in streams, and in ponds.

Instar: In animals such as insects, crustaceans, and other arthropods that go through several molts, or stages, as they develop, each stage is known as an instar. Thus the first stage after hatching is the first larval instar, the next is the second instar, and so on. Depending on the kind of animal, there may be just a few instars — some water beetle larvae, for example, go through only four instars before they pupate and transform into adults — or there may be many; some damselfly nymphs, for example, may go through more than eleven instars as they develop.

Iteroparous: Reproducing more than once.

Kettle, kettle hole: A geologic feature in glaciated areas, consisting of a depression in glacial deposits, originally produced when an ice block left behind by the retreating glacier melted. Kettle holes that intersect the groundwater table may be intermittently or permanently flooded, forming vernal pools or perennial lakes and ponds.

K-selected: One of two extremes in theoretical models of population growth and reproductive fitness, which describes species with a long life span, slow growth, and the production of relatively few offspring. K-selected species usually live in fairly stable or predictable environments. Their population growth is regulated by the carrying capacity of the environment, or the ability of the environment to provide food, shelter, and other resources needed for survival and reproduction. K-selected species include animals such as tortoises and humans. (See r-selected for the alternative extreme.)

Lagg: The moat between a wetland (usually a bog or fen) and the upland.

Larva: In salamanders, the gilled, aquatic, juvenile stage that hatches from eggs. In insects, an immature stage that is very different from the adult and that undergoes a dramatic transformation into the adult form (see definition of metamorphosis below) through a process known as pupation. (Larvae is the plural of larva.)

Lentic: Still water, as in lakes and ponds (contrasted with lotic: flowing water, as in streams and rivers).

Littoral: In shallow water near shore.

Lymnaeid: A freshwater snail in the family Lymnaeidae.

Macroalga: An alga that is large and looks superficially like a higher plant.

Macrophytes: Aquatic plants, such as pondweeds, water lilies, coontail, duckweed, etc.

Macropterous: With fully developed, functional, wings.

Mantle: In molluscs, a specialized tissue on the surface of the body which surrounds the internal organs, secretes the shell and, in pulmonate snails, forms a cavity in which oxygen diffuses from the air into the blood.

Metamorphosis: In amphibians, the transformation of aquatic tadpoles or larvae into terrestrial juvenile frogs, toads, or salamanders. The process involves physical changes, such as the development of limbs and loss of the tail in frogs and toads and resorption of the gills and loss of the keeled tailfin in salamanders, and physiological changes that adapt the animals to a terrestrial, rather than aquatic, existence.

In insects, the process whereby the juvenile (larva or nymph) transforms into an adult. In groups of insects with larvae, metamorphosis occurs during pupation. In insects with nymphs, metamorphosis is gradual, with the nymph becoming more like the adult with each successive molt.

Metanauplius: A larval stage in some crustaceans. It represents a more complex level of development than the nauplius larva (see definition, below). In some crustaceans, eggs hatch as nauplii and the larvae then molt, developing into the metanauplius stage. In some other species, the early developmental stages are passed in the egg and the larva hatches as a metanauplius. (Metanauplii is the plural of metanauplius.)

Metapopulation: Metapopulation theory relates to species in which regular connections between distinct breeding populations are important for the long-term continuation of the species. In metapopulations, local breeding populations are connected genetically to other breeding populations by the migration of individuals. Some local breeding populations, described as "source" populations, regularly raise more young than the local habitat can support, and these individuals colonize new sites when they are ready to breed. Other local breeding populations, known as "sink" populations, are not able to produce enough young to replace the adults that die and are maintained only by the immigration of individuals from source populations. Some populations that breed in variable habitats may be sources in some years and sinks in others.

Metazoan: A multicellular animal (as opposed to a plant or single-celled organism such as a bacterium or protozoan).

Miracidium: The first larval stage of parasitic trematode worms, which hatches from the egg and seeks out and infects a snail. (Miracidia is the plural of miricidium.)

Molt, molting: In insects and crustaceans, the act or process of shedding the hard external skin and replacing it with a new one, to allow the animal to grow.

Moraine: A geologic feature that contains a poorly sorted mixture of large and small rocks, gravels, sands, silts, and clays and was created and emplaced by an advancing or melting glacier. A terminal or end moraine was deposited along the edge of the glacier's farthest advance, a lateral moraine sloughed off the side of the glacier, and a ground moraine was dropped on the surface of the underlying land as the glacier melted. The poorly sorted sediment making up moraines is known as till.

Morainal: Derived from, or part of, a moraine.

Multivoltine: With several generations per year.

Myriapod: A many-legged arthropod such as a centipede or millipede.

Nauplius: An early, relatively simple, larval stage in many crustaceans. (Nauplii is the plural of nauplius.)

Neoteny: In some populations of salamanders, a condition in which adults become reproductively mature while retaining the physical characteristics of the larvae, such as gills and keeled tails.

Nymph: In arthropods such as water mites and some insects, a juvenile stage that broadly resembles the adult in overall body form and gradually develops adult structures over time, transforming into an adult (for example, with genitalia, wings, adult color and pattern) at the final molt.

Obligate: Required, or necessary, for completion of the life cycle. For example, some species have an obligate diapause and do not mature until they have completed it.

Omnivore: An animal that eats a variety of foods including plant, animal, and dead material.

Omnivorous: Describing or pertaining to an omnivore.

Order of magnitude: Ten times. A population of 10,000 individuals is an order of magnitude larger than one with 1,000 individuals. Two orders of magnitude equals one hundred, and three orders of magnitude equals one thousand.

Outwash: Geologic deposits of well-sorted sands and gravels that were carried and deposited by streams flowing off the melting glaciers.

Ovisac: Egg sack.

Palustrine: A term used in the US Fish and Wildlife Service's wetland classification system to refer to inland freshwater wetlands that do not have flowing water, such as marshes, fens, bogs, and swamps.

Parthenogenesis: A process wherein eggs develop without being fertilized.

Perched: In terms of hydrology, located above the groundwater table on impermeable bedrock, soils, or peat. Perched pools receive all of their water from precipitation and surface runoff.

Periphyton: Algae growing upon the surfaces of submerged objects, including sediment, leaves, plant stems, rocks, and woody debris.

Photosynthesis: The process through which some bacteria and protozoans, algae, mosses, ferns, and higher plants use the sun's energy to transform carbon dioxide plus water into glucose and oxygen.

Physid: A freshwater snail in the family Physidae.

Phytoplankton: Small, floating or weakly swimming organisms, such as some algae and protozoans, which are able to produce food through photosynthesis in water bodies.

Planorbid: A freshwater snail in the family Planorbidae.

Plastron: The hard, lower half of a turtle's shell, protecting the body from below.

Pond: A shallow, non-flowing, continuously flooded water body.

Predator: An animal that actively captures and feeds on living animals.

Predaceous: Describing or pertaining to a predator.

Primary production: The creation by plants and algae of living matter and stored food from carbon dioxide, oxygen, and inorganic minerals through photosynthesis.

Protist: A member of the Kingdom Protista, which includes single-celled microorganisms such as ciliates, flagellates, sporozoans, amoebas and their relatives, and slime molds.

Pulmonate: Any of several kinds of air-breathing snails, in which the mantle forms a chamber from which oxygen is absorbed into the blood.

Pupa: An insect during the stage of the life cycle in which a larva transforms into an adult; an insect that is undergoing pupation. The chrysalis of butterflies and moths is a pupa, as is the "tumbler" of mosquitoes. (Pupae is the plural of pupa.)

Pupation: The stage in the life cycle of some kinds of insects during which the larval form is completely reorganized into the adult form. Also, the active process of this reorganization, for example, the physiological process by which a caterpillar transforms into a butterfly or moth.

Resorb: To take back into the body. For example, deciduous trees take back sugars and nutrients from their leaves before autumn leaf-fall, retaining the nutrients for future growth. Similarly, as amphibian larvae and tadpoles grow and prepare to transform into terrestrial juveniles, parts of the body specialized for an aquatic existence, such as swimming tails in tadpoles and gills and tail keels in salamander larvae, are broken down and the proteins and other materials in them are taken back into the bloodstream, where they are available for re-use as the animals grow.

Rhizoids: Root-like structures in macro-algae such as *Chara*.

Richness: The number of kinds of organisms (plants, animals, microbes, or a combination of taxa) present in a particular location.

r-selected: One extreme in theoretical models of population growth and reproductive fitness, which describes species that usually have short life spans and produce very large numbers of offspring because they occur in highly variable, unpredictable, or risky environments where the probability that the offspring will die is very high. For members of r-selected species, the production of very large numbers of young helps ensure that at least one will live to reproduce. Bacteria and many kinds of plankton are typical examples of r-selected species. (See K-selection for the opposite extreme.)

Saprophyte: An organism, usually a plant, fungus, or bacterium, that feeds on dead plants or animals.

Saprophytic: Describing or pertaining to a saprophyte.

Scavenger: An animal that feeds on a variety of materials including dead plant and animal remains.

Scute: One of the large scale- or tile-like structures that cover a turtle's shell. Patterns of growth and coloring on the scutes can be helpful in identifying different species of turtles.

Semelparous: Reproducing only once.

Semi-permanent vernal pool: A shallow, isolated, non-flowing woodland water body that retains water year-round for several consecutive years and dries periodically, often on a five-to-ten-year cycle, and in which the biological community lacks fish and includes species requiring the absence of fish predation and adapted to seasonal drying.

Shrub swamp: A wetland in which the dominant vegetation consists of shrubs.

Species richness: The number of species of a taxon or group of taxa present at a site or location.

Sphaeriid: A fingernail clam in the family Sphaeriidae.

Sporocyst: A stage in the life cycle of parasitic trematode worms and passed in the body of host snails. Sporocysts develop from miracidia larvae and, depending on the species, produce either daughter sporocysts or redia larvae. These leave the mother sporocyst and enter the host's liver or gonad, where they produce cercaria larvae.

Stream order: A numbering system that indicates the relative position of a stream within a watershed, and the number of smaller streams occurring upstream of it. River systems start as small streams that eventually flow together forming larger streams. The most upstream channels are identified as first-order streams, and when two first-order streams join they form a second-order stream. Two second-order streams flowing into each other form a third-order stream; two third-order streams form a fourth-order stream, and so on. A very large river might have a stream order of eight or higher.

Swale: A low wet area, often narrow, and often carrying water after storms.

Swamp: A wetland dominated by woody vegetation, such as trees or shrubs.

Sympatric: Occurring together.

Tadpole: The initial, aquatic developmental stage that hatches from the eggs of frogs, toads, and treefrogs, characterized by a large, combined head-body and a keeled tail adapted for swimming. The tadpole gradually develops limbs and resorbs the tail, eventually transforming into a juvenile miniature of the adult.

Taxon: A group of genetically related plants, animals, or other organisms that are classified together. A species represents a single taxon; a genus, family, or phylum each represents a taxon that usually includes several taxa. (Taxa is the plural of taxon.)

Temporary pond or pool: A shallow, non-flowing water body that alternates between being flooded for a period of time and dry. Usually as used it is synonymous with annual vernal pool, but it may also refer to ephemeral pools, rainpools that fill at any time of year in response to heavy rainfall, and semi-permanent pools.

Tetraploid: With four sets of chromosomes rather than two; a condition occurring, for example, in some hybrid salamanders.

Thallus: The flat, ribbon- or plate-like growth form of some liverworts (other liverworts have a leafy growth form and look a lot like mosses).

Thallose: Describing or pertaining to a thallus.

Till: Poorly sorted and poorly drained sediments that consist of a mixture of rock, gravel, sand, silt, and clay and were deposited by glaciers.

Tip-up: When trees blow over, particularly in forested wetlands, the hole left by the dislodged roots forms a shallow pool with a mound of earth adjacent to it.

Tree-hole: A miniature, ephemeral aquatic habitat with a unique biological community, formed when rainwater or melted snow collects between the base of a branch and the tree trunk, or in another depression in or on a tree.

Triploid: With three sets of chromosomes rather than two; seen in some hybrid salamanders.

Trophic: Relating to the mode by which an organism gets its energy. Autotrophs — "self-feeders" — include plants and algae that produce their own food by photosynthesis. They are also known as primary producers, because they produce the food that supports the entire community. Heterotrophs — "other feeders" — obtain their energy by eating other organisms. They are also known as consumers. There are many trophic levels, from the primary producers, to primary consumers such as herbivores and detritivores that eat living or dead algae and plants, to secondary consumers that eat the herbivores, to tertiary consumers that eat the secondary consumers, and so on. Trophic structure refers to how a biological community is organized in terms of the sources of food (for example, algae, plants, detritus from decaying tree leaves) and how the animal community is organized in terms of functional feeding groups, or "who eats what."

Type locality: The collection location for the "type" specimen — the specimen on which the taxonomic description of an organism was based, and which serves as a reference for future investigators.

Unisexual: With only one sex present in the population, as in many hybrid salamanders in which all individuals are females.

Univoltine: A life cycle with only one generation per year.

Vernal pool: A shallow, isolated, non-flowing woodland water body that attains its maximum depth and volume in spring, remains flooded for a minimum of two months, and periodically loses all or most of its water volume and surface area, and in which the biological community lacks fish and includes species requiring the absence of fish predation and adapted to seasonal drying.

Watershed: The catchment area contributing water, either through surface runoff or groundwater discharge, to a water body such as a stream, lake, or vernal pool. All of the precipitation that falls within a watershed and that is not used by plants and animals, evaporated back into the atmosphere, or retained in the surface soils, eventually flows overland or underground to surface waters. Also, the dividing line between two catchments. Precipitation that falls on one side of the watershed will flow to one water body, and precipitation that falls on the other side will flow to a different water body.

Appendix

An Annotated List of the Fauna of Vernal Pools and Seasonal Ponds in and near the Glaciated Northeast

This appendix summarizes information on macroscopic animals that have been reported from vernal pools of the glaciated northeast. Invertebrates are presented first, arranged alphabetically by phylum and then by class, order, family, genus, and species. Vertebrates are listed at the end. For vertebrates, detailed information provided in the text is not reproduced in the appendix; instead, readers are referred to chapters 11 and 12. An addendum on pages 356 and 357 contains a small number of species that were added to the appendix immediately before this book went to press.

Differing amounts of information are available for different taxa. Because the focus here is on habitats and life histories of pool animals, the citations are drawn largely from studies of temporary ponds and aquatic communities, from ecological research and life history investigations, and from broad natural history surveys of major groups of animals. Relatively few citations of detailed systematic studies are included. Some reports of taxa in vernal pools contradict other distributional and habitat data, and I have not attempted to verify the identification of those taxa.

The classification of taxa included in this appendix is based on the following authorities: Borkent 1981 (Chaoboridae); Bousquet 1991 (Coleoptera); Brigham 1996 (Haliplidae); Clarke 1981 (snails and fingernail clams); Jokinen 1992 (snails); Eugen Kempf and Koen Martens, personal communications (Ostracoda); Larson et al 2000 (Dytiscidae); Morse 2001 (Trichoptera); Pennak 1989 (non-insect invertebrates); Pesce 2003 (Copepoda); Poole 1997 (Hemiptera: Heteroptera); Poole and Gentili 1996a (Coleoptera); Poole and Lewis 1996 (Diptera); Roughley 2004 (Dytiscidae); I. M. Smith 1987, 1997 (water mites); and US Department of Agriculture 2002 (animals not covered above).

Higher Taxa	Genus/Species	Vernal Pool Study [1]	State or Province Studied [2]	Hydrologic Class [3], Setting [4], and Bottom [5]
Phylum ANNELIDA				
Class Clitellata, Subclass Hirudinea		S&F, EAC	MA, WI	la/w/l, sp/w/l, sp/wo/l, sp/ox
Order Arhynchobdellilda				
Erpobdellidae				
	Dina — see *Mooreobdella*			
	Erpobdella punctata (Leidy 1870)	W, K	MI, ON	sa/o/h, la/o/h, p(sp?)/o/h
	Mooreobdella bucera Moore 1949 (as *Dina*)	K	MI	sa/o/h, p(sp?)/o/h
Order Rhynchobdellida				
Glossiphoniidae		EAC, W	MA, ON	ss/w/l, ls/wo/l, sp/wo/l
	Batracobdella — see *Desserobdella*			
	Desserobdella picta (Verrill 1872) (as *Batracobdella* or *Placobdella*)	W, K, Wilbur	MI, ON	ss/w/l, sa/o/h, la/o/h, temp/w, p(sp?)/o/h
	Helobdella fusca (Castle 1900)	K	MI	sa/o/h
	Helobdella stagnalis (Linnaeus 1758)	W, K, DDW	MI, ON	ss/w/l, ss/o/h, sa/o/h, la/o/h, p(sp?)/o/h
	Marvinmeyeria lucida (Moore 1954)	M&M79	ON	la/o/hm
	Placobdella ornata (Verrill 1872) (as *P. rugosa*)	K	MI	sa/o/h
	Placobdella parasitica (Say 1824)	K	MI	p(sp?)/o/h
	Placobdella picta — see *Desserobdella*			
Class Clitellata, Subclass Oligochaeta		W	ON	
Order Haplotaxida				
Enchytraeidae		W, DDW	ON	ss/w/l, ss/o/h
Lumbricidae		W, EAC	MA, ON	ls/w/h, sp/wo/hm
	Eiseniella tetraedra (Savigny 1826)	W	ON	ss/w/l
Naididae		K	MI	sa/o/h, p(sp?)/oh
	Chaetogaster diaphanus (Gruithuisen 1828)	K	MI	pfish/h
	Chaetogaster limnaei von Baer 1827	K	MI	p(sp?)/o/h
	Dero sp.	K, W	MI, ON	

[1] Studies listed in this column report the occurrence of the respective taxa in vernal pools. The references are identified at the end of the appendix on page 358.

[2] This column identifies the geographic area covered by the cited studies. This area rarely encompasses the full distributional range of the respective taxa.

[3] ss = short-cycle (early summer drying), spring flood; ls = long-cycle (late summer drying), spring flood; sa = short-cycle (early summer drying), fall flood; la = long-cycle (late summer drying), fall flood; sp = semi-permanent; p = permanent water; fp = floodplain pool; sw = swale/intermittent stream; temp = reported as temporary water, no additional information available; pfish = permanently flooded, with fish present

[4] w = woodland pool; wo = woodland setting, basin open to sky; o = open, e.g., in field or edge of woods; d = ditch; x = altered by human activity

Life History A[6]	Life History B[7]	Life History C[8]	Drought-Resisting Strategy or Stage	Comments
		1	unknown, probably aestivation	found in Italian temporary pools (Baz)
1?	2?,4?	1?		common May-June in MI (K)
2?				abundant in spring, known only from MI (K)
1		1		family in Australian temporary pools (Lake)
1?	3?,4?	1		common in June in MI (K), parasitic on amphibians
	3?			found in May in MI (K)
1	2,3?	1		
1		1	within aestivating molluscs, particularly *Sphaerium occidentale*	
	3?			May in MI (K); genus in European pools (Baz)
	3?			In April in MI (K)
1	1	1		
1	1	1	aestivation	in Italian temporary pools (Baz)
1		1		known from European vernal pools (Baz)
1		1	aestivation	in Italian temporary pools (Baz)
1	2?	1		
	3			feeds on chadocerans, chydorids, and other small crustaceans
	3?			
1		1		

(continued)

[5] h = bottom with grasses and/or emergent herbaceous plants; m = aquatic macrophytes; D = *Decodon;* R = red maple swamp pool; s = shrub swamp; l = litter; S = *Sphagnum*

[6] After Wiggins et al. 1980: 1 = overwintering residents; 2 = overwintering spring recruits; 3 = overwintering summer recruits; 4 = non-wintering spring migrants (see Chapter 6). Data are provided only where life history is known.

[7] After Williams 1983: 1 = present year-round; 2 = present soon after flooding; 3 = not present until 3–4 weeks after flooding; 4 = present 2-3 weeks before pool dries; 5 = present only in dry pool

[8] After this book: 1 = permanent resident; 2 = migratory breeder, early spring; 3 = migratory breeder, spring-summer; 4 = migratory breeder, fall; 5 = migratory non-breeder

Higher Taxa	Genus/Species	Vernal Pool Study [1]	State or Province Studied [2]	Hydrologic Class [3], Setting [4], and Bottom [5]
	Dero digitata (Müller 1773) (as *D. limosa* Leidy)	K	MI	ss/w, sa/o/h, p(sp?)/o/h
	Nais sp.	K, W	MI, ON	sa/o/h, p(sp?)/o/h
	Pristina longiseta Ehrenberg 1828 (also recorded as *P. leidyi* Smith 1896)	K	MI	ss/w/l, sa/o/h
	Slavinia appendiculata (d'Udekem 1855)	K, W	MI, ON	sa/o/h, pfish/h
Tubificidae		K	MI	ss/w/l, sa/o/h, p(sp?)/o/h
Order Lumbriculida				
Lumbriculidae		W	ON	
	Lumbriculus sp.	DDW	ON	ss/o/h
	Lumbriculus variegatus (Müller 1774)	W, K	MI, ON	ss/w/l, sa/o/h, p(sp?)/o/h
Class Polychaeta (Scolecida)				
Aeolosomatidae		W lit.	MI	temp
	Aeolosoma spp.	K (two species)	MI	sa/o/h
Phylum ARTHROPODA				
Class Arachnida				
Order Araneae				
Microphantidae, unidentified species		K	MI	ss/w
Pisauridae				
	Dolomedes triton (Walckenaer 1837) (as *Dolomedes triton sexpunctatus* Hentz)	K	MI	ss/w/l, la/o/h, p(sp?)/o/h
Class Arachnida, Subclass Acarina				
Order Trombidiformes				
Arrenuridae				
	Arrenurus sp.	K, W, EAC	MA, MI, ON	ss/w/l, sa/o/h, p/o
	Arrenurus (A.) planus Marshall 1908	W, K	MI, ON	ss/w/l, sa/o/h, temp
	Arrenurus (Megaluracarus) neobirgei Cook 1954	S97	ON, QB	ss/, temp, ponds
	Arrenurus (Megaluracarus) neomamillanus Cook 1954	S97	ON, QB	ss/, temp, ponds
	Arrenurus (Megaluracarus) rotundus Marshall 1908	S97	ON, QB	temp, p, bog
	Arrenurus (Micruracarus) pseudosetiger (Marshall 1921)	S97	ON, QB	temp, p, bog
	Arrenurus (Truncaturus) acuminatus Mullen 1976	S97	ON, QB	temp
	Arrenurus (Truncaturus) angustilimbatus Mullen 1976	S97	ON, QB	temp
	Arrenurus (Truncaturus) danbyensis Mullen 1976	S97	ON, QB	temp, p, bog

Life History			Drought-Resisting Strategy or Stage	Comments
A[6]	B[7]	C[8]		
	3?,4?			common May-June in MI (K); found in European vernal pools (Baz)
1	1	1		
1	2	1		
1	3	1		
				family in Italian temporary pools (Baz)
	1	1		
1	2?	1	encystment of fragments?	known from temporary waters worldwide (W)
1		1		
1	2	1	aestivation	
				May in MI (K)
				May-June in MI (K)
	2			
		1,2,3		
2				
2			in most, adult overwinters in pool basin	April-May (MI), genus present in Australian temporary waters (Lake)
2		1	nymphochrysalis over-winters in pool basin	widely distributed, restricted to vernal pools, larvae are parasitic on odonates, adults and deutonymphs prey on ostracods and cladocerans (S87), occurs April-May in MI (K), complex life cycle, see Chapter 10
				northeastern, larval hosts unknown (S97)
				northeastern, larvae parasitize chironomids (S97)
				Great Lakes basin, common in small ponds, larvae parasitize chironomids (S97)
				boreal, larval host unknown (S97)
				Great Lakes basin, larvae parasitize culicids (S97)
				Great Lakes basin, larvae parasitize culicids (S97)
				Great Lakes basin, larvae parasitize culicids (S97)

(continued)

Higher Taxa	Genus/Species	Vernal Pool Study [1]	State or Province Studied [2]	Hydrologic Class [3], Setting [4], and Bottom [5]
	Arrenurus (Truncaturus) kenki Marshall 1944	W, K, S97	MI, ON, QB	sa/, sa/o/h, temp, p
	Arrenurus (Truncaturus) lacrimatus Cook 1955	W, S97	ON, QB	sa?, temp, p
	Arrenurus (Truncaturus) palustris Mullen 1976	S97	ON, QB	temp, p, bog
	Arrenurus (Truncaturus) ringwoodi Mullen 1976	S97	ON, QB	temp
	Arrenurus (Truncaturus) rubropyriformis Habeeb 1954	S97	ON, QB	temp, bog
Eylaidae[a]				
	Eylais sp.	EAC, W, K, S97, Lanciani	MA, MI, ON, NY	ss/w/l, sa/o/h, sp/wo/l, p(sp?)/o/h, temp
	Eylais harmani Lanciani 1970	W lit., Lanciani	NY	temp
Hydrachnidae				
	Hydrachna sp.	EAC, K	MA, MI	ss/w/l, sa/o/h, sp/wo/l
	Hydrachna baculoscutata Crowell 1960 and sp. nr. *comosa* (Koenike 1895)	W, S97	ON, QB	ss/w/l, temp
	Hydrachna conjecta Koenike 1895	K	MI	sa/o/h
	Hydrachna crenulata Marshall 1930	K	MI	sa/o/h
	Hydrachna magniscutata Marshall 1927	K	MI	sa/o/h, p(sp?)/o/h
	Hydrachna militaria Berlese 1888	K	MI	sa/o/h
	Hydrachna rotunda Marshall 1930	K	MI	sa/o/h
	Hydrachna stipata Lundblad 1934	K	MI	sa/o/h
Hydryphantidae		EAC, W	MA, ON	ss/w/l, sp/wo/l
	Euthyas truncata (Neuman 1875)	W, DDW	ON	ss/o/h
	Hydryphantes sp.	EAC, DDW	MA, ON	ss/o/h, temp
	Hydryphantes ruber (DeGeer 1778)	W, K	MI, ON	ss/w/l, la(sp?)/o/h, p(sp?)/o/h

[a] See also pages 356–357.

Life History A[6]	B[7]	C[8]	Drought-Resisting Strategy or Stage	Comments
2		1		boreal, larvae parasitize culicids (S97)
2		1		boreal, larvae parasitize culicids (S97)
				boreal, larvae parasitize culicids (S97)
				Great Lakes basin, larvae parasitize culicids (S97)
				northeastern, larvae parasitize culicids (S97)
4,2?				
4,2?	3	1?,2,3	larvae carried by hosts to permanent waters, where they overwinter, and back to pools in spring	April-June in MI (K), also occurs in Australian temporary waters (Lake), adults eat ostracods, larvae parasitize water bugs and water beetles, about 15 species are specialized for life in temporary waters (S87,97)
		1?		larvae parasitize resident *Hydroporus* and may overwinter in pools on their hosts (W lit., Lanciani)
4		1	larvae remain on hosts during pool drawdown	more than 15 species in northeastern vernal pools, found April-June in MI (K), genus also occurs in Australian and European temporary waters (Baz, Lake), adults and deutonymphs feed on aquatic insects' eggs, larvae parasitize water bugs and water beetles (S97)
4		1		only found in Great Lakes basin, larvae parasitic on backswimmers (S97)
4		1		May in MI (K)
4		1		May in MI (K)
4		1		May in MI (K)
4		1		May in MI (K)
4		1		widespread distribution, found in May in MI (K), larvae parasitic on water scorpions and giant water bugs (S97)
4		1		May in MI (K)
2		1	inactivity of deutonymphs or adults; passive resistance to drought conditions	
2	2	1	adult	rare, restricted to temporary ponds (Pennak)
	2	1		temporary pond specialists (S97), larvae of most species parasitize pupae of ephydrid midges or other dipterans, adults prey on insect eggs (S87)
2		1	deutonymph or adult overwinters on pool bottom	April-May in MI (K), larvae parasitic on ephydrids (W lit., S87)

(continued)

Higher Taxa	Genus/Species	Vernal Pool Study[1]	State or Province Studied[2]	Hydrologic Class[3], Setting[4], and Bottom[5]
	Thyas sp.	K, Pennak	MI	ss/w/l
	Thyas barbigera Viets 1908	W, DDW	ON	ss/o/h, ss/w/l, la(sp?)/o/h
	Thyas stolli Koenike 1895	W lit.	MI	temp
	Thyasides sphagnorum Habeeb 1958	W lit.	MI	temp
	Thyopsis sp.	EAC	MA	ss/w/l
	Thyopsis cancellata (Protz 1896)	W lit.	MI	temp
	Zschokkea bruzelii Lundblad 1926	W lit., DDW (as *Thyas*)	MI, ON	ss/o/h, temp
Ixodidae				
	Ixodes sp.	EAC	MA	ss/w/s, ls/o/hm, ss/w/R, temp
Limnesiidae				
	Limnesia sp.	EAC	MA	ss/w/R, ls/o/h, sp/wo/hm
	Limnesia maculata Müller 1776	K	MI	s/o/h, p(sp?)/o/h
	Limnochares sp.	EAC	MA	sp/wo/hm
Mideopsidae				
	Mideopsis — see *Xystonotus*			
	Xystonotus robustus Habeeb 1954 (as *Mideopsis robusta*)	S97	ON, QB	temp, p
Oribatidae		EAC, S97	MA, ON, QB	ss/w/s, ls/o/h, ls/wo/hD, ls/wo/h, sp/wo/D, sp/wo/hD
Oxidae				
	Oxus connatus Marshall 1929	K	MI	ss/w/l, sa/o/h
Piersigiidae				
	Piersigia sp.	S&C, Newell	ON, QB	temp
Pionidae		EAC	MA	ss/w/s, ls/o/h, ls/wo/hD, ls/wo/h, sp/wo/D, sp/wo/hD
	Acercus — see *Tiphys*			
	Piona sp.	EAC, K, S97	MA, MI, ON, QB	ss/w/l, ss/w/R, sa/o/h, sp/wo/hm

Life History			Drought-Resisting Strategy or Stage	Comments
A[6]	B[7]	C[8]		
2		1		January-June in MI (K), temporary pond specialists (Pennak, S97), larvae feed on pupae of culicids, tipulids, or chironomids (W lit.)
2	2	1	adult overwinters in pool	most common mite collected in ON pools (W)
			adult overwinters in pool	
2		1	adult overwinters in pool	larvae parasitic on culicids (S87)
				temporary pond specialists (Pennak, S97, W)
2		1	adult overwinters in pool	larvae feed on tipulids (S87)
2	2	1	adult overwinters in pool	temporary pond specialists (S97, W)
				Ixodes are parasitic on vertebrates and probably occur only incidentally on vegetation along the edges of vernal pools, adults feed on aquatic insects' eggs and chironomid larvae
				genus in Australian temporary waters (Lake)
				May in MI (K)
				larvae of *L. americana* parasitize adult odonates, those of *L. aquatica* feed on gerrids and hydrometrids (S87), genus in Australian temporary waters (Lake)
				boreal, parasitizes chironomids (S97)
				oribatid mites are important decomposers in forest soils, 10 families with >30 species that are aquatic or semiaquatic (S97), found in large numbers in pools in spring in MA
				boreal, with southern limit in southern ON, QB, and the upper Great Lakes states, larvae parasitize chironomids (S97)
			unknown; adults have been found in pools in early spring	larvae parasitize water beetles, adults and deutonymphs prey on ostracods
			deutonymph overwinters in pool	larvae parasitize chironomids, adults feed on cladocerans and insect larvae (S87)
		1	deutonymph overwinters in pool	five species restricted to vernal pools in southern ON and QB (S97), larvae parasitize chironomids (W), found December-April in MI (K)

(continued)

Higher Taxa	Genus/Species	Vernal Pool Study [1]	State or Province Studied [2]	Hydrologic Class [3], Setting [4], and Bottom [5]
	Piona carnea (Koch 1836)	K	MI	la(sp?)/o/h, p(sp?)/o/h
	Piona clavicornis (Müller 1776)	W	ON	la/o/h, la(sp?)/o/h, p(sp?)/o/h
	Piona constricta (Wolcott 1902)	?W, K	MI, ON	la(sp?)/o/h
	Piona mitchelli Cook 1960	W	ON	temp
	Piona napio Crowell 1960	W	ON	ss/w/l, la(sp?)/o/h
	Piona neumani (Koenike 1883)	K	MI	la(sp?)/o/h
	Pionopsis lutescens paludis Habeeb 1954 (as *P. paludis*)	S97	ON, QB	temp
	Tiphys sp.	EAC, W, S97	MA, ON, QB	la/wo/h, temp
	Tiphys americanus (Marshall 1937) (as *Acercus*)	W, K	MI, ON	ss/w/l, sa/o/h, la(sp?)/o/h
	Tiphys brevipes Habeeb 1954	W	ON	ss/w/l, la(sp?)/o/h
	Tiphys simulans (Marshall 1924) (as *Acercus*)	K	MI	ss/w/l
	Tiphys vernalis (Habeeb 1954)	W	ON	temp
Class Apterygota				
Order Collembola				
	Isotomidae	DDW, EAC	MA, ON	ss/o/h, ls/wo/D, ls/wo/h, sp/wo/hD
	Isotomurus palustris (Müller 1776)	K	MI	sa/o/h
	Poduridae	EAC	MA	ss/w/R, ls/wo/hD, sp/wo/D
	Podura aquatica (Linnaeus 1758)	EAC	MA	
	Sminthuridae	DDW, EAC	MA, ON	ss/o/h/ ls/w/h, sp/wo/hD, sp/wo/hm
	Bourletiella sp.	EAC	MA	ss/w/R, ss/wo/hD, ls/wo/h, sp/wo/h
Subphylum Crustacea				
Class Branchiopoda				
Order Anomola				
	Chydoridae	DDW, W lit., EAC	MA, ON	ss/o/h, ss/w/l, ls/wo/D, ls/wo/hD, sp/wo/hD, sp/wo/hm
	Alona guttata Sars 1862	W lit, EAC	MA	temp, sp/wo/hm
	Alona rectangula Sars 1861	K	MI	sa/o/h, p(sp?)/o/h
	Alonella sp.	S&F	WI	sp/w

Life History A[6]	Life History B[7]	Life History C[8]	Drought-Resisting Strategy or Stage	Comments
		1	deutonymph overwinters in pool	
		1	deutonymph overwinters in pool	in pools with much water fluctuation (W)
2		1	deutonymph overwinters in pool	in pools with much water fluctuation (W), April-May in MI (K)
2		1	deutonymph overwinters in pool	in pools with much water fluctuation (W)
2		1	deutonymph overwinters in pool	in pools with much water fluctuation (W)
			deutonymph overwinters in pool	
				northeastern distribution (S97)
2		1	deutonymph embeds mouthparts in mosses and overwinters in pool	eight species in temporary ponds in southern ON and QB (S97), parasitize chironomids (W)
2		1		found April-June in MI (K), occur in pools with much water fluctuation (W)
2		1		in pools with much water fluctuation (W)
2		1		May in MI (K)
2		1		in pools with much water fluctuation (W)
	1			
	2			
	4?			
				these are terrestrial springtails usually found in leaf litter, and probably only incidentally associated with vernal pools
		1		
2		1		
1		1		genus in Australian and CA contemporary pools (Lake, E&B, Simovich, Z)
	3?			April-June in MI (K), known from German temporary pools (Maier)

(continued)

Higher Taxa	Genus/Species	Vernal Pool Study [1]	State or Province Studied [2]	Hydrologic Class [3], Setting [4], and Bottom [5]
	Chydorus sp.	S&F	WI	sp/w
	Chydorus sphaericus (Müller 1785)	W lit.		temp
	Kurzia latissima (Kurz 1875)	K	MI	sa/o/h, p(sp?)/o/h
	Pleuroxus denticulatus Birge 1879	K	MI	sa/o/h
Daphniidae				
	Ceriodaphnia reticulata (Jurine 1820)	W lit., H&M01, K, EAC	MA, MI, ON	sa/o/h, sp/wo/hm, la/o/h, p(sp?)/o/h, temp
	Ceriodaphnia quadrangula (Müller 1785)	K	MI	ls/w/h, sa/o/h
	Daphnia sp.	K, S&F, EAC	MA, MI, WI	ss/w, ss/w/l, ls/w, ls/wo/h, ls/w/l, sa/w, la/w, sp, sp/o/h, sp/wo/hm, sp/wo/hD, p/w, pfish/h
	Daphnia ephemeralis (Schwartz and Hebert 1985)	H&M01, R&H, S&H, Innes, T&W	MI, ON	sa/w/l, la/w/l
	Daphnia obtusa Kurz 1874	N&J	IL	ss/w/l
	Daphnia pulex Leydig 1860	EAC, W, K, DDW, H&M01, T&W	MA, MI, ON	ss/w/l, ss/o/h, sa/o/h, p(sp?)/o/h
	Scapholeberis kingi Sars 1903	W, K, S&F	MI, ON, WI	ss/w/l, ls/w, sp/w
	Simocephalus exspinosus (Koch 1841)	?W, H&M01, EAC	MA, MI, ON	ss/w/l
	Simocephalus serrulatus (Koch 1841)	K, EAC	MA, MI	ss/w/l, ls/w/h, ls/wo/D, sp/wo/hD, sp/wo/hm, p(sp?)/o/h
	Simocephalus vetulus (Müller 1776)	?W, K, EAC	MA, MI, ON	ss/w/l, ls/w/h, ls/w/D, sa/o/h, la(sp?)/o/h, sp/wo/hD, sp/wo/hm, p(sp?)/o/h
Moinidae				
	Moina spp.	W lit.		temp
	Moina brachiata (Jurine 1820)	K	MI	sa/o/h

Life History A[6]	Life History B[7]	Life History C[8]	Drought-Resisting Strategy or Stage	Comments
				genus in Australian and CA temporary pools (Lake, Simovich)
1		1		species occurs in Australian, European, and CA temporary pools (Lake, Maier, Simovich)
	4?			rare, present in June in MI (K)
	3?			common April-June in MI (K)
			ephippial eggs	
	2			present December, abundant May-June, in semi-permanent water in MI (K), genus in Australian pools (Lake), species in CA and European pools (Maier, Simovich)
	3	1		common April-June in MI (K)
1	2	1	resting eggs	genus in Australian and European temporary pools (Lake, Maier)
1	2	1	resting eggs	northern deciduous forest of Eastern NA: ON, southern New England, west to OH, MI, IA, occurs in pools with high levels of dissolved humic acids and tannins, eggs hatch in late autumn, are present through winter, disappear by early spring (S&H)
1	2	1	resting eggs	found in fishless permanent and temporary ponds throughout the glaciated northeast, found in vernal pools April-June in MI (K), occurs in CA pools (Simovich)
1	4?	1		as *Scapholeberis mucronata* (Müller), common in June (K), known from Europe (Maier)
1		1		known from European and CA temporary waters (C&M, E&B, Z), associated with vegetation in permanent waters
	2	1		present December-May in MI (K), known from European temporary waters (Baz), associated with vegetation in permanent waters
1		1		December-June, MI (K), known from Australian, European, and CA temporary pools (Lake, Maier, Simovich)
1		1		
1		1		genus in CA and European vernal pools (E&B, Z, Maier, Simovich)
	2	1		rare, found in January in MI (K), species in German temporary ponds (Maier)

(continued)

Higher Taxa	Genus/Species	Vernal Pool Study [1]	State or Province Studied [2]	Hydrologic Class [3], Setting [4], and Bottom [5]
Polyphemidae				
	Polyphemus pediculus (Linnaeus 1761)	EAC	MA	sp/wo/hm, sp/wo/hD
Sididae				
	Diaphanosoma brachyurum (Lie'ven 1848)	K	MI	p(sp?)/o/h
Order Anostraca				
Branchinectidae				
	Branchinecta paludosa (Müller 1788)	Dexter 53, 59	LB, NS, QB	temp
Chirocephalidae		see Chapter 8		ss/w/l, ss/o/h, ls/w/l, fp/o/h, fp/w/l, sa/wo/l, la/w/l, sp/w/l, sp/wo/sh, temp
	Eubranchipus sp.	S&F	WI	ss/w, ls/w, sp/w
	Eubranchipus bundyi Forbes 1876	see Chapter 8	IL, IN, MI, MN, NH, NY, OH, ON, QB, WI	temp, temp/w/l, a/o/h, ls/w/l
	Eubranchipus holmani (Ryder 1879)	see Chapter 8	CT, IL, NJ, NY, OH	temp, fp
	Eubranchipus intricatus Hartland-Rowe 1967	EAC, Smith	MA	ss/w/l, ls(la?)/w/l
	Eubranchipus neglectus Garman 1926	see Chapter 8, Belk et al. 1998	IL, IN, MI, OH, ON	ss/w/l, ss/w/R, temp, a/o/h, ls/w/l, ls/w/h, sa/o/h, sp/wo/h, sp/wo/hD, sp/wo/D, sp/wo/s, p(sp?)/o/h
	Eubranchipus ornatus Holmes 1910	see Chapter 8	MN, WI	temp
	Eubranchipus vernalis (Verrill 1869)	see Chapter 8, Dexter, Belk et al. 1998	CT, MA, NJ, NY, PA, RI	ss/w/l, ss/w/R, fp/o/h, fp/w/l, ls/w/l, ls/w/h, sa/o/h, sa/w/l, sp/wo/h, sp/wo/hD, sp/wo/D, sp/wo/s
Streptocephalidae				
	Streptocephalus seali Ryder 1879	Dexter 1953	IL, MN, NJ	temp, temp/o, d

Life History A[6]	B[7]	C[8]	Drought-Resisting Strategy or Stage	Comments
				predatory on rotifers, protozoans, and other small animals, common in marshes and weedy pond shores (Brooks)
1	2	1		occurs in the far north, genus in CA vernal pools (E&B, Simovich, Z)
1	2	1	drought-resistant eggs (cysts)	December-April, family in Italian and German temporary pools (Baz, Maier, Mura)
1		1	drought-resistant eggs (cysts)	
1	2	1	drought-resistant eggs (cysts)	most common species in northern half of the glaciated northeast
1	2	1	drought-resistant eggs (cysts)	at the northern edge of its range in southern parts of the glaciated northeast
1	2,(3?)	1	drought-resistant eggs (cysts)	appears later than *E. vernalis,* seems only to occur in high-water years following dry summers (EAC)
1	2	1	drought-resistant eggs (cysts)	most common in southern half of glaciated northeast west of Appalachians, found December-May in MI (K), apparently intolerant of turbid conditions, in many collections in the Midwest, incorrectly identified as *E. vernalis*
1	2	1	drought-resistant eggs (cysts)	a species of the prairies and Great Plains, rather than of northeastern woodlands
1	2	1	drought-resistant eggs (cysts)	most common species in southern half of glaciated northeast east of Appalachians, tolerant of a wide range of conditions
1		1	drought-resistant eggs (cysts)	widely distributed except in New England and the Maritimes, occurs later in the season than *Eubranchipus*

(continued)

Higher Taxa	Genus/Species	Vernal Pool Study [1]	State or Province Studied [2]	Hydrologic Class [3], Setting [4], and Bottom [5]
Order Laevicaudata				
Lynceidae		see Chapter 8		
	Lynceus brachyurus Müller 1776	W, EAC, K, DDW, Dexter & Kuehnle 1951, S&F, W&M	MA, MI, OH, ON, WI	ss/w, ss/w/l, ss/o/h, ls/w, ls/w/lh, sa/o/h, sp, sp/wo/hm, sp/o/h
Order Spinicaudata				
Cyzicidae				
	Caenestheriella gynecia Mattox 1950	S&G, W&M	OH, MA, PA	temp
Class Copepoda				
Order Calanoida		K	MI	ss/w/l, sa/o/h, p(sp?)o/h
Diaptomidae				
	Diaptomus sp.	S&F, EAC	MA, WI	sp/w, ls/wo/sS
	Diaptomus stagnalis Forbes 1882	W lit., Cole	IL	a
Order Cyclopoida				
Cyclopidae		W, S&F, K, EAC	MA, MI, ON, WI	ss/w/l, ls/w, la(sp?)/o/h, sp, p, temp
	Cyclops spp. — see also *Diacyclops, Megacyclops, Microcyclops*			
	Cyclops haueri Kiefer 1931	Yeatman	CT, OH	temp
	Cyclops vernalis Fischer 1853	W, K, S&J	MI, ON, IL	ss/w/l, sa/o/h, p(sp?)/o/h
	Diacyclops bicuspidatus (cf. thomasi) (Forbes 1882) (as *Cyclops*)	W, K, DDW	MI, ON	ss/o/h, ss/w/l, sa/o/h, p(sp?)/o/h
	Diacyclops navus (Herrick 1882) (as *Cyclops*)	K	MI	ss/w/l
	Eucyclops speratus (Lilljeborg 1901)	K	MI	p(sp?)/o/h
	Macrocyclops albidus (Jurine 1820)	K	MI	p(sp?)/o/h
	Megacyclops latipes (Lowndes 1927) (as *Cyclops*)	W, K	MI, ON	ss/w/l
	Mesocyclops leukarti (Claus 1857)	K	MI	sa/o/h, p(sp?)/o/h
	Microcyclops varicans rubellus (Lilljeborg 1901) (aka *Microcyclops varicans* (Sars 1863) and *M. rubellus* (Lilljeborg 1901)) (as *Cyclops*)	W, K	MI, ON	ss/w/l, sa/o/h, p(sp?)/o/h
	Paracyclops fimbriatus (Fischer 1853) (as *Platycyclops*)	K	MI	p(sp?)/o/h
	Platycyclops fimbriatus — see *Paracyclops*			

Life History A[6]	Life History B[7]	Life History C[8]	Drought-Resisting Strategy or Stage	Comments
1		1	drought-resistant eggs (cysts)	
1		1	eggs	
1	3	1	eggs	April-June in MI (K), appears after *Eubranchipus,* found in CA vernal pools (Simovich)
		1	eggs	rare, occurs sporadically
1?	4?			
		1	resistant eggs	family in German temporary pools (Maier)
1	2	1		
1	2	1		occurs in Carolina Bays in the southeastern US (TWM)
				family in French and German vernal pools (Maier, V&A)
			copepodite encystment	can stay dormant 3+ years, family in French and German vernal pools (Maier, V&A)
				rare
1		1	protective slime? resistant eggs?	widely distributed, abundant December-March in MI, common April-July (K), found in CA temporary pools (E&B, Z)
	1	1		February-May, November-May, MI (K), found in European temporary pools (Mastrantuono, V&A)
1		1		abundant May-June in MI (K)
				found in CA temporary pools (E&B, Z)
1	3?	1		February-June in MI (K), genus in German temporary pools (Maier)
				rare June-July in MI (K), genus in Italian and Australian temporary pools (Mastrantuono, Lake)
1		1		common May-June in MI (K), genus in Australian temporary pools (Lake)

(continued)

Higher Taxa	Genus/Species	Vernal Pool Study [1]	State or Province Studied [2]	Hydrologic Class [3], Setting [4], and Bottom [5]
Order Harpacticoida				
	Canthocamptidae	EAC	MA	ls/wo/D
	Attheyella americana (Herrick 1884) (as *A. northumbrica americana*)	K	MI	ss/w/l, sa/o/h, p(sp?)/o/h
	Bryocamptus (minutus complex*)* (Claus 1863)	W	ON	ss/w/l
	Canthocamptus sp.	S&J	IL	temp
	Canthocamptus staphylinoides Pearse 1905	W, K, DDW	MI, ON	ss/o/h, ss/w/l, sa/o/h, p(sp?)/o/h
Class Malacostraca				
Order Amphipoda		W, B&S	NY, MA, ON	temp, a, sp
	Crangonyctidae	EAC	MA	
	Crangonyx sp.	EAC	MA	temp/w
	Crangonyx gracilis Smith 1871	K	MI	ss/w/l, sa/o/h, p(sp?)/o/h
	Crangonyx obliquus (Hubricht and Mackin 1941)	K	MI	p(sp?)/o/h
	Crangonyx pseudogracilis Bousfield 1958	D&B	NY	la/w/l
	Crangonyx rivularis Bousfield 1958	W, EAC	MA, ON	ss/w/l, sa/o/h
	Crangonyx shoemakeri (Hulbricht and Mackin 1940)	K, EAC	MA, MI	ss/w/l, p(sp?)/o/h, temp
	Talitridae			
	Hyalella azteca (Saussure 1858)	EAC, W, K	MA, MI, ON	ss/w/l, la(sp?)/o/h, pfish/h, p(sp?)/o/h
Order Decapoda				
	Astacidae			
	Cambarus spp. — see also *Orconectes*			
	Cambarus fodiens (Cottle 1863)	W, K	MI, ON	ss/w/l, sa/o/h
	Orconectes immunis (Hagen 1870) (as *Cambarus*)	K	MI	p(sp?)/o/h
Order Isopoda				
	Asellidae	W, B&S		
	Asellus — see *Caecidotea, Conasellus*			
	Caecidotea communis — see *Conasellus*			
	Caecidotea forbesi (Williams 1970) (as *Asellus*)	W	ON	ss/w/l
	Caecidotea intermedius (Forbes 1876) (as *Asellus*) (= *A. militaris* Hay)	K	MI	ss/w/l, sa/o/h, p(sp?)/o/h
	Caecidotea racovitzai (Williams 1970)	B&S	NY	la/w/l
	Conasellus communis (Say 1818) (aka *Caecidotea*)	EAC, K	MA, MI	pfish

Life History			Drought-Resisting Strategy or Stage	Comments
A[6]	B[7]	C[8]		
			resting eggs, encystment	order in Australian temporary pools (Lake)
1,(2?)		1	cyst or eggs	December-June in MI (K)
1		1		
	2	1	encystment or slime layer, possibly eggs	common November-May, MI (K), genus in CA vernal pools (Simovich)
	1	1	generally not considered to be drought-resistant	
1				
	2,3			January-May in MI (K)
1		1	burrow into wet sediment	
				April in MI (K)
1			colonize — not drought resistant?	
1			remain in burrows at water table during dry phase	May-June in MI (K)
			generally not considered able to withstand drying	
1			in crayfish burrows? hypogean (in groundwater)?	
	2,3		aestivation	found in pools December-June in MI (K)

(continued)

Higher Taxa	Genus/Species	Vernal Pool Study [1]	State or Province Studied [2]	Hydrologic Class [3], Setting [4], and Bottom [5]
Class Ostracoda				
Order Podocopida				
Candonidae				
	Candona candida (Müller 1776)	Tressler	MA	temp
	Candona crogmaniana Turner 1894	K, Tressler	MI	p(sp?)/o/h, temp, p
	Candona decora Furtos 1933 (= *C. fossulensis* Hoff 1942)	W, K, DDW, Hoff	IL, MI, ON	ss/o/h, ss/w/l, temp
	Candona distincta Furtos 1933	Hoff	IL	temp
	Candona inopinata Furtos 1933 (= *C. indigena* Hoff 1942)	Hoff	IL, MI, TN	temp
	Candona rostrata Brady and Norman 1889 (= *C. limbata* Sars 1890, *Eucandona marchica* Hartwig 1889)	D70c		temp
	Candona suburbana Hoff 1942	Hoff	IL	temp
	Candona truncata Furtos 1933	K	MI	ss/w/l, p(sp?)/o/h
	Pseudocandona albicans (Brady 1864)	W, Hoff	IL, ON	ss/w/l, temp
	Pseudocandona hartwigi (Müller 1900)	D70c	ON, QB	temp
Cyclocyprididae				
	Cyclocypris laevis (Müller 1776)	W	ON	ss/w/l
	Cypria exculpta Fischer 1855	K	MI	ss/w/l, sa/o/h, p(sp?)/o/h
	Cypria ophthalmica (Jurine 1820)	W, DDW, K, Hoff, D70b	IL, MI, ON, NS	ss/o/h, la(sp?)/o/h, temp
	Cypria palustera Furtos 1935	D70b, Tressler	MA, Canada	temp
Cyprididae				
	Bradleycypris tincta (Furtos 1933) (as *Cypricercus*)	W, K, Tressler	MI, OH, ON	ss/w/l, sa/o/h, temp
	Bradleystrandesia fuscata (Jurine 1820) (as *Eucypris fuscata, Cypricercus fuscatus, C. reticulatus* (Zaddach 1884))	Tressler, W, Hoff, K	IL, IN, MA, MI, NY, OH, NJ, ON	ss/w/l, sa/o/h, p(sp?)/o/h, temp
	Bradleystrandesia splendida (Furtos 1933) (as *Cypricercus*)	Tressler	MA, OH	temp
	Cypricercus spp. — see *Bradleycypris, Bradleystrandesia*			
	Cypriconcha princeps Sars 1898 (as *Megalocypris*)	D70a	Canada	temp
	Cypridopsis vidua (Müller 1776) (= *Cypria obesa* Sharpe 1897)	W, Fe, K, D70a	MI, ON, Canada	ss/w/l, sa/o/h, p(sp?)/o/h, temp, p
	Cypris — see also *Eucypris*			
	Cypris pubera (Müller 1776)	W	ON	ss/w/l

Life History			Drought-Resisting Strategy or Stage	Comments
A[6]	B[7]	C[8]		
			torpor and eggs	eggs can hatch after 20 years (Pennak), usually univoltine
				also in MT and OR
1	2	1	torpor	January-June in MI (K), typical temporary pond species (D70c)
				typical temporary pond species (D70c)
				holarctic, occurs in the mixedwood and boreal zones in Canada (D70c)
		1		
		1		January-May in MI (K)
1		1		common in temporary pools all the way to boreal zone (D70c), genus in German pools (Maier)
				typical temporary pond species (D70c)
1		1		
		1		November-June in MI (K)
1	2	1	torpor in dry basins, or burrowing into wet sediment	several generations per year
		1		
1	4	1		April-May, MI (K), small, grassy temporary pools (Tressler), genus in CA vernal pools (Z)
1	3			found in April and May in MI (K)
				genus in Australian temporary ponds (Lake), genus not in North America
1	3	1	eggs	2 to 3 generations per year, most common Canadian ostracod (D70a), January-June in MI (K)
1	4	1		

(continued)

Higher Taxa	Genus/Species	Vernal Pool Study [1]	State or Province Studied [2]	Hydrologic Class [3], Setting [4], and Bottom [5]
	Cypris subglobosa Sowerby 1840	K	MI	ss/w/l, p(sp?)/o/h
	Eucypris crassa (Müller 1785)	W, K, D70a	MI, ON	ss/w/l, p(sp?)/o/h
	Eucypris hystrix Furtos 1933	Tressler	OH	temp
	Sarscypridopsis aculeata (Liljeborg 1853) (as *Cypridopsis*)	D70a, Tressler	ON, Canada	temp
	Spirocypris horridus (Sars 1895) (as *Cypricercus*)	W	ON	la(sp?)/o/h
Notodromadidae				
	Cyprois marginata (Strauss 1821)	W, K, Tressler	IL, MI, OH, ON	ss/w/l
	Notodromas monacha (Müller 1776)	W	ON	sp/o/h
Subphylum Uniramia				
Class Insecta				
Order Coleoptera				
Bruchidae		EAC	MA	sp/o/h, sp/wo/sS
Curculionidae		EAC	MA	ls/wo/D, ls/wo/h, sp/wo/hm
	Listronotus sp.	K	MI	ss/w/l, p(sp?)/o/h
	Lixellus sp.	EAC	MA	la/o/h, sp/o/h
	Lixus sp.	EAC	MA	la/o/h, sp/o/h
	Steremnius sp.	EAC	MA	la/o/h, sp/o/h
Dytiscidae				
	Acilius sp.	DDW, W, S&F, K, EAC	MA, MI, ON, WI	ss/w/l, ss/o/h, ls/w, sa/o/h, sp/w, sp/wo/l, p(sp?)/o/h
	Acilius fraternus (Harris 1828)	EAC	MA	ls/wo/h, ls/wo/D, ls/w/h, sp/wo/D
	Acilius mediatus (Say 1823)	LAR	glaciated northeast	small forest pools
	Acilius semisulcatus Aubé 1838	EAC, W, J66, J67, J69, K, S&F	MA, MI, ON, WI	ss/w/l, ss/o, p
	Acilius sylvanus Hilsenhoff 1975	EAC, S&F	MA, WI	ls/w/l, sp/w/l
	Agabetes acuductus (Harris 1828)	L28, EAC	MA, NY	temp/w, sp/w/h, sp/wo/hm

Life History A[6]	Life History B[7]	Life History C[8]	Drought-Resisting Strategy or Stage	Comments
				common May-June in MI (K)
1	4	1		present June in MI (K), an early-spring species (D70a), genus in CA and Australian temporary pools (E&B, Z, Lake)
				small weedy temporary pools in woods
				genus in German pools (Maier)
1		1		
1	3	1		typically found in temporary ponds in prairie (D70d)
1		1		boreal, rare in North America (D70d)
				family in Australian temporary pools (Lake)
4	2	4	migratory adults	in Italian and British temporary pools (Baz, Davy-Bowker)
				in temporary ponds in DE (FCF), and in shaded woodland pools in the southeastern US (Matta and Michael), found in the southern two-thirds of the glaciated northeast, not in Canada
				found throughout glaciated northeast (LAR) in permanent ponds and not vernal pools in ON (J67)
4		4	adults overwinter in permanent water	complex life cycle, see Chapter 9, found through the glaciated northeast, typical of vernal pools, breed in vernal pools and not in permanent ponds (J67)
				found through the glaciated northeast, except the Canadian Maritime Provinces (LAR)
				"usually in woodland pools" (L28), in temporary ponds and rarely permanent ponds in DE (FCF), larvae in fishless ponds in southeastern PA (FFM)

(continued)

Higher Taxa	Genus/Species	Vernal Pool Study [1]	State or Province Studied [2]	Hydrologic Class [3], Setting [4], and Bottom [5]
	Agabus sp.	DDW, S&F, J66	ON, MI	ss/w, ss/o/h, ss/w/R, ls/wo/D, ls/w, ls/wo/h, sp/o, sp/w, sp/w/h, sp/wo/hm, p/w, p/ox
	Agabus anthracinus Mannherheim 1852	W, J67, J69, S&F	ON WI	ss/w/l, sp
	Agabus bifarius (Kirby 1837)	J67, J69, K	MI, ON	ss/w/l
	Agabus canadensis Fall 1922	J69	ON	
	Agabus confinis (Gyllenhal 1808)	W, J69	ON	ss/w/l
	Agabus disintegratus (Crotch 1873)	G&H	IN	temp
	Agabus erichsoni Gemminger and Harold 1868	EAC, J61, J66, J69, W, S&F	MA, ON WI	ss/w/l, la/o/h, sp/o, sp/w
	Agabus falli (Zimmerman 1934) (as *A. sharpi* Fall 1922)	J67, J69	ON	ss/w
	Agabus gagates Aubé 1838	EAC	MA	ls/w/h
	Agabus phaeopterus (Kirby 1837)	J66, J67, J69	ON	ss/w, sp/w, sp/o
	Agabus semipunctatus (Kirby 1837)	L28, J66	NY, ON	ss/w, ss/o/h
	Agabus wasastjernae (Sahlberg 1824) (as *A. kenaiensis* Fall 1926)	J69	ON	ss/w
	Agoporus — see *Laccornis*			
	Celina sp.	EAC	MA	ls/w/h, sp/wo/hm
	Colymbetes sp.	S&F	WI	sp/w
	Colymbetes sculptilis Harris 1829	J61, J67, J69, W	ON	ss/w/l, p(sp?)/o/h, sp, p
	Coptotomus sp.	EAC, S&F	MA, WI	ls/w/h, sp/w, sp/wo/hm
	Deronectes — see *Stictotarsus*			
	Desmopachria sp.	K, S&F	MI, WI	ss/w, sa/o/h, sp/w, p(sp?)/o/h

Life History A[6]	Life History B[7]	Life History C[8]	Drought-Resisting Strategy or Stage	Comments
2	2			genus present in European and CA temporary pools (Baz, Davy-Bowker, Z)
2			eggs	occurs throughout the glaciated northeast, reported to be a permanent pond species (LAR)
	3		nonbreeding migrant	distributed across the glaciated northeast (LAR)
			nonbreeding migrant	possibly misidentified; this species occurs in grassland pools in the prairies and other western grassland areas (LAR)
2		5	nonbreeding migrant	found in Canada, New England, and the upper Great Lakes states (LAR)
		4?	adults aestivate in CA, overwinter in permanent waters in IN	a widespread species of temporary ponds, found throughout glaciated northeastern US except ME and MN, and in southern ON (LAR)
2,4	2	4	eggs, adults	distributed throughout the glaciated northeast (LAR)
		5	nonbreeding migrant	known from central New England, MI, and a few sites in southern ON, QB, and NB, absent from Canadian Shield (LAR)
				in temporary ponds in DE (FCF), occurs in US portion of glaciated northeast and possibly southern Canada (LAR)
		4?,5?	nonbreeding migrant(?), migratory breeder?	known from vernal pools in Canada and from New England, NY, and the upper Great Lakes states (LAR)
				found throughout the glaciated northeast
		5	nonbreeding migrant	northern, found in the upper Great Lakes to the Gaspé Peninsula, and in ME (LAR)
				genus in Italian and English temporary pools (Baz, Davy-Bowker)
2?,4?		1?,4?	eggs? (larval identification questionable) or adult	associated with emergent vegetation, found throughout glaciated northeast (LAR), found in temporary and permanent ponds in ON but breed only in temporary water (J67)
				C. interrogatus in ON, found in permanent ponds but not vernal pools (J67)

(continued)

Higher Taxa	Genus/Species	Vernal Pool Study [1]	State or Province Studied [2]	Hydrologic Class [3], Setting [4], and Bottom [5]
	Desmopachria convexa (Aubé 1838)	EAC, K, W	MA, MI, ON	ss/w, ls/wo/D, sp/wo/h
	Dytiscus sp.	DDW, K, S&F, H&K, J66	MI, ON, WI	ss/w/l, ss/w/R, ls/w, ls/w/h, sa/w, la/w, sa/o/h, sp/wo/hm, sp/w, p/w
	Dytiscus fasciventris Say 1824	EAC, W, J61, J67, J69	MA, ON	ss/w/l, la/o/h, sp
	Dytiscus verticalis Say 1823	F&B81	ON	temp, p
	Graphoderus sp.	EAC, K, W, S&F	MA, MI, ON, WI	ss/w/l, ls/w/h, sa/o/h, sp/w, sp/wo/hm
	Graphoderus fascicollis (Harris 1828)	K, J67	MI, ON	sa/o/h, p
	Graphoderus liberus (Say 1825)	K, S&F	MI, WI	p(sp?)/o/h, sp/w
	Hydaticus sp.	EAC	MA	ls/wo/hD, sp/w/hm, sp/w/h
	Hydaticus modestus Sharp 1882	J69, DDW	ON	ss/w, ss/o/h
	Hydroporus sp.	EAC, K, DDW	MA, ON, MI	ss/o/h, ss/w/R, ss/w/l, ls/wo/hD, ls/wo/h, sp/w/hm, sp/wo/D, p(sp?)/o/h
	Hydroporus americanus Aubé 1838	EAC, LAR	MA, NJ	la/o/h, sp/w/l, small forest pools
	Hydroporus appalachius Sherman 1913 (as *H. signatus* Sharp 1882)	K	MI	ss/w/m, sa/o/h
	Hydroporus dentellus Fall 1917	J69, EAC	MA, ON	ss/w/l
	Hydroporus despectus Sharp 1882	J69, W	ON	ss/w/l
	Hydroporus fuscipennis Schaum 1868	J69, W	ON	ss/w/l
	Hydroporus niger Say 1823 (= *H. modestus* Aubé 1838)	K, J61, J67, J69, EAC	MA, MI, ON	ss/w/l, ls/wo/D, sa/o/h, p(sp?)/o/h
	Hydroporus notabilis LeConte 1850	W	ON	ss/w/l
	Hydroporus striola Gyllenhal 1808	J69, S&F	ON, WI	ss/w/l
	Hydroporus tenebrosus LeConte 1850	W lit., J69, S&F, EAC	MA, ON WI	ss/w/l, ls/wo/D
	Hydroporus tristis (Paykull 1798)	EAC, J69, S&F	MA, ON, WI	ss/wo/hD, ss/w/R, ls/wo/D, sp
	Hydrovatus sp.	EAC, DDW	MA, ON	ss/o/h
	Hydrovatus pustulatus (Melsheimer 1844)	EAC, K	MA, MI	ls/wo/h, sa/o/h, sp/wo/hm, p(sp?)/o/h

Life History A[6]	Life History B[7]	Life History C[8]	Drought-Resisting Strategy or Stage	Comments
		5	nonbreeding migrant	in DE, found in temporary and permanent ponds early in the season, and in temporary ponds later (FCF), occurs throughout the glaciated northeast (LAR)
(4)	2			genus in European temporary ponds (Baz, Davy-Bowker, Tejedo), in ON, *D. circumcinctus, D. harrissii,* and *D. verticalis* found in permanent, not temporary ponds (J67)
4	4	4	possible larval overwintering, eggs, adults overwinter in permanent water	adults in permanent and temporary water in ON, breed in temporary pools (J67)
4	3			breeds in MI vernal pools (K)
				March in MI (K), found through glaciated northeast (LAR), in ON, in permanent water and not in nearby vernal pools (J67)
				February-April in MI (K), occurs throughout the glaciated northeast, although localized in distribution (LAR)
4				
	3		nonbreeding migrant	
2	1			genus present in European temporary pools (Baz)
				infrequently collected, known from MA, NJ, IL, MI (LAR)
			nonbreeding migrant	associated with dense vegetation (LAR)
2			nonbreeding migrant	
2			nonbreeding migrant	widespread, more typical of grassland than forest pools, in dense vegetation (LAR)
			nonbreeding migrant	found throughout glaciated northeast
			eggs	found in small, barren peat pools throughout glaciated northeast (LAR)
			nonbreeding migrant	ubiquitous in pools with dense emergent vegetation (LAR)
	2			in temporary ponds in DE (FCF), found throughout glaciated northeast (LAR)
			nonbreeding migrant	in temporary and permanent ponds in DE (FCF)
2	3			
				January-May in MI (K)

(continued)

Higher Taxa	Genus/Species	Vernal Pool Study [1]	State or Province Studied [2]	Hydrologic Class [3], Setting [4], and Bottom [5]
	Hygrotus sp.	S&F, EAC	MA, WI	ls/w/h, sp
	Hygrotus impressopunctatus (Schaller 1783)	J69, K, L28	ON, MI, NY	ss/w/l
	Hygrotus laccophilinus (LeConte 1878)	J69, L28, S&F	NY, ON, WI	ss/w/l, ss/w, p
	Hygrotus nubilis (LeConte 1855)	EAC, W, L28	MA, NY, ON	ss/w/l
	Hygrotus sayi Balfour-Browne 1944	J69, K, S&F	MI, ON WI	ss/w/l, sa/o/h, p(sp?)/oh, sp/w
	Hygrotus sylvanus (Fall 1917)	L28	NY	woodland pools
	Hygrotus turbidus (LeConte 1855)	J69, K	MI, ON	ss/w/l
	Ilybius sp.	EAC, W	MA, ON	ss/w, sp/w
	Ilybius biguttulus (Germar 1824)	J69	ON	ss/w/l
	Ilybius discedens Sharp 1882	J66, J69	ON	ss/w/l, sp/w, sp/o
	Ilybius ignarus (LeConte 1862)	J69, L28	NY, ON	ss/w/l
	Laccophilus sp.	EAC, W DDW, S&F	MA, ON, WI	ss/o/h, ss/w/l, sp/w, sp/wo/hm, p/ox
	Laccophilus biguttatus Kirby 1837	J69	ON	ss/w/l
	Laccophilus fasciatus Aubé 1838	Z60, Z70	IN	la/o/l, temp
	Laccophilus maculosus Say 1823	W, J69, K	MI, ON	ss/w/l, p(sp?)/o/h, pfish/h
	Laccophilus proximus Say 1823	Z60, Z70	IN	temp, p
	Laccophilus undatus Aubé 1838	Z70	glaciated northeast	woodland pools

Life History			Drought-Resisting Strategy or Stage	Comments
A[6]	B[7]	C[8]		
			nonbreeding migrant	as *H. similis* Kirby 1837, in May in MI (K), reportedly prefers permanent water, found throughout the glaciated northeast (LAR)
			nonbreeding migrant	found throughout the glaciated northeast, appears to prefer long-hydroperiod ponds (LAR)
			nonbreeding migrant	found throughout the glaciated northeast (LAR), in temporary and permanent ponds in DE (FCF)
			nonbreeding migrant	uncommon in temporary habitats, found throughout glaciated northeast (LAR)
				rarely collected, specimens from NY, MA, WI, MN, ON and QB (LAR), long believed to be extinct (Anderson)
			nonbreeding migrant	reportedly a species of well-vegetated, permanent ponds, found throughout the glaciated northeast (LAR)
			nonbreeding migrant	genus present in European temporary pools (Baz, Davy-Bowker)
			nonbreeding migrant	found throughout the glaciated northeast (LAR)
			nonbreeding migrant	a boreal species that occurs from the northern limit of trees into northern New England and the upper Midwest (LAR)
			nonbreeding migrant	found in New England, NY, the upper Midwest, IL, southern ON and QB, and NS (LAR)
4				larvae in ss/w/l June in ON (W), genus in Italian temporary pools (Baz)
			nonbreeding migrant	found in the northern portions of the glaciated northeast including New England and the upper Great Lakes states, usually in permanent water (LAR)
4			nonbreeding migrant	larvae in June in MI floodplain pool (K), found in permanent and temporary waters, colonizes newly flooded areas (Z60, Z70), occurs throughout the glaciated northeast, usually in permanent waters (LAR)
				occurs throughout the glaciated northeast (LAR), readily colonizes newly flooded pools in unglaciated areas of IN (Z70)
			migratory breeder?	not known from Canada but may occur in southern ON (LAR), not found in anthropogenic pools (Z60), a common species in temporary forest pools (Z70)

(continued)

Higher Taxa	Genus/Species	Vernal Pool Study [1]	State or Province Studied [2]	Hydrologic Class [3], Setting [4], and Bottom [5]
	Laccornis sp.	EAC, S&F	MA, WI	sp/w, la/o/h
	Laccornis difformis (LeConte 1855) (as *Agoporus*)	L28	NY	ss/w/l
	Liodessus sp.	EAC, S&F	MA, WI	ss/w/l, sp/w
	Liodessus affinis (Say 1823)	J69	ON	ss/w/l
	Liodessus fuscatus (Crotch 1873)	EAC, W, J61, L28	MA,ON, NY	ss/w/l, ls/wo/hD, sp/wo/hD
	Matus bicarinatus (Say 1823)	L28	NY	ss/w
	Neoporus lobatus (Sharp 1882) (as *Hydroporus*)	K	MI	sa/o/h, p(sp?)/o/h
	Neoporus undulatus (Say 1823) (= *Hydroporus consimilis* LeConte 1850)	W, EAC, K, J69	MA, MI, ON	ss/w, sa/o/h, sp/wo/hm, p(sp?)/o/h
	Neoscutopterus sp.	DDW	ON	ss/o/h
	Rhantus sp.	K, EAC, W, S&F, DDW, J66	MA, MI, ON, NY	ss/w/l, ss/o/h, sa/o/h, p(sp?)/o/h, p/ox, sp
	Rhantus binotatus (Harris 1828)	J67, J69	ON	ss/w/l, p
	Rhantus consimilis Motschulsky 1859 (in J, K as *R. tostus* LeConte 1866)	J67, K, W, EAC, S&F	MA, MI, ON, WI	ss/w/l, p/ox, sp, p
	Stictotarsus griseostriatus (DeGeer 1774) (as *Deronectes*)	K, W?	MI, ON	ss/w, sa/o/h
	Thermonectus basillaris (Harris 1829)	EAC	MA	ss/w/l
	Uvarus sp.	EAC	MA	ss/w/l, sp/wo/hm
	Uvarus suburbanus (Fall 1917)	J61	ON	ss/w/l
Elmidae				
	Stenelmis sp.	DDW	ON	ss/o/h
Georyssidae		DDW	ON	ss/o/h
	Georyssus sp.	EAC	MA	ss/fp/l, sp/o/sh
Gyrinidae				
	Dineutus sp.	EAC, S&F	MA, WI	sp/w, sp/wo/hm
	Dineutus assimilis Kirby 1837	K	MI	sa/o/h, p(sp?)/o/h
	Dineutus nigrior Roberts 1895	K, EAC	MA, MI	sa/o/h, sp/wo/l, pfish/h
	Gyrinus sp.	K, EAC, S&F, DDW	MA, MI, ON, WI	ss/w/l, ss/o/h, sa/w, ls/w, la/w, sp/w, p/w

Life History A[6]	Life History B[7]	Life History C[8]	Drought-Resisting Strategy or Stage	Comments
				"Occurs in woodland pools among rotted leaves and is very sluggish in its movements" (L28)
				genus in Australian temporary pools (Lake), related genus *Bidessus* present in European temporary pools (Baz)
			nonbreeding migrant	
			nonbreeding migrant	in DE, found in temporary and permanent ponds early in the season and in temporary ponds later in the season (FCF)
2			nonbreeding migrant	
	2			
2	3		eggs?	genus in Australian and Italian temporary pools (Lake, Baz)
				occurs throughout the glaciated northeast, mostly in small seeps and springs (LAR)
2			eggs or adult, possibly nonbreeder in some sites	found in ON and QB in Canada, and in the northern parts of the eastern US, but absent from the lower Midwest, typically in grasslands not forests (LAR)
2?				widely distributed through the glaciated northeast on mineral substrates in physically disturbed habitats (LAR), genus in CA vernal pools (Z)
				common in clear temporary ponds (LAR)
				type locality in woodland pools (as *Deronectes*)
	4			late in the season, rare
			migrants	
4				
				May in MI (K)
		2,3,4		

(continued)

Higher Taxa	Genus/Species	Vernal Pool Study [1]	State or Province Studied [2]	Hydrologic Class [3], Setting [4], and Bottom [5]
	Gyrinus affinis Aubé 1838	J61, J69, EAC, S&F	MA, ON, WI	ss/w/l, sp/w
	Gyrinus borealis Aubé 1838	EAC	MA	ls/wo/l, sp/w/l, p/ox
	Gyrinus lecontei Fall 1922	J61, J66, J69 S&F, K, W	MI, ON WI	ss/w/l, sa/o/h, sp, p(sp?)/o/h, sp/w
Haliplidae				
	Haliplus sp.	W, S&F, DDW, K	MI, ON, WI	ss/w/l, sa/o/h, ls/w, sp/w
	Haliplus blanchardi Roberts 1913	EAC, K	MA, MI	sa/o/h, p(sp?)/o/h
	Haliplus borealis LeConte 1850	EAC	MA	sp/wo/l
	Haliplus connexus Matheson 1912	K	MI	p(sp?)/o/h
	Haliplus immaculicollis Harris 1828	EAC, K, W J69, S&F	MA, MI, ON, WI	ss/w/l, ls/w/h, ls/wo/lsh, p(sp?)/o/h, sp/w
	Haliplus longulus LeConte 1850	J69, W , K, S&F	MI, ON, WI	ss/w/l, p(sp?)/o/h, sp/w
	Haliplus ohioensis Wallis 1933	W lit.	IN	temp
	Haliplus pantherinus Aubé 1838	K	MI	p(sp?)/o/h
	Haliplus subguttatus (Fabricius 1801)	K	MI	sa/o/h, sp/wo/l
	Peltodytes sp.	W	ON	ss/w/l
	Peltodytes edentulus (LeConte 1863)	EAC, K	MA, MI	ss/w/l, p(sp?)/o/h, p
	Peltodytes muticus (LeConte 1863)	EAC	MA	ls/wo/h, sp/w/hm
	Peltodytes pedunculatus (Blatchley 1910)	K	MI	ss/w/l, p(sp?)/o/h
	Peltodytes tortulosus Roberts 1913	EAC, W	MA, ON	ss/wo/l, ls/w/h
Helophoridae				
	Helophorus sp.	EAC, W, S&F	MA, ON, WI	ss/w/l, ss/o/h
	Helophorus aquaticus (Linnaeus 1758)	W, K	MI, ON	ss/w/l
	Helophorus grandis (Illiger 1798)	DDW	ON	ss/o/h
	Helophorus lacustris LeConte 1850	W	ON	ss/w/l
	Helophorus lineatus Say 1823	K?, W	MI, ON	ss/w/l
	Helophorus orientalis Motschulsky 1860	DDW	ON	ss/o/h
Hydraenidae				
	Hydraena sp.	EAC	MA	p/w/x

Life History A[6]	Life History B[7]	Life History C[8]	Drought-Resisting Strategy or Stage	Comments
4				
4				
	4,5		egg, larva, pupa, or adult	larvae, June, January in MI (K), genus present in European and Australian temporary waters (Baz, Lake)
				common April-May in MI, widespread, CT, MA, MO, MN, NJ, NY, QB, WI (Brigham)
				distributed throughout glaciated northeast (Brigham)
				widespread through glaciated northeast (Brigham)
2			aestivation	occurs April-May in MI (K), is widespread throughout glaciated northeast (Brigham)
				abundant April-June in MI (K)
				occurs in IL, IN, MO, OH
				distributed throughout glaciated northeast (Brigham)
				found in May in MI (K); in glaciated northeast occurs in southern Canada and upper Midwest (Brigham)
2	1			genus known from CA temporary pools (Z)
				January, April-May in MI (K), widely distributed in glaciated northeast
				widespread through glaciated northeast (Brigham)
				widespread through glaciated northeast (Brigham)
				northern, in glaciated northeast found in ME, MI, ON, QB, WI (Brigham)
2	1			genus present in European and CA temporary pools (Baz, Z)
2			terrestrial adults	April-June, MI (K)
	4		terrestrial adults	
2			terrestrial adults	in DE, found late season in open temporary ponds and permanent ponds (FCF)
2			terrestrial adults	April-June in MI (K)
	4,5			
				genus present in Australian temporary pools (Lake)

(continued)

Higher Taxa	Genus/Species	Vernal Pool Study [1]	State or Province Studied [2]	Hydrologic Class [3], Setting [4], and Bottom [5]
Hydrochidae				
	Hydrochus sp.	S&F, EAC	MA, WI	ss/w/h, sp/wo/hm, sp/w, sp/wo/D
	Hydrochus currani Brown 1929	K, EAC	MI, MA	ss/w/l, ss/w/h, sp/wo/hm
	Hydrochus squamifer LeConte 1855	K, S&F	MI, WI	ss/w/l, sp/w
Hydrophilidae				
	Anacaena sp.	DDW, S&F	ON, WI	ss/o/h, sp
	Anacaena limbata (Fabricius 1792)	EAC, W, DDW, J69	MA, ON	ss/w/l, ss/o/h, ls/wo/lhs
	Berosus sp.	EAC, J69	MA, ON	ss/w/l, p/o/x
	Berosus fraternus LeConte 1855	DDW	ON	ss/o/h
	Berosus striatus (Say 1825)	K	MI	ss/w, sa/o/h
	Cercyon sp.	EAC, K	MA, MI	ss/w/l, sl/w/h, sp
	Cymbiodyta sp.	W, J69, EAC	MA, ON	ss/w/l
	Enochrus sp.	K, EAC	MI, MA	ss/w/l, p(sp?)/o/h, sp/o/h
	Enochrus cinctus (Say 1824)	J69	ON	ss/w/l
	Enochrus hamiltoni (Horn 1890)	EAC, W	MA, ON	ss/w/l
	Enochrus ochraceus (Melsheimer 1846)	W, EAC, K	MA, MI, ON	ss/w/l, ss/wo/h, ls/wo/D, sp/wo/hD, sp/wo/h
	Helocombus bifidus (LeConte 1855)	EAC	MA	ls/wo/hD, sp/w/hD
	Hydrobius sp.	EAC, S&F	MA, WI	ss/w, ls/wo/D, ls/wo/hD, sp/w/l, sp/w/h
	Hydrobius fuscipes (Linnaeus 1758)	J69, W	ON	ss/w/l
	Hydrochara sp.	W, EAC, DDW, K, S&F, J69	MA, MI, ON, WI	ss/w/l, ss/o/h, ls/wo/lhs, sp/w
	Hydrochara obtusatus (Say 1823)	J61, J67	ON	ss/w
	Hydrophilus sp.	K, S&F	MI, WI	sa/o/h, sp/w
	Paracymus subcupreus (Say 1825)	EAC	MA	ls/wo/hD, sp/wo/hD, sp/wo/h
	Tropisternus sp.	S&F, K	MI, WI	ss/w, ls/w, sa/o/h, sp/w
	Tropisternus blatchleyi modestus d'Orchymont 1922	EAC	MA	ss/w/h, ls/o/D, ls/o/h, sp/o/hm
	Tropisternus mixtus (LeConte 1855)	EAC, W, J69, K	MA, MI, ON	ss/w/l, sa/o/h, ls/w, sp/w
	Tropisternus natator d'Orchymont 1938	J69, EAC	MA, ON	ss/w/l, ls/w/h, sp/wo/hm, sp/wo/h

Life History A[6]	Life History B[7]	Life History C[8]	Drought-Resisting Strategy or Stage	Comments
				genus present in Australian and Italian temporary pools (Lake, Baz), in temporary and permanent pools in DE (FCF)
				April-June in MI (K)
				April-June in MI (K)
	2			genus present in European temporary pools (Baz)
2	2	5	nonbreeder, overwinters as adult	
				genus present in Australian, Italian, and CA temporary pools (Lake, Baz, Z)
	4			
4				*C. minima* and *C. vindicata* in temporary pools in DE (FCF), genus in Italian temporary pools (Baz)
				genus present in Australian temporary pools (Lake)
		5	nonbreeding migrant	
4				
4				
				genus present in European temporary pools (Baz)
2		5	nonbreeding migrant	
4	4			
		4?		
			adults overwinter (Matheson)	larvae in June in MI (K), genus in European temporary pools (Baz)
				genus present in Australian temporary pools (Lake)
				genus present in CA temporary pools (Z)
4		5	adults overwinter in permanent water	nonbreeder in vernal pools
		5	adults overwinter in permanent water	nonbreeder in vernal pools

(continued)

Higher Taxa / Genus/Species	Vernal Pool Study [1]	State or Province Studied [2]	Hydrologic Class [3], Setting [4], and Bottom [5]
Noteridae			
Hydrocanthus iricolor Say 1823	K, EAC	MA, MI	la/w/h, sp/wo/hm, sp/wo/h, p(sp?)/o/h
Scirtidae (formerly **Helodidae**)	EAC	MA	ls/wo/h, sp/wo/h, la/wo/D, sp/wo/hD, sp/wo/hm
Cyphon sp.	W, EAC, DDW	ON, MA	ss/w/l, ss/o/h, la/w/hl, la/o/sS
Prionocyphon limbatus LeConte 1866	EAC	MA	la/o/sS
Scirtes sp.	EAC	MA	la/o/sS
Order Diptera			
Anthomyidae?	DDW	ON	ss/o/h
Ceratopogonidae			
Alluaudomyia sp.	W	ON	ss/w/l
Atrichopogon sp.	EAC	MA	p/w/x
Atrichopogon cf. *geminus* Boesel 1973	W	ON	ss/w/l
Bezzia/Palpomyia sp.	K	MI	ss/w/l, sa/o/h, p(sp?)/o/h
Bezzia sp.	EAC, W	MA, ON	ss/w/l
Palpomyia sp.	EAC	MA	ss/w/l
Palpomyia lineata (Meigen 1818) group	W	ON	ss/w/l
Chaoboridae			
Chaoborus sp.	DDW, EAC	MA, ON	ss/o/h, ss/w/l, sp/wo/hm, sp/wo/D, sp/wo/hS
Chaoborus americanus (Johannsen 1903)	W, EAC, K	MA, MI, ON	ss/w/l, p(sp?)/o/h, p(sp?)/w/l, sp/wo/hD, sp/wo/hm
Chaoborus flavicans (Meigen 1830)	W	ON	ss/w/l, la(sp?)/w
Chaoborus trivittatus (Loew 1862)	W	ON	ss/w/l, sa(sp?)/o/h
Eucorythra underwoodi Underwood 1903	EAC, Borkent	MA, glaciated northeast	ss/w/l, ls/wo/l
Mochlonyx sp.	EAC, W, S&F	MA, ON, WI	ss/w/l, ls/w, sp/w, ls/w/h, la/o/h, sp/wo/D, sp/wo/hS, p/ox
Mochlonyx cinctipes (Coquillett 1903)	EAC, W, DDW	MA, ON	ss/w/l, ss/o/h, la(sp?)/wo/h
Mochlonyx fuliginosus (Felt 1905)	Borkent	glaciated northeast	temp
Mochlonyx velutinus (Ruthe 1831)	W, EAC, J66	ON, MA	ss/w/l, sp/w, sp/o

Life History A[6]	Life History B[7]	Life History C[8]	Drought-Resisting Strategy or Stage	Comments
				largely in permanent ponds in DE (FCF)
2	2		terrestrial adults?	
	2			family in CA vernal pools (Z)
			eggs?, larvae	
2				
2				
2			larvae	
				in European temporary pools (Baz)
2			larvae	
				known from Australian vernal pools (Lake)
2			larvae?	
	4		most *Chaoborus* overwinter as third-instar larvae in permanent waters, western species *C. cooki* Saether 1970 overwinters in egg stage in Alberta (W lit.)	genus in CA temporary pools (Z)
4				
4			possibly overwintering eggs?	also in European temporary pools (Baz)
4				
				restricted to small woodland ponds (Borkent)
3		1	eggs	
3	2	1	eggs laid on bottom in spring, undergo partial development in summer, hatch when flooded	
		1	eggs	
3		1	eggs	

(continued)

Higher Taxa	Genus/Species	Vernal Pool Study [1]	State or Province Studied [2]	Hydrologic Class [3], Setting [4], and Bottom [5]
Chironomidae				
Chironominae				
	Chironomus sp.	EAC, W	MA, ON	ss/w/l
	Chironomus decorus Johanssen 1905	K	MI	ss/w/l
	Chironomus stigmaterus Say 1823	?W	ON	ss/w/l
	Chironomus tuxis Curran 1930	W	ON	ss/w/l
	Cryptochironomus sp.	K	MI	p(sp?)/o/h
	Einfeldia? dorsalis (Meigen 1818)	DDW	ON	ss/o/h
	Endochironomus nigricans (Johannsen 1905)	W	ON	ss/w/l
	Glyptotendipes sp.	W	ON	ss/w/l
	Limnochironomus sp.	K	MI	ss/w/l, sa/o/h
	Microtendipes pedella (DeGeer 1776)	K	MI	p(sp?)/o/h
	Parachironomus sp.	W, DDW	ON	ss/w/l, ss/o/h
	Paratanytarsus spp.	W	ON	ss/w/l
	Phaenopsectra sp.	W, DDW	ON	ss/w/l, ss/o/h
	Polypedilum spp. including *P. fallax* (Johannsen 1905) group	W, DDW	ON	ss/o/h
	Pseudosmittia sp.	DDW	ON	ss/o/h
	Tanytarsus sp.	W	ON	ss/w/l
Diamesinae				
	Prodiamesa sp.	W	ON	ss/w/l
Orthocladiinae		K	MI	ss/w/l, sa/o/h, p(sp?)/o/h
	Acricotopus nitidellus (Malloch 1915)	W	ON	ss/w/l
	Corynoneura sp.	DDW	ON	ss/o/h
	Cricotopus spp. including *C. sylvestris* (Fabricius 1794)	W, K, DDW	MI, ON	ss/w/l, ss/o/h, sa/o/h, p(sp?)/o/h
	Eukiefferella sp.	DDW, W	ON	ss/o/h, ss/w/l
	Hydrobaenus spp. including *H. pilipes* (Malloch 1915)	W	ON	ss/w/l
	Limnophyes sp.	W	ON	ss/w/l
	Paraphenocladius sp.	W	ON	ss/w/l
	Psectrocladius spp.	W	ON	ss/w/l
	Smittia sp.	W	ON	ss/w/l
	Trissocladius sp.	DDW	ON	ss/o/h
Tanypodinae				
	Ablabesmyia sp.	EAC, DDW	MA, ON	ss/o/h, la/w/l, p/wx
	Ablabesmyia monilis (Linnaeus 1758)	W	ON	ss/w/l

Life History A[6]	Life History B[7]	Life History C[8]	Drought-Resisting Strategy or Stage	Comments
			no egg diapause, larvae overwinter in cocoons	family in CA vernal pools (Z), family with many species in European temporary pools (Baz)
2		1		genus in Australian and European temporary waters (Lake, Baz)
2			larvae	
2			larvae	
	1			
2				
2				
				also occurs in Italian pools (Baz)
2	2		larvae	genus in Australian temporary waters (Lake)
2			larvae	genus in Italian pools (Baz)
2	2		larvae	
2	2		larvae	genus in European temporary pools (Baz)
	1			
2				genus in Australian temporary waters (Lake)
2			larvae	
				subfamily present in Australian temporary pools (Lake)
2			larvae	
	4			
2	4		larvae	genus in Australian and Italian temporary waters (Lake, Baz)
2				genus in Australian temporary waters (Lake)
2			larvae	
3			larvae	
3			larvae	
2			larvae	genus in Italian temporary pools (Baz)
3			larvae	
	2			
			most presumed to develop from eggs laid by adults from permanent waters (W)	
	4	4	larvae overwinter in flooded pools, do not resist pod drying (W)	
4		4		

(continued)

Higher Taxa	Genus/Species	Vernal Pool Study [1]	State or Province Studied [2]	Hydrologic Class [3], Setting [4], and Bottom [5]
	Guttipelopia sp.	W	ON	ss/w/l
	Procladius sp.	W	ON	ss/w/l, sa(sp?)/o/h, la(sp?)/o/l
	Psectrotanypus sp.	DDW	ON	ss/o/h
	Psectrotanypus dyari (Coquillett 1902)	W	ON	ss/w/l
Tanytarsinae				
	Micropsectra sp.	DDW	ON	ss/o/h
Corethrellidae				
	Corethrella brakeleyi (Coquillett 1902)	EAC	MA	sp/wo/hS
Culicidae				
	Aedes sp.	S&F, K, W	MI, ON, WI	ss/w, ls/w, sa/w, la/w, sp/w, p(sp?)/o/h, p/w
	Aedes abserratus (Felt and Young 1904)	E&C	MA	s?s/w/S
	Aedes aurifer (Coquillet 1903)	E&C	MA	s?s/o
	Aedes canadensis (Theobald 1901)	E&C, W lit., EAC, Worth	MA, NJ, ON	ss/w/l
	Aedes cinereus Meigen 1818	E&C	MA	s?s/w/o
	Aedes communis (De Geer 1776)	E&C, J66, M&M	MA, ON	sp, s?s/w/l
	Aedes diantaeus Howard, Dyar and Knab 1913	E&C, J66	MA, ON	s?s/w/l, sp
	Aedes euedes Walker 1913	J66 (as *Ae. barri*), W	ON	ss/w/l, sa(sp?)/o/h, sp
	Aedes excrucians (Walker 1856)	E&C, W, J61, J66, EAC, K	MA, MI, ON	ss/w/l, sa(sp?)/o/h, ls/wo/hD, la(sp?)/o/h, sp/wo/hS, sp
	Aedes fitchii (Felt and Young 1904)	W, DDW, E&C, K, J66	MA, MI, ON	ss/w/l, ss/o/h, sa/o/h, sa(sp?)/o/h, la(sp?)/o/l, sp
	Aedes flavescens (Müller 1764)	Wood et al.	Canada	temp
	Aedes grossbecki Dyar and Knab 1906	Worth	NJ	s?s/w
	Aedes impiger (Walker 1848)	K	MI	ss/w/l, sa/o/h
	Aedes implicatus Vockeroth 1954	Rodney, E&C	MA, MN	ss/w/l
	Aedes intrudens Dyar 1919	J66, E&C	MA, ON	ss/w/l, sp
	Aedes provocans (Walker 1848)	W, E&C (as *Ae. trichurus* Dyar), J61, J66, DDW	MA, ON	ss/w/l, ss/o/h, sa(sp?)/o/h, sp
	Aedes punctor (Kirby 1837)	E&C, Horsfall	MA, ON	ss/w/l, s?s/w/S

Life History			Drought-Resisting Strategy or Stage	Comments
A[6]	B[7]	C[8]		
2			larvae	
4			not drought tolerant (W)	genus in Australian and European temporary waters (Lake, Baz)
	4		not drought tolerant (W)	genus in European temporary waters (Baz)
4				
	4			genus in European temporary waters (Baz)
3			eggs	genus in Australian temporary waters (Lake)
			eggs	vernal pools in coniferous forest, especially in *Sphagnum* bogs (E&C)
			eggs	vernal pools in cranberry bogs (E&C)
3			eggs	vernal pools in mixed forest, especially in *Sphagnum* bogs (E&C)
			eggs	vernal pools in mixed forest especially in *Sphagnum* bogs (E&C)
			eggs	vernal pools in deciduous forest (E&C)
			eggs	vernal pools in deciduous forest (E&C)
3		1	eggs on dry pool bottom	
3		1	eggs on dry pool bottom	vernal pools in mixed forest and grassy swamps (E&C)
3	2	1	eggs on dry pool bottom	vernal pools in mixed forest and grassy swamps (E&C)
				found through glaciated northeast (Darsie)
				coastal shore woodland pool (Worth)
				deciduous forest pools (E&C)
				mixed forest (E&C)
3	2	1	eggs on dry pool bottom	pools in deciduous forest (E&C)
				coniferous forest, especially *Sphagnum* bogs (E&C) and sloping unvegetated depressions (Horsfall)

(continued)

Higher Taxa	Genus/Species	Vernal Pool Study [1]	State or Province Studied [2]	Hydrologic Class [3], Setting [4], and Bottom [5]
	Aedes sticticus (Meigen 1838)	E&C, DDW, J66	MA, ON	ss/o/h, a/w, sp
	Aedes stimulans (Walker 1848)	W, J61, E&C, D&B	MA, ON, NY	ss/w/l, sa(sp?)/o/h, la(sp?)/o/l, la/w/l, sp
	Aedes trivittatus (Coquillet 1902)	J66	ON	temp
	Aedes vexans (Meigen 1830)	W, E&C	MA, ON	ss/w/l, ss/o/h
	Anophleles sp.	S&F	WI	sp/w
	Anopheles earlei Vargas 1943	W	ON	sa(sp?)/o/h
	Anopheles punctipennis (Say 1823)	W	ON	sa(sp?)/o/h
	Anopheles quadrimaculatus Say 1824	E&C	MA	sp, p
	Anopheles walkeri Theobald 1901	W	ON	la/o/h, a(sp?)/o/h, sp/o/h
	Culiseta melanura (Coquillet 1902)	DPH	MA	p, (temp)
	Culiseta morsitans (Theobald 1901)	W (as *Culex*), J66	ON	la/w/l, sp/w, s?f, sp
	Culex sp.	EAC, S&F	MA, WI	ss/w, sp/w
	Culex apicalis Adams 1903	K	MI	ss/w/l
	Culex territans Walker 1856	W, E&C	MA, ON	la(sp?)/o/h, sp/wo/hS
	Psorophora sp.	W lit., EAC	MA	ls/w/h
	Uranotaenia sapphirina (Osten Sacken 1868)	DPH, EAC	MA	ls/w/h
Dixidae				
	Dixa sp.	K, W, EAC	MA, MI, ON	ss/w/l, ss/wo/sS
	Dixella sp.	W	ON	ss/w/l
Dolichopodidae		W, K	MI, ON	ss/w/l
Ephydridae		W, K	MI, ON	ss/w/l, sa/o/h
	Hydrellia sp.	W	ON	ss/w/l
	Lycoria sp.	K	MI	ss/w/l
Ptychopteridae		EAC	MA	fp/wl, p/wx
Sciomyzidae		K, W	MI, ON	ss/w/l
	Anticheta melanosoma Melander 1920	W lit.		temp, s
	Atrichomelina pubera (Loew 1862)	W	ON	sa(sp?)/o/h
	Dictya sp.	W	ON	la(sp?)/o/h, a

Life History A[6]	Life History B[7]	Life History C[8]	Drought-Resisting Strategy or Stage	Comments
	2	1	eggs on dry pool bottom	reflood species, especially in ground pools in wooded or semiwooded floodplains (E&C)
3	3	1	eggs on dry pool bottom	vernal pools in mixed forest and grassy swamps (E&C)
			eggs	found in rainpools
3		1	eggs on dry pool bottom	mostly in ruts and shallow grassy depressions (E&C)
				genus in Australian and Italian temporary pools (Baz, Lake)
3		2	adults overwinter on land	
3		2	adults overwinter on land	
3		1	eggs in water	
			not drought tolerant	overwinters in water as larva, prefers permanent wetland pools (DPH)
3		2	adults overwinter, eggs in water	genus known from CA temporary waters (Z)
		2	adults overwinter on land	genus known from Australian and CA temporary waters (Lake, Z)
3		2	winter adults, eggs in water, no diapause	
3		2	eggs	
3?,4?				
3,4?				
				members of this group are predators on molluscs, family found in Australian temporary pools (Lake)
3			winter puparium diapause	vernal pool specialist (W)
3				
3			reproduce year-round, no diapause, adults presumed to migrate from permanent water	

(continued)

Higher Taxa	Genus/Species	Vernal Pool Study [1]	State or Province Studied [2]	Hydrologic Class [3], Setting [4], and Bottom [5]
	Hedria mixta Steyskal 1954	W lit.		temp, s
	Pherbellia sp.	W	ON	sa(sp?)/o/h
	Renocera sp.	W lit.		
	Sepedon sp.	K	MI	ss/w/l, sa/o/h
	Sepedon fuscipennis Loew 1859	W	ON	ss/w/l
	Tetanocera loewi Steyskal 1959	W	ON	ss/w/l, sa(sp?)/o/h, a
	Tetanocera vicina Macquart 1843	W	ON	a(sp?)/o/h, a
Stratiomyiidae		EAC	MA	ss/w/l, la/w/l
	Hedriodiscus truquii (Bellardi 1859) (as *Eulalia*)	K	MI	ss/w/l
	Hedriodiscus vertebratus (Say 1824) (as *Eulalia*)	K	MI	ss/w/l, sa/o/h, p(sp?)/o/h, pfish
	Odontomyia sp.	EAC, W, DDW	MA, ON	ss/w/l, ss/o/h, la/o/h
	Odontomyia cincta Olivier 1812 (as *Eulalia*)	K	MI	ss/w/l, p(sp?)/o/h
	Odontomyia virgo (Wiedemann 1830)	W	ON	ss/w/l
Tabanidae				
	Chrysops sp.	K, EAC	MA, MI	sp/wo/hS, p(sp?)/o/h
	Tabanus sp.	K, EAC	MA, MI	ls/w/h, sa/o/h
	Tabanus similis Macquart 1850	W	ON	ss/w/l
Tipulidae		EAC, K	MA, MI	ss/w/l, ls/w/h, sa/o/h, sp/wo/hm, sp/wo/hS, sp/wo/hsS, pfish/h
	Limnophila sp.	K	MI	ss/w/l, pfish/h
	Nephrotoma sp.	W	ON	ss/w/l
	Tipula sp.	DDW	ON	ss/o/h
	Tipula (Pterelachisus) spp.	W	ON	ss/w/l
	Tipula (Yamatotipula) sp.	W	ON	ss/w/l
Order Ephemeroptera				
Arthropleidae				
	Arthroplea bipunctata (McDunnough 1924)	B&G, EAC	MA, ME	la/w//lR, la/w/SR
Baetidae		DDW	ON	ss/o/h
	Callibaetis sp.	K, EAC	MA, MI	p(sp?)/o/h, sp/w/hD, sp/wo/hm
	Callibaetis ferrugineus (Walsh 1862)	W	ON	ss/w/l, la(sp?)/o/h
	Callibaetis fluctuans (Walsh 1862)	EAC	MA	sp/w/hm
	Centroptilum sp.	EAC	MA	sp/wo/hS
	Cloeon sp.	DDW, EAC	MA, ON	ss/o/h, ls/w/h
	Procloeon sp.	EAC	MA	ls/w/D, ls/w/h, ls/o/h, ls/w/sS, sp/w/hS, sp/wo/hm, temp

Life History A[6]	Life History B[7]	Life History C[8]	Drought-Resisting Strategy or Stage	Comments
3		1	egg diapause	species is confined to vernal pools (W)
3		1	puparium	feed on stranded snails (larvae drown in water)
3		1	pupal diapause	prey on sphaeriids, 2 species in temporary pools (W)
3		2	multivoltine, go to other, permanent habitats in summer, reproductive diapause winter	
3		2?	adult reproductive diapause, eggs	
3		2?	adult reproductive diapause	
3				family in Australian temporary pools (Lake)
2	4		larva?	
			larva?	
			larva?	
				genus in European temporary pools (Baz)
2		2	larvae	
?				many in family are terrestrial, hard to determine life histories (W)
	4			genus known from CA temporary pools (Z)
	4			
				November-April in MI (K), genus in CA temporary pools (Z)
4				
4				
	4			genus in Italian temporary pools (Baz)

(continued)

Higher Taxa	Genus/Species	Vernal Pool Study [1]	State or Province Studied [2]	Hydrologic Class [3], Setting [4], and Bottom [5]
Caenidae				
	Caenis sp.	EAC, K	MA, MI	p(sp?)/o/h, p/xo, pfish/h
Ephemerellidae				
	Eurylophella funeralis (McDunnough 1925)	EAC	MA	sa/w/l
Heptageniidae				
	Stenonema sp.	DDW	ON	sa/o/h
Leptophlebiidae				
	Leptophlebeia sp.	W	ON	sf(sp?)/o/h
	Leptophlebeia cupida (Say 1823)	Burks	IL	temp
	Leptophlebeia nebulosa (Walker 1853)	Burks	IL	temp
	Paraleptophlebeia praepedita (Eaton 1884)	W	ON	ss/w/l
	Paraleptophlebeia volitans (McDunnough 1924)	EAC	MA	sa/w/l
Siphlonuridae				
	Siphlonurus spp.	W, EAC	MA, ON	fp/o/h, a, sa/o/h, sp/o/l, temp
Order Hemiptera				
Belostomatidae		DDW	ON	ss/o/h
	Belostoma sp.	K, S&F	MI, WI	ss/w/l, sa/o/h, p(sp?)/o/h, sp/w
	Belostoma flumineum Say 1832	EAC, W, K	MA, MI, ON	ss/w/l, sa/o/h, a(sp?)/o/h
	Lethocerus americanus (Leidy 1847)	EAC, W, S&F	MA, ON, WI	ss/w/l, a/(sp?)/o/h, sp/w/D, sp/w
Corixidae				
	unidentified nymphs	K	MI	ss/w/l, sa/o/h, p(sp?)/o/h
	Callicorixa sp.	S&F	WI	ss/w, sp/w
	Callicorixa alaskensis Hungerford 1926	W	ON	ss/w/l
	Callicorixa audeni Hungerford 1928	W, J66	ON	ss/w/l, sp/o
	Hesperocorixa sp.	EAC, S&F	MA, WI	ss/w/l, ls/w/D, ls/w/h, ls/w, sp/w/D, sp/wo/l, sp/wo/hS, sp/w/hm, sp/w, p/w
	Hesperocorixa atopodonta (Hungerford 1927)	W	ON	ss/w/l
	Hesperocorixa kennicottii (Uhler 1897)	W	ON	a(sp?)/o/h
	Hesperocorixa michiganensis (Hungerford 1926)	W	ON	ss/w/l
	Hesperocorixa vulgaris (Hungerford 1925)	W, K, EAC	MA, MI, ON	ss/w/l, sa/o/h, sp/w/hm, p(sp?)/o/h
	Palmacorixa sp.	EAC	MA	ss/w/l, sp/w/hm
	Palmacorixa nana Walley 1930	W	ON	a(sp?)/o/h
	Sigara sp.	K, S&F	MI, WI	ss/w/l, ls/w, sp/w, sp/w/hm
	Sigara alternata (Say 1825)	EAC, W, K	MA, MI, ON	ss/w/l, ls/w/D, p(sp?)/o/h, sp/w/D
	Sigara decoratella (Hungerford 1926)	W	ON	ss/w/l

Life History A[6]	Life History B[7]	Life History C[8]	Drought-Resisting Strategy or Stage	Comments
	4			
2				
2				
2			diapausing egg	
2			egg?	
2				
	4			family in CA vernal pools (Z)
				nymphs in June in MI (K)
4			adult	May-June in MI (K)
4				
		3	overwinter as adults in permanent water	family in Italian and CA pools (Baz, Z)
				May-June in MI (K)
4				
4				
4				
4				
4				
4				May in MI (K) (as *Sigara*)
4				
				June in MI (K), genus in Australian, European, and CA temporary waters (Lake, Baz, Z)
4				January in MI (K)
4				

(continued)

Higher Taxa	Genus/Species	Vernal Pool Study [1]	State or Province Studied [2]	Hydrologic Class [3], Setting [4], and Bottom [5]
	Sigara grossolineata Hungerford 1948	W, DDW	ON	ss/o/h, ss/w/l
	Sigara hubbelli (Hungerford 1928)	K	MI	p(sp?)/o/h
	Sigara knighti Hungerford 1948	W	ON	ss/w/l
	Sigara mathesoni Hungerford 1948	W	ON	a(sp?)/o/h
	Sigara modesta (Abbott 1916)	W	ON	ss/w/l
	Sigara solensis (Hungerford 1926)	W	ON	ss/w/l
Gerridae				
	unidentified gerrids	K	MI	sa/o/h
	Gerris sp.	EAC, S&F	MA, WI	ss/w, sa/o/h, ls/w, ls/wo/D, l, s/w/h, sp/w/hm, sp/w
	Gerris buenoi Kirkaldy 1911	W, J66	MI, ON	ss/w/l, sa/o/h, sp/w, sp/o
	Gerris comatus Drake and Harris 1925	W	ON	ss/w/l
	Gerris incurvatus Drake and Harris 1925	K	MI	p(sp?)/o/h
	Gerris marginatus Say 1832	EAC, W, K	MA, MI, ON	ss/w/l, sa/o/h, ls/w/h, ls/wo/hD, sp/w/hD, sp/w/hm
	Gerris remigis Say 1832	EAC, DDW	MA, ON	ss/o/h, ls/o/h, ls/wo/h
	Limnoporus dissortis (Drake and Harris 1930) (as *Gerris*)	W, WSW	ON	ss/w/l
Hebridae		EAC	MA	la/w/sS
Hydrometridae		EAC	MA	ls/w/h
	Hydrometra sp.	W lit., K, EAC	MA, MI, NY	sp/wo/hm, p(sp?)/o/h, temp
Mesoveliidae				
	Mesovelia mulsanti White 1879	EAC, W lit.	MA, NY	ls/wo/hS, ls/wo/D, sp/wo/D, sp/wo/hm, sp/wo/hS, temp
Naucoridae				
	Pelocoris sp.	EAC	MA	sp/wo/hS, sp/wo/hm
Nepidae				
	Ranatra sp.	EAC, S&F	MA, WI	sa/o/h, sp/wo/h, sp/wo/D, sp/wo/hm, sp/w/hS, sp/w, p(sp?)/o/h
	Ranatra fusca Palisot de Beauvois 1820	W, K	MI, ON	ss/w/l, sa/o/h
Notonectidae				
	unidentified notonectids	K, EAC	MA, MI	ss/w/l, sa/o/h, ls/w/h, ls/wo/hD, sp/w/hD, sp/w/hm
	Buenoa confusa Truxal 1953	EAC	MA	la/wo/h, sp/wo/hS, sp/wo/hm
	Notonecta sp.	S&F	WI	sp/w
	Notonecta borealis Hussey 1919	W	ON	a(sp?)/o/h

Life History A[6]	Life History B[7]	Life History C[8]	Drought-Resisting Strategy or Stage	Comments
4	2			
				February in MI (K)
4				
4				
4				
4				
4		3	adults typically overwinter in vegetation adjacent to permanent water	
				May-June in MI (K)
4				genus in CA temporary pools (Z)
4				
4				
4				April-June in MI (K)
	2			
4				vernal pool species entirely macropterous (W)
4				genus occurs in Australian temporary waters (Lake)
4				genus occurs in Australian temporary waters (Lake)
				family in Australian temporary waters (Lake)
4				
				common in June in MI and through summer in MA (K, EAC)
				genus in CA and Italian temporary pools (Z, Baz)
			eggs?	

(continued)

Higher Taxa	Genus/Species	Vernal Pool Study [1]	State or Province Studied [2]	Hydrologic Class [3], Setting [4], and Bottom [5]
	Notonecta insulata Kirby 1867	EAC, W	MA, ON	a(sp?)/o/h, sp/wo/hm
	Notonecta irrorata Uhler 1879	EAC	MA	ls/w/h, sp/wo/hS, sp/wo/hm
	Notonecta kirbyi Hungerford 1925	DDW	ON	ss/o/h
	Notonecta undulata Say 1832	K, EAC, W	MA, MI, ON	ss/w/l, ls/w/h, ls/wo/D, sa/o/h, la(sp?)/o/h, sp/w/sS, sp/wo/hm, p(sp?)/o/h
Pleidae				
	Plea sp.	EAC	MA	ss/w/l, p/ox
	Plea striola Fieber 1844	K	MI	p(sp?)/o/h
Saldidae		W lit.	NY	temp
Veliidae				
	Microvelia sp.	EAC, W	MA, NY	ls/w/D, ls/w/h, ls/o/h, ls/w/sS, sp/w/hS, sp/wo/hm, temp
Order Lepidoptera		K	MI	p(sp?)/o/h
Order Megaloptera				
Corydalidae				
	Chauliodes sp.	EAC	MA	ss/w/sS, ls/wo/h, sp/wo/hS, sp/wo/hm
Order Neuroptera		K	MI	p(sp?)/o/h
Order Odonata — Anisoptera				
Aeshnidae				
	Aeshna sp.	S&F	WI	sp/w
	Aeshna interrupta Walker 1908	EAC	MA	sp/w/l, sp/wo/hm
	Aeshna mutata Hagen 1861	EAC	MA	sp/wo/hm
	Aeshna tuberculifera Walker 1908	EAC	MA	ls/w/h, sp/wo/hS
	Aeshna umbrosa Walker 1908	W	ON	la(sp?)/o/l
	Anax junius (Drury 1770)	EAC, K, W	MA, MI, ON	ls/w/l, sp/wo/hm, sp/wo/D, sp/wo/hS, pfish/h
	Boyeria vinosa (Say 1839)	EAC	MA	ls/w/h, ls/o/h, sp/wo/hS, sp/wo/hm
	Epiaeshna heros (Fabricius 1798)	EAC	MA	ls/w/l, la/w/lR
Gomphidae				
	Arigomphus villosipes (Selys 1854) (as *Gomphus*)	K	MI	p(sp?)/o/h
Libellulidae				
	Erythemis simplicicollis (Say 1839)	EAC	MA	ss/o/h, sp/w/shl, sp/o/h
	Leucorrhinia sp.	S&F	WI	sp/w

Life History A[6]	Life History B[7]	Life History C[8]	Drought-Resisting Strategy or Stage	Comments
			adults overwinter in permanent water	
	2			
4			adults overwinter in permanent water	most common vernal pool species in North America (W), April-June in MI (K)
				genus known from Australian and Italian temporary waters (Lake, Baz)
4				
4				
4				genus known from Australian temporary waters (Lake)
				family in Australian temporary waters (Lake)
	3			
		2,3		genus in Australian temporary waters (Lake)
3				
4			overwinter as adults, migrate to pools in spring	residents use permanent or semi-permanent pools, migrants use vernal pools and breed early, known from CA vernal pools (Z)
				April in MI (K)
				family in Australian temporary waters (Lake)
				found in short-cycle pools during unusually wet years, normally in semi-permanent pools

(continued)

Higher Taxa	Genus/Species	Vernal Pool Study [1]	State or Province Studied [2]	Hydrologic Class [3], Setting [4], and Bottom [5]
	Leucorrhinia intacta (Hagen 1861)	K, EAC	MA, MI	ls/w/hD, p(sp?)/o/h, sp/wo/hS
	Libellula sp.	K, EAC, S&F	MA, MI, WI	ss/w/l, ls/o/h, sp/wo/hm, sp/w
	Libellula pulchella Drury 1770	K	MI	p(sp?)/o/h
	Libellula quadrimaculata Linnaeus 1758	W	ON	la(sp?)/o/l
	Pachydiplax longipennis (Burmeister 1839)	K	MI	ls/w/h, sa/o/h, p(sp?)/o/h, sp/w/hm, sp/wo/hD
	Pantala flavescens (Fabricius 1798)	W lit.	QB	ls/w/l?
	Somatochlora sp.	S&F	WI	sp/w
	Sympetrum ambiguum (Rambur 1842)	EAC, W&C	MA, ON	la/w/h, la/w/l, temp
	Sympetrum costiferum (Hagen 1861)	EAC	MA	ls/w/l, sp/w/l
	Sympetrum danae (Sulzer 1776)	EAC	MA	sp/wx/l
	Sympetrum internum Montgomery 1943	W lit.	ON	a
	Sympetrum obtrusum (Hagen 1867)	K, EAC, W, W lit.	MA, MI, ON, IN	ss/w/l, sa/o/h, p(sp?)/o/h
	Sympetrum rubicundulum (Say 1839)	EAC, W	MA, ON	ss/w/l, la(sp?)/o/h
	Sympetrum semicinctum (Say 1839)	W lit.	IN	p
	Sympetrum vicinum (Hagen 1861)	W lit., EAC	IN, MA	sp/w/hS, sp/w/hm, sp/o/h, p
	Tramea carolina (Linnaeus 1763)	EAC	MA	sp/o/h
	Tramea lacerata Hagen 1861	W lit.	ON	ls/w/l?
Order Odonata — Zygoptera				
Coenagrionidae				
	Coenagrion resolutum (Hagen 1876)	W, EAC	MA, ON	la(sp?)/o/h, sp/wo/hm
	Enallagma sp.	S&F	WI	sp/w, sp/wo/hS
	Enallagma boreale Selys 1875	W	ON	la(sp?)/o/h
	Enallagma civile (Hagen 1861) or *E. hageni* (Walsh 1863)	K	MI	p(sp?)/o/h, pfish/h
	Enallagma ebrium (Hagen 1861) or *E. carunculatum* Morse 1895	K	MI	p(sp?)/o/h
	Enallagma hageni (Walsh 1863) and/or *E. ebrium* (Hagen 1861)	W, EAC	MA, ON	ls/w/h, la(sp?)/o/h, sp/w/hS, sp/wo/hm
	Ischnura posita (Hagen 1861)	W, EAC	MA, ON	la(sp?)/o/h, sp/wo/hS
	Ischnura verticalis (Say 1839)	K	MI	p(sp?)/o/h, pfish/h
	Nehalennia irene (Hagen 1861)	W	ON	ls/wo/sS, la(sp?)/o/h
	Nehalennia gracilis Morse 1895	EAC	MA	sp/o/shS, sp/o/h

Life History			Drought-Resisting Strategy or Stage	Comments
A[6]	B[7]	C[8]		
				May in MI (K), genus in Italian temporary pools (Baz)
				December-May in MI (K)
3				
4		2,3	similar to *A. junius*, migrants oviposit early	
				genus in Italian temporary pools (Baz)
3		3	probably egg, terrestrial oviposition	
3		3	egg diapause	April-June in MI (K)
3	3	3	egg diapause	
4		2,3	similar to *A. junius*, migrants oviposit early	with *Anax junius*, an important invertebrate predator in fishless ponds
4		2,3	similar to *A. junius*, migrants oviposit early	
				group as a whole not adapted to drawdown
3				
3				
				May in MI (K)
3				
3				genus in Australian and European temporary waters (Lake, Baz)
				present in winter in MI (K)
3				

(continued)

Higher Taxa	Genus/Species	Vernal Pool Study [1]	State or Province Studied [2]	Hydrologic Class [3], Setting [4], and Bottom [5]
Lestidae				
	Lestes sp.	DDW, S&F	ON, WI	ss/o/h, ls/o/h, ls/w/h, ls/wo/hD, sp/wo/hS, sp/wo/hm, sp/w
	Lestes congener Hagen 1861	W	ON	sa/o/h, la(sp?)/o/h
	Lestes dryas Kirby 1890	EAC, W, K	MA, MI, ON	ss/w/l, sa/o/h, la/o/h, p(sp?)/o/h
	Lestes eurinus Say 1839	EAC	MA	sp/o/h
	Lestes forcipatus Rambur 1842	EAC	MA	ls/w/h, ls/o/h, ls/o/Dh, sp/o/h
	Lestes rectangularis Say 1839 (possibly *L. forcipatus* Rambur 1842)	K	MI	ss/w/l, sa/o/h, pfish/h
	Lestes unguiculatus Hagen 1861	W	ON	ss/w/l, la(sp?)/o/h
Order Plecoptera				
Leuctridae		EAC	MA	fp/w/l
Nemouridae		EAC	MA	fp/w/l, la/wo/ls
	Prostoia similis (Hagen 1861)	EAC	MA	fp/w/l, la/wo/ls
Order Trichoptera				
Lepidostomatidae				
	Lepidostoma sp.	EAC	MA	fp/w/l
Leptoceridae				
	unidentified larvae	K	MI	sa/o/h, p(sp?)/o/h
	Oecetis sp.	EAC, W, S&F	MA, ON, WI	la(sp?)/o/h, sp/wo/hm, sp/w
	Triaenodes sp.	K	MI	sa/o/h, p(sp?)/o/h
	Triaenodes aba Milne 1935	W	ON	la(sp?)/o/h
	Triaenodes nox Ross 1941	EAC	MA	la/wo/hS, sp/wo/hm, sp/wo/hs
Limnephilidae		K	MI	sa/o/h, p(sp?)/o/h
Dicosmoecinae				
	Ironoquia sp.	DDW, W	ON	ss/o/h, la(sp?)/o/h
	Ironoquia parvula (Banks 1900)	EAC, F60	MA, ON	ss/w/l, ss/o/lh, la/o/h, sp/wo/l
	Ironoquia punctatissima (Walker 1852)	EAC, W	MA, ON	ss/w/l, la(sp?)/o/h
Limnephilinae				
	Anabolia sp.	S&F, EAC	WI, MA	ss/w/l, ls/w, sp/w
	Anabolia bimaculatus (Walker 1852)	W, F60, F&M	ON	ss/w/l, sa(sp?)/o/h
	Anabolia sordida Hagen 1861	EAC, F60	MA, ON	ss/w/l, sp/w/l
	Asynarchus sp.	J66	ON	sp/o, sp/w
	Asynarchus batchawanus (Denning 1949) (as *A. curtus* (Banks))	EAC, F&M	MA, NB, NF, ON, QB	ss/w/l, ss/o/lh, temp
	Lenarchus crassus (Banks 1920)	F&M	QB	temp

Life History A[6]	Life History B[7]	Life History C[8]	Drought-Resisting Strategy or Stage	Comments
			egg diapause, eggs laid in plants	family in Australian temporary waters (Lake)
	2			
3		3		
		3		
		3		
		3		
3		3		
				may be associated with overbank flooding from streams or intermittent stream flows
3?				
			pupal aestivation	
3			aestivate as pupa ready to emerge in fall, egg overwinters	
		4		
3			egg	

(continued)

Higher Taxa	Genus/Species	Vernal Pool Study [1]	State or Province Studied [2]	Hydrologic Class [3], Setting [4], and Bottom [5]
	Limnephilus sp.	S&F	WI	ss/w, ls/w, sp/w
	Limnephilus indivisus Walker 1852	EAC, W, K, DDW, J66	MA, MI, ON	ss/w/l, ss/o/h, sa/o/h, la(sp?)/o/h, sp/w/l, sp/w/s, p(sp?)/o/h, sp/o
	Limnephilus sericeus (Say 1824)	EAC, W	MA, ON	ss/w/l, ss/o/lh, la(sp?)/o/h
	Limnephilus submonilifer Walker 1852	W	ON	ss/w/l, la(sp?)/o/h
	Nemotaulius hostilis (Hagen 1873)	EAC, W	MA, ON	ss/w/h, la(sp?)/o/h, p/wox/l
	Phanocelia canadensis (Banks 1924)	C&G	MA	la/o/sS, sp/o/hS
	Platycentropus radiatus (Say 1824)	S&S	WV	temp
	Pycnopsyche divergens (Walker 1852)	EAC	MA	sw/w/lh
Phryganeidae		K	MI	pfish/h
	Banksiola sp.	EAC	MA	ls/w/sS, sp/wo/hS, sp/wo/hD, sp/wo/hm
	Banksiola crotchi Banks 1943	W	ON	ls(sp?)/o/h
	Banksiola dossuaria (Say 1828)	S&S	WV	p
	Ptilostomis ocellifera (Walker 1852)	EAC, W	MA, ON	ss/w/R, ss/w/l, la(sp?)/o/h, ss/w/sS, ls/wo/h, sp/w/hm, sp/wo/hD
	Ptilostomis postica (Walker 1852)	Rowe	PA	temp
Polycentropodidae				
	Holocentropus flavus Banks 1908 (as *Polycentropus*)	W73	ON	ss
	Plectrocnemia aureola Banks 1930 (as *Polycentropus*)	W73	ON	ss
	Plectrocnemia crassicornis (Walker 1852) (as *Polycentropus*)	W, F&G	ON	ss/w/l
	(Polycentropus) sp.	EAC	MA	ss/w/l, ls/w/sS, ls/wo/hD, sp/wo/hD, sp/wo/hS, sp/wo/hm
Phylum BRYOZOA				
Plumatellidae				
	Fredericella sp.	W	ON	la/o/h
	Plumatella sp.	K, M	MI	temp
Phylum CNIDARIA				
Hydridae				
	Hydra viridissima Pallas 1776 (as *Chlorohydra*)	K	MI	p(sp?)/o/h
	Hydra oligactus Pallas 1766 (as *Pelmatohydra*)	K	MI	p(sp?)/o/h
	Pelmatohydra — see *Hydra*			

Life History A[6]	Life History B[7]	Life History C[8]	Drought-Resisting Strategy or Stage	Comments
1		1,4	egg, larvae ready to hatch upon flooding	found in vernal pools throughout the glaciated northeast
3				
3		1,4		
			do not appear to withstand drawdown	
		(1,4)	appear to survive as last-instar larvae in damp *Sphagnum*	found in dense *Sphagnum*, adults emerge in late fall
			intolerant of drawdown	
				prey on amphibian eggs and larvae
3		4	adult ovarian diapause, oviposit on dry basin, eggs overwinter	
				prey on amphibian eggs and larvae
				throughout glaciated northeast including IL, MI, MA, NH, and ON
				northern distribution, including NH and MI
2		1	oviposit in water on emergence, eggs overwinter in mud, hatch in spring	also known from KY, MA, MI, NH
2				due to taxonomic revisions in this group, genus is unknown; keys as *Polycentropus*
		1	statoblasts	
1			statoblasts	appears to require high moisture during dry phase
		1		

(continued)

Higher Taxa	Genus/Species	Vernal Pool Study [1]	State or Province Studied [2]	Hydrologic Class [3], Setting [4], and Bottom [5]
Phylum MOLLUSCA				
Class Bivalvia				
Order Veneroidea				
Sphaeriidae		S&F	WI	ss/w, sp/w
	Musculium partumeium (Say 1822)	S, Way et al., Hornbach et al.	MA, OH	ls/s, temp/w, p/o/h
	Musculium securis (Prime 1852)	S, Clarke, M&Q74, M&M80, M&M83	MA, ON	ss/w/l, temp/w, p
	Sphaerium occidentale (Prime 1853)	W, K, D&B, S, M&M83	MA, MI, NY, ON	ss/w/l, la/w/l, p(sp?)/o/h, la/o/h
	Sphaerium truncatum (Linsley)	K	MI	ss/w/l, sa/o/h, p(sp?)/o/h
	Pisidium sp.	DDW, Clarke	ON, Canada	ss/o/h, temp
	Pisidium casertanum (Poli 1795)	S, Clarke, K, D&B	MA, MI, NY, Canada	temp, la/w/l, pfish/h
	Pisidium rotundatum Prime 1852	K, Clarke	MI, Canada	ss/w/l, p
Class Gastropoda				
Order Basomotophora				
Ancylidae				
	Ferrissia sp.	K	MI	p(sp?)/o/h
	Ferrissia walkeri (Pilsbry and Ferris 1907)	Jo92	NY	temp
Lymnaeidae		S&F, EAC 1975	WI	sp/w, sp/w/h, fp/w, fp/o/h
	Fossaria modicella (Say 1825) (= *F. humilis* and *Lymnaea humilis*)	Jo92, EAC 1975, DDW	NY, WI, ON	ss/o/h, sp/w/h
	Fossaria parva (Lea 1841)	Jo92	NY	temp
	Pseudosuccinea columella (Say 1817)	Jo92, K&S	NY, CT	temp, p
	Stagnicola caperata (Say 1829)	Jo92, Clarke	NY, Canada	fp, temp/w
	Stagnicola elodes (Say 1821) (= *Lymnaea palustris, L. elodes*)	Jo92, W, EAC, K	MA, MI, NY, ON	ss/w/l, ss/o/h, sa/o/h, fp/w, fp/o/h, p(sp?)/o/h, temp
	Stagnicola reflexa (Say 1821)	Clarke	Canada	temp, p
Physidae				
	Aplexa elongata (Say 1821) (often identified as *A. hypnorum* (Linnaeus 1758))	Jo92, W, EAC, K	MA, NY, MI, ON	ss/w/l
	Physa sp.	EAC	MA	
	Physa gyrina (Say 1821)	Jo92, K	MI, NY, WI	ss/w/l, sa/o/h, sp/w/h, p(sp?)/o/h, temp, p
	Physa heterostropha (Say 1816)	Jo92	NY	temp

Life History A[6]	Life History B[7]	Life History C[8]	Drought-Resisting Strategy or Stage	Comments
		1	aestivation	
			juveniles aestivate	
			newly released juveniles aestivate	genus in CA vernal pools (Z)
1			all stages aestivate, live 2–3 years	restricted to temporary waters
	2			
				= *P. abditum* Haldemann, found in CA vernal pools (Z)
	1	1	aestivation	
				genus in Australian temporary pools (Lake)
		1		family in CA vernal pools (Z)
1		1	aestivate as juveniles in mud	
		1	aestivate 9–11 months	
1	4	1	survive as juveniles	juveniles leave pools, climb trees, later return to pools (Jo78, 92)
1		1		family in European temporary pools (Baz)
1		1	survive as juvenile burrowing into mud	primarily in temporary waters, usually hard water (Jo92)
1		1		
1		1		
		1		

(continued)

Higher Taxa	Genus/Species	Vernal Pool Study [1]	State or Province Studied [2]	Hydrologic Class [3], Setting [4], and Bottom [5]
	Physa vernalis Taylor and Jokinen 1984	Jo92	NY	temp
	Physella ancillaria (Say 1825) (as *Physa*)	Jo92	NY	temp
Planorbidae				
	Gyraulus circumstriatus (Tryon 1866)	Jo92	NY	temp
	Gyraulus crista (Linnaeus 1758) (= *Armiger crista*)	Jo92, K	MI, NY	ss/w, temp
	Gyraulus parvus (Say 1817)	W, Jo92, K	MA, MI, NY, ON, WI	ss/w/l, sa/o/h, ls/wo/l, p(sp?)/o/h, temp la/o/h
	Micromenetus dilatus (Gould 1841)	Jo92	NY	temp
	Planorbula armigera (Say 1821)	Jo92, EAC, K	MA, MI, NY	ss/w/l, sa/o/h, p(sp?)/o/h, pfish/h, temp, p
	Promenetus exacuous (Say 1821)	Jo92	NY	temp
	Helisoma sp.	S&F	WI	sp
	Helisoma trivolvis (Say 1817)	W	ON	la/o/h
Phylum NEMATODA		K	MI	ss/w/l, sa/o/h, p(sp?)/o/h
Phylum PLATYHELMINTHES				
Class Turbellaria				
Order Catenulida				
Catenulidae				
	Catenula lemnae Duges 1832	K	MI	sa/o/h
Stenostomidae				
	Stenostomum spp.	K	MI	sa/o/h
Order Lecithoepitheliata				
Prorhynchidae				
	Geocentrophora sp.	Pennak	cosmo-politan	temp
	Prorhynchus stagnalis Schultze 1851	K, Pennak	MI	ss/w/l
Order Macrostomida				
Macrostomidae				
	Macrostomum sp.	K	MI	pfish/h
Order Neorhabdocoela				
Dalyelliidae				
	Dalyellia viridis (Shaw 1791)	K, W, EAC	MA, MI, ON	ss/w/l, sa/o/h, sp/o/h, p(sp?)/x/l
Typhloplanidae				
	Bothromesostoma sp.	K, (Pennak)	MI	ss/w/l, sa/o/h, p(sp?)/o/h
	Mesostoma sp.	Kolasa		s
	Mesostoma ehrenbergi (Focke 1836), form *wardi* Husted and Rue 1940	K	MI	sa/o/h
	Phaenocora sp.	K	MI	pfish/h
	Rhynchomesostoma sp. (*R. rostratum* (Muller 1774) is only North American species in genus)	Kolasa		temp

Life History A[6]	B[7]	C[8]	Drought-Resisting Strategy or Stage	Comments
				genus in Italian pools (Baz)
1		1		family in European temporary pools (Baz)
				genus in Australian temporary pools (Lake)
1	2	1	burrow in sediment	survival related to moisture content of sediment
		1		
1		1	aestivate in sediment	*Planorbula jenksii* (Carpenter 1871) is a synonym
1		1		
1				
			in moist microhabitats	
		1		present in temporary waters in Italy (Baz)
		1	cysts and eggs	
	2		encystment	
	2		encystment	most common microturbellarian genus (Pennak)
		1		
				locally common in temporary habitats
	3	1	encyst or follow water table into sediment	locally common in temporary habitats, known from Europe
	3			
1	3	1	eggs	common April-June in MI temporary ponds (K), known from Europe
	3			genus in CA vernal pools (Z)
		1	eggs	
	2	1	diapausing egg	diapause up to seven months (Pennak)
				widely distributed in Northern Hemisphere

(continued)

Higher Taxa	Genus/Species	Vernal Pool Study [1]	State or Province Studied [2]	Hydrologic Class [3], Setting [4], and Bottom [5]
	unidentified rhabdocoel	K	MI	ss/w/l, sa/o/h, p(sp?)/o/h, pfish/h
Order Tricladida				
Planariidae				
	Hymanella retenuova Castle 1941 (= *Phagocata vernalis* of Kenk 1944)	W, K, DDW	MI, ON	ss/w/l, ss/o/h, sa/o/h, la/w/l
	Phagocata velata Stringer 1909	W, K	MI, ON	ss/w/l, sa/o/h, p(sp?)/o/h
Phylum PORIFERA				
Spongillidae				
	Eunapius fragilis (Leidy 1851)	S	MA	sp, p
Phylum ROTIFERA				
Class Bdelloidea				
Order Bdelloida				
Philodinidae				
	Philodina roseola Ehrenberg 1832	Masters	glaciated northeast	temp
	Rotaria citrina (Ehrenberg 1838)	Masters	glaciated northeast	temp
Class Monogononta				
Order Ploimida				
Brachionidae				
	Epiphanes senta (Müller 1773)	Masters	glaciated northeast	temp
Phylum CHORDATA				
Subphylum Vertebrata				
Class Amphibia				
Order Anura: literature is extensive, see Chapter 11				
Bufonidae				
	Bufo americanus Holbrook 1836	-- See Chapter 11 --		
	Bufo woodhousei fowleri (Hinkley 1882)	-- See Chapter 11 --		
Hylidae		EAC, W	MA, ON	s, a, sp
	Hyla versicolor LeConte 1823	-- See Chapter 11 --		
	Pseudacris crucifer (Wied-Neuwied 1824)	-- See Chapter 11 --		ss/w/ls, ls/w/ls, sa/w/ls, la/w/ls, sp/wo/ls
	Pseudacris triseriata triseriata (Weid-Neuwied 1838)	WSL	ON	temp
Pelobatidae		-- See Chapter 11 --		
	Scaphiopus holbrookii holbrookii (Harlan 1835)	EAC, D&R	MA, CT, RI, NJ, NY	ephemeral, ss/o/h, temp
Ranidae		-- See Chapter 11 --		
	Rana catesbeiana Shaw 1802	-- See Chapter 11 --		sp/o/h, p/o/h
	Rana clamitans Latreille 1801	See Chapter 11	glaciated northeast	ls/w/l, ls/o/h, sp/w/l, sp/o/h, p

Life History A[6]	Life History B[7]	Life History C[8]	Drought-Resisting Strategy or Stage	Comments
				in temporary waters in Italy (Baz)
1	2,4	1	eggs in cocoon	winter in MI (K)
1	2	1	fragments form cysts	
		1	gemmules	widespread throughout glaciated northeast, in acid and hard waters, apparently tolerant of mild pollution (S)
4			adults and metamorphs are terrestrial	
4	3	3		
4	3	3		
4	2		adults and metamorphs are terrestrial	family in CA vernal pools (Z)
4	3	3		
4	2	2		
4	2	2		
			rapid growth of tadpoles, metamorphs and adults burrow into soils in uplands	breed spring to fall when ephemeral pools flood; not restricted to vernal pools
4	2,3			
			not drought-tolerant	
4	3	3	not drought tolerant	

(continued)

Higher Taxa	Genus/Species	Vernal Pool Study [1]	State or Province Studied [2]	Hydrologic Class [3], Setting [4], and Bottom [5]
	Rana sylvatica LeConte 1825	See Chapter 11	glaciated northeast	ss/w/l, ss/o/h, sp/wo/l
Order Caudata: literature is extensive, see Chapter 11				
Ambystomatidae		-- See Chapter 11 --		ss/w/l, ls/wx/l, la/w/l, sp, temp
	Ambystoma jeffersonianum (Green 1827)	-- See Chapter 11 --		ss/w/l, ls/wx/l, la/w/l, sp, temp
	Ambystoma laterale Hallowell 1856	-- See Chapter 11 --		ss/w/l, sp/wo/sl, p/xo, temp
	triploid hybrids *A. jeff* x *A. jeff* x *A. laterale* (*A. platineum*), *A. lat* x *A. lat* x *A. jeff* (*A. tremblayi*)	-- See Chapter 11 --		ss/w/l, sa/w/l, ls/w/l, sp/w/sl, p/x, temp
	Ambystoma maculatum (Shaw 1802)	-- See Chapter 11 --		ss/w/l, ls/w/l, sa/w/l, la/w/l, sp/wo/hm, sp/wo/hS
	Ambystoma opacum (Gravenhorst 1807)	-- See Chapter 11 --		ss/w/l, ls/w/l, sa/w/l, la/w/l
	Ambystoma texanum (Matthes 1855)	-- See Chapter 11 --		temp
	Ambystoma tigrinum (Green 1825)	-- See Chapter 11 --		a, sp, temp
Salamandridae				
	Notophthalmus viridescens (Rafinesque 1820)	-- See Chapter 11 --		temp, sp
Class Reptilia				
Order Testudines				
Chelydridae				
	Chelydra serpentina (Linnaeus 1758)	-- See Chapter 12 --		temp, sp
Emydidae				
	Chrysemys picta (Schneider 1783)	-- See Chapter 12 --		temp, sp
	Clemmys guttata (Schneider 1792)	-- See Chapter 12 --		temp, sp
	Emydoidea blandingii (Holbrook 1838)	-- See Chapter 12 --		temp, sp
	Terrapene carolina carolina (Linnaeus 1758)	-- See Chapter 12 --		temp, sp

Addendum

Higher Taxa	Genus/Species	Vernal Pool Study	State or Province Studied	Hydrologic Class, Setting, and Bottom
Class Arachnida, Subclass Acarina[b]				
Order Trombidiformes				
Eylaidae				
	Eylais discreta Koenike 1897	Lanciani	NY	temp
	Eylais falcata Koenike 1897	Lianciani	NY	temp, p
	Eylais mitchelli Lanciani 1970	Lanciani	NY	temp
	Eylais ovaliporus Lanciani 1970	Lanciani	NY	temp, p
	Eylais vernalis Lanciani 1970	Lanciani	NY	temp

[b] See also pages 296–303 for Class Arachnida, pages 298–299 for Family Eylaidae.

Life History A[6]	Life History B[7]	Life History C[8]	Drought-Resisting Strategy or Stage	Comments
4	2	2	adults and metamorphs are terrestrial	
4	2	2,4	adults and post-metamorphic juveniles are terrestrial	
4	2,3	2		
4	2,3	2		
4	2,3	2		
4	2,3	2		
(3)	5+2	4		
4	2,3	2		
4	2,3	2		
			adults aestivate on land	adults and larvae aquatic, juveniles terrestrial
4		5		
4		5		
4		5		move to pools in spring to feed
4		5		move to pools in spring to feed
4		5		move to pools in summer to hydrate
			larva overwinters on host in permanent water	parasitic on corixids (Lanciani)
			larva overwinters on host in permanent water	parasitic on hydrophilids (Lanciani)
			larva overwinters on host in permanent water	parasitic on Deronectes in temporary ponds (Lanciani)
			larva overwinters on host in permanent water	parasitic on hydrophilids (Lanciani)
			larva overwinters on host in permanent water	parasitic on dytiscids (Lanciani)

(continued)

Abbreviations and References for Vernal Pool Studies Cited in Appendix

Abbreviations: Baz — Bazzanti et al. 1996; **B&G** — Burian and Gibbs 1991; **B&S** — Batzer and Sion 1998; **C&G** — Colburn and Garretson Clapp 2004; **C&M** — Crosetti and Margaritora 1987; **D70a-d** — Delorme 1970a, b, c, d; **D71** — Delorme 1971; **D&L** — Dodson and Lillie 2001; **D&R** — DeGraaf and Rudis 1983; **DDW** — Williams 1983; **DPH** — Massachusetts Department of Public Health 1940; **EAC** — Colburn unpublished data; **E&B** — Ebert and Balko 1987; **E&C** — Edman and Clark 1988; **F60** — Flint 1960; **FCF** — Fairchild et al. 2003; **FFM** — Fairchild et al. 2000; **F&B** — Formanowicz and Brodie 1981; **F&M** — Flannagan and MacDonald 1987; **Fe** — Ferguson 1944; **G&H** — Garcia and Hagen 1987; **H&K** — Herreid and Kinney 1966; **H&M98, H&M01** — Higgins and Merritt 1998, 2001; **J61, J66, J67, J69** — James 1961, 1966, 1967, 1969; **Jo78, Jo92** — Jokinen 1978, 1992; **K** — Kenk 1949; **K&S** — Kiesecker and Skelly 2001; **Lake** — Lake et al. 1989; **L28** — Leonard 1928; **LAR** — Larson et al. 2000; **M&M79, M&M80, M&M83** — McKee and Mackie 1979, 1980, 1983; **M&Q** — Mackie and Qadri 1974; **Maier** — Maier et al. 1998; **Mura** — Mura et al. 2003; **N&J** — Nix and Jenkins 2000; **R&H** — Rowe and Hebert 1999; **S&C** — I. M. Smith and Cook 1991; **S&F** — Schneider and Frost 1996; **S&G** — Smith and Gola 2001; **S&H** — Schwartz and Hebert 1987; **S&J** — Stevens and Jenkins 2000; **S87, S97** — I. M. Smith 1987, 1997; **S** — D. G. Smith 1992; **T&W** — Tessier and Woodruff 2002; **TWM** — Taylor et al. 1990; **V&A** — Vallières and Aubin 1986; **W** — Wiggins et al. 1980 ("W genus" means that the taxon is covered only at the generic level in the reference); **W lit.** — literature cited in Wiggins et al. 1980; **W73** — Wiggins 1973; **W&C** — Walker and Corbet 1978; **W&M** — Weeks and Marcus 1997; **WSL** — Weller et al. 1976; **WSW** — Wellborn et al. 1996; **Z** — Zedler 1987; **Z60, Z70** — Zimmerman 1960, 1970. A "?" means the author's identification of the taxon was tentative.

References: Batzer and Sion 1998; Bazzanti et al. 1996; Brigham 1996; Brooks 1959; Burian and Gibbs 1991; Burks 1953; Clarke 1981; Colburn 1975, unpublished data; Colburn and Garretson Clapp 2004; Cole 1959; Cook 1981; Cory and Manion 1953; Crosetti and Margaritora 1987; Darsie 1981; Davies 1991; DeGraaf and Rudis 1983; Delorme 1970a, 1970b, 1970c, 1970d, 1971; Dexter 1953, 1959; Dexter and Kuehnle 1951; Dodson and Lillie 2001; Ebert and Balko 1987; Edman and Clark 1988; Fairchild et al. 2000, 2003; Ferguson 1944; Flannagan and MacDonald 1987; Flint 1960; Floyd 1995; Formanowicz and Brodie 1981; Garcia and Hagen 1987; Herreid and Kinney 1966; Higgins and Merritt 1998, 2001; Hoff 1942; Hornbach et al. 1980; Horsfall 1963; Innes 1997; James 1961, 1966, 1967, 1969; James and Redner 1965; Jeffries 1996; Jokinen 1978, 1992; Jokinen and Morrison 1998; Kenk 1944, 1949, 1972; Kiesecker and Skelly 2001; Kolasa 1991; Lake et al. 1989; Lanciani 1970; Leonard 1928; Mackie and Qadri 1974; Maier 1992, 1993; Maier et al. 1998; Massachusetts Department of Public Health 1940; Masters 1968; Mastrantuono 1994; Matta and Michael 1976; McKee and Mackie 1979, 1980, 1983; Mura et al. 2003; Newell 1959; Nimmo 1971; Nix and Jenkins 2000; Pennak 1978, 1989; Ross and Merkley 1952; Rowe and Hebert 1999; Rowe et al. 1994; Schneider and Frost 1996; Schwartz and Hebert 1987; Simovich 1998; D. G. Smith 1992; D. G. Smith and Gola 2001; I. M. Smith and Cook 1991; I. M. Smith 1987, 1997; Stevens and Jenkins 2000; Stout and Stout 1992; Taylor et al. 1990; Tejedo 1993; Tessier and Woodruff 2002; Tressler 1959; Vallières and Aubin 1986; Walker and Corbet 1978; Way et al. 1980; Weeks and Marcus 1997; Wellborn et al. 1996; Weller et al. 1976; Wiggins 1973; Wiggins et al. 1980; Wilbur 1971; Williams 1983; Wolfe 1980; Wood et al. 1979; Worth 1972; Yeatman 1959; Zedler 1987; Zimmerman 1960, 1970.

Bibliography

Abernethy, Y., and R. E. Turner. 1987. "US forested wetlands: 1940–1980." *Bioscience* 37: 721–727.

Acker, P. M., K. C. Kruse, and E. B. Krehbiel. 1986. "Aging *Bufo americanus* by skeletochronology." *Journal of Herpetology* 20: 570–574.

Aerts, R., R. van Logtestijn, M. van Staalduiner, and S. Toet. 1995. "Nitrogen supply effects on productivity and potential leaf decay of *Carex* species from peatlands differing in nutrient limitation." *Oecologia* 104: 447–453.

Alarie, Y., P. J. Spangler, and W. E. Steiner, Jr. 2002. "Larval morphology of *Agabetes* Crotch (Coleoptera: Adephaga: Dytiscidae): The hypothesis of sister-group relationship with the subfamily Laccophilinae revisited." *The Coleopterists Bulletin* 56: 547–567.

Albers, P. H., and R. M. Prouty. 1987. "Survival of spotted salamander eggs in temporary woodland ponds of coastal Maryland." *Environmental Pollution* 46: 45–61.

Ali, A. 1980. "Nuisance chironomids and their control: A review." *Bulletin of the Entomological Society of America* 26: 3–16.

Aliberti, M. A., S. R. Bradt, S. Greene, Dr. J. Haney, J. E. Morrissey, and J. L. Nowak. 2003. *An Image-based Key to the Macrozooplankton of the Northeast (USA).* Version 1, January, 2003. Durham, NH: Center for Freshwater Biology, Department of Zoology, University of New Hampshire. Available at *http://cfb.unh.edu*.

Altig, R. A. 1970. "A key to the tadpoles of the continental United States and Canada." *Herpetologica* 26: 180–207.

Altig, R. A. 1998. "A key to the anuran tadpoles of the United States and Canada." Available at *http://www.pwrc.usgs.gov/tadpole/.*

Altig, R. A., and P. H. Ireland. 1984. "A key to salamander larvae and larviform adults of the United States and Canada." *Herpetologica* 40: 212–218.

Altig, R. A., R. W. McDiarmid, K. A. Nichols, and P. C. Ustach. 1999. *Tadpoles of the United States and Canada: A Tutorial and Key.* Washington, DC: US Geological Survey. Available at *http://www.prc.usgs.gov/tadpole/.*

Altig, R. A., and R. Lohoefener. 1980. *A Bibliography of Larval and Neotenic Salamander Biology.* Smithsonian Herpetological Reference Service 47. Washington, DC: National Museum of Natural History, Smithsonian Institution.

Anderson, J. D., and R. V. Giacosie. 1967. "*Ambystoma laterale* in New Jersey." *Herpetologica* 23: 108–111.

Anderson, J. D., and R. E. Graham. 1967. "Vertical migration and stratification of larval *Ambystoma*." *Copeia* 1967: 148–153.

Anderson, J. D., D. D. Hassinger, and G. H. Dalrymple. 1971. "Natural mortality of eggs and larvae of *Ambystoma t. tigrinum*." *Ecology* 52: 1107–1112.

Anderson, R. D. 1976. "A revision of the Nearctic species of *Hygrotus* Groups II and III (Coleoptera: Dytiscidae)." *Annals of the Entomological Society of America* 69: 577–584.

Andren, C., M. Marden, and G. Nilson. 1989. "Tolerance to low pH in a population of moor frogs, *Rana arvalis*, from an acid and a neutral environment: A possible case of rapid evolutionary response to acidification." *Oikos* 56: 215–223.

Andren, H. 1994. "Effects of habitat fragmentation on birds and mammals in landscapes with different proportions of suitable habitat: A review." *Oikos* 71: 355–366.

Arnold, S. J., and R. J. Wassersug. 1978. "Differential predation on metamorphic anurans by garter snakes (*Thamnophis*) and social behavior as a possible defense." *Ecology* 59: 1014–1022.

Ash, A. N. 1988. "Disappearance of salamanders from clearcut plots." *Journal of the Elisha Mitchell Scientific Society* 104: 116–122.

Ash, A. N., and R. C. Bruce. 1994. "Impacts of timber harvesting on salamanders." *Conservation Biology* 8: 300–301.

Atlantic Canada Conservation Data Centre. 2004. "Four-toed salamander *(Hemidactylium scutatum)*." Webpage. *http://www.accdc.com/products/profiles/salamander.html.*

Austin, N. E., and J. P. Bogart. 1982. "Erythrocyte area and ploidy determination in the salamanders of the *Ambystoma jeffersonianum* complex." *Copeia* 1982: 485–488.

Avery, J. L. 1939. "Effect of drying on the viability of fairy shrimp eggs." *Transactions of the American Microscopical Society* 58: 356.

Babcock, H. L. 1938. *Field Guide to New England Turtles*. Natural History Guides #2. Boston, MA: Boston Society of Natural History.

Bachmann, M. D. 1988. "Habitat use differences among aquatic larvae of three sympatric salamanders." *Verhandlungen Internationale Vereinigung für Theoretische und Angewandte Limnologie* 23: 2209.

Bachmann, M. D., and R. W. Bachmann. 1994. "Larval salamander growth in acidic, low oxygen temporary ponds." Abstract, *Verhandlungen Internationale Vereinigung für Theoretische und Angewandte Limnologie* 25: 2490.

Baldwin, W. F., H. G. James, and H. E. Welch. 1955. "A study of predators of mosquito larvae and pupae with a radioactive tracer." *Canadian Entomologist* 87: 350–356.

Ball, I. R., N. Gourbault, and R. Kenk. 1981. *The Planarians (Turbellaria) of Temporary Waters in Eastern North America*. Life Science Contribution 127. Toronto, ON: Royal Ontario Museum.

Ball, J. C. 1999. "A survey of breeding salamanders in a twenty hectare Michigan wood lot." *Michigan Academician* 31: 467–481.

Bärlocher, F., and B. Kendrick. 1974. "Dynamics of the fungal population on leaves in a stream." *Journal of Ecology* 62: 761–791.

Bärlocher, F., R. J. Mackay, and G. B. Wiggins. 1978. "Detritus processing in a temporary vernal pool in southern Ontario." *Archiv für Hydrobiologie* 81: 269–295.

Barnes, R. D. 1968. *Invertebrate Zoology.* Philadelphia, PA: W. B. Saunders Company.

Batzer, D., and K. Sion. 1999. "Autumnal woodland pools of western New York: Temporary habitats that support permanent water invertebrates." Pp. 319–332 *in:* D. P. Batzer, R. B. Rader, and S. A. Wissinger, (eds.). *Invertebrates in Freshwater Wetlands of North America.* New York, NY: J. Wiley and Sons.

Bazzanti, M., S. Baldoni, and M. Seminara. 1996. "Invertebrate macrofauna of a temporary pond in Central Italy: Composition, community parameters, and temporal succession." *Archiv für Hydrobiologie* 137: 77–94.

Beatini, S. J. 2003. Using DNA fingerprinting to assess the genetic structure of the vernal pool amphibian *Rana sylvatica.* MS Thesis, Biology, Worcester Polytechnic Institute, Worcester, MA.

Belk, D. 1972. "The biology and ecology of *Eulimnadia antlei* Mackin (Conchostraca)." *The Southwestern Naturalist* 16: 297–305.

Belk, D. 1975. "A key to the Anostraca (fairy shrimps) of North America." *Southwestern Naturalist* 20: 91–103.

Belk, D. 1984. "Antennal appendages and reproductive success in the Anostraca." *Journal of Crustacean Biology* 4: 66–71.

Belk, D. 1998. "Global status and trends in ephemeral pool invertebrate conservation: Implications for Californian fairy shrimp." Pp. 147–150 *in*: C. W. Witham, E. T. Bauder, D. Belk, W. R. Ferren, Jr., and R. Ornduff, (eds.). *Ecology, Conservation, and Management of Vernal Pool Ecosystems — Proceedings from a 1996 Conference.* Sacramento, CA: California Native Plant Society.

Belk, D., and J. Brtek. 1995. "Checklist of the Anostraca." *Hydrobiologia* 298: 315–353.

Belk, D., and G. A. Cole. 1975. "Adaptational biology of desert temporary-pond inhabitants." Pp. 207–226 *in*: N. F. Hadley, (ed.). *Environmental Physiology of Desert Organisms.* Stroudsburg, PA: Dowden, Hutchinson and Ross, Inc.

Belk, D., G. Mura, and S. C. Weeks. 1998. "Untangling confusion between *Eubranchipus vernalis* and *Eubranchipus neglectus* (Branchiopoda: Anostraca)." *Journal of Crustacean Biology* 18:147–152.

Belk, D., and T. S. Nelson. 1995. "Observations on the effects of incubation at inhibitory temperature on subsequent hatching of anostracan cysts." *Hydrobiologia* 298: 179–181.

Bellis, E. D. 1959. "A study of movement of American toads in a Minnesota bog." *Copeia* 1959: 173–174.

Bellis, E. D. 1961. "Growth of the wood frog, *Rana sylvatica*." *Copeia* 1961: 74–77.

Bellis, E. D. 1965. "Home range and movements of the wood frog in a northern bog." *Ecology* 46: 90–98.

Benfield, E. F. 1996. "Leaf breakdown in stream ecosystems." Pp. 579–590 *in*: F. R. Hauer and G. A. Lamberti, (eds.). *Methods in Stream Ecology*. Boston, MA: Academic Press.

Benke, A. C., and S. S. Benke. 1975. "Comparative dynamics and life histories of coexisting dragonfly populations." *Ecology* 56: 302–317.

Bennett, S. H., J. W. Gibbons, and J. Glanville. 1980. "Terrestrial activity, abundance, and diversity of amphibians in differently managed forest types." *American Midland Naturalist* 1–3: 412–616.

Berger, L., R. Speare, P. Daszak, D. E. Green, A. A. Cunningham, C. L. Goggin, R. Slocombe, M. A. Ragan, A. D. Hyatt, K. R. McDonald, H. B. Hines, K. R. Lips, G. Marantelli, and H. Parkes. 1998. "Chytridiomycosis causes amphibian mortality associated with population declines in the rain forests of Australia and Central America." *Proceedings of the National Academy of Sciences, USA* 95: 9031–9036.

Bernot, R. J. 2003. "Trematode infection alters the antipredator behavior of a pulmonate snail." *Journal of the North American Benthological Society* 22: 241–248.

Berven, K. A. 1988. "Factors affecting variation in reproductive traits within a population of wood frogs (*Rana sylvatica*)." *Copeia* 1988: 605–615.

Berven, K. A. 1990. "Factors affecting population fluctuations in larval and adult stages of the wood frog (*Rana sylvatica*)." *Ecology* 71: 1599–1608.

Berven, K. A., and T. A. Grudzien. 1990. "Dispersal in the wood frog (*Rana sylvatica*): Implications for genetic population structure." *Evolution* 44: 2047–2056.

Biebighauser, T. R. 2003. *A Guide to Creating Vernal Ponds*. Morehead, KY: USDA Forest Service. Available at *http://www.southernregion.fs.fed.us/boone/vernform.pdf*.

Biernbaum, C. K. 1989. "Distribution and seasonality of branchiopod and malacostracan crustaceans at the Santee National Wildlife Refuge, South Carolina." *Brimleyana* 15: 7–30.

Bishop, S. C. 1941. *The Salamanders of New York*. Bulletin 324. Albany, NY: The New York State Museum.

Bishop, S. 1943. *Handbook of Salamanders*. Ithaca, NY: Comstock Publishing Associates.

Blaustein, L., B. P. Kotler, and D. Ward. 1995. "Direct and indirect effects of a predatory backswimmer (*Notonecta maculata*) on community structure of desert temporary pools." *Ecological Entomology* 20: 311–318.

Bogart, J. P. 1982. "Ploidy and genetic diversity in Ontario salamanders of the *Ambystoma jeffersonianum* complex revealed through electrophoretic examination of larvae." *Canadian Journal of Zoology* 60: 848–855.

Bogart, J. P., L. E. Licht, M. J. Oldham, and S. J. Darbyshire. 1985. "Electrophoretic identification of *Ambystoma laterale* and *Ambystoma texanum* as well as their diploid and triploid interspecific hybrids (Amphibia: Caudata) in Pelee Island, Ontario." *Canadian Journal of Zoology* 63: 340–347.

Bolek, M. G., and J. R. Cobbins. 2003. "Helminth community structure of sympatric eastern American toad, *Bufo americanus*, northern leopard frog, *Rana pipiens*, and blue-spotted salamander, *Ambystoma laterale*, from southeastern Wisconsin." *Journal of Parasitology* 89: 673–680.

Bollinger, T. T., J. Mao, D. Schock, R. M. Brigham, and V. G. Chinchar. 1999. "Pathology, isolation, and preliminary molecular characterization of a novel iridovirus from tiger salamanders in Saskatchewan." *Journal of Wildlife Disease* 35: 413–429.

Boobar, L. R., P. J. Spangler, K. E. Gibbs, J. R. Longcore, and K. M. Hopkins. 1998. "Predaceous diving beetles in Maine: Faunal list and keys to subfamilies." *Northeastern Naturalist* 5: 1–20.

Borkent, A. 1979. "Systematics and bionomics of the species of the subgenus *Schadonophasma* Dyar and Shannon (*Chaoborus,* Chaoboridae, Diptera)." *Quaestiones Entomologicae* 15: 122–255.

Borkent, A. 1980. "The potential use of larvae of *Chaoborus cooki* Saether (Diptera: Chaoboridae) as a biological control of mosquito larvae." *Mosquito News* 40: 634–635.

Borkent, A. 1981. "The distribution and habitat preferences of the Chaoboridae (Culicomorpha: Diptera) of the Holarctic region." *Canadian Journal of Zoology* 59: 122–133.

Bormann, F. H., and G. E. Likens. 1979. *Pattern and Process in a Forested Ecosystem*. New York, NY: Springer–Verlag.

Bousquet, Y., (ed.). 1991. *Checklist of Beetles of Canada and Alaska*. Publication 1861/E. Ottawa, ON: Research Branch, Agriculture Canada.

Bouvet, Y. 1971. "La diapause des Trichoptères cavernicoles." *Bulletin de la Société Zoologique de France* 96: 375–384.

Boyce, S. 2003. "The Conchostraca." Webpage: *http://atiniui.nhm.org/peet/conchostraca/.*

Boycott, A. E. 1936. "The habits of freshwater Mollusca in Britain." *Journal of Animal Ecology* 5: 116–186.

Brame, A. H., Jr. 1959. "Status of the salamander *Ambystoma tremblayi* Comeau." *Herpetologica* 15: 20.

Brandon, R. A. 1977. "Interspecific hybridization among Mexican and United States salamanders of the genus *Ambystoma* under laboratory conditions." *Herpetologica* 33: 133–152.

Brewer, R. H. 1964. "The phenology of *Diaptomus stagnalis* (Copepoda: Calanoida): The development and the hatching of the egg stage." *Physiological Zoology* 37: 1–20.

Brigham, W. U. 1996. "Haliplidae. Crawling Water Beetles." Webpage: *http://www.inhs.uiuc.edu/biod/waterbeetles/haliplidae/family.html.*

Brinkhurst, R. O., and S. R. Gelder. 1991. "Annelida: Oligochaeta and Branchiobdellida." Pp. 401–435 *in*: J. H. Thorp and A. P. Covich, (eds.). *Ecology and Classification of North American Freshwater Invertebrates*. New York, NY: Academic Press.

Brinson, M. M. 1993. *A Hydrogeomorphic Classification for Wetlands*. Technical Report WRP-DE-4. Vicksburg, MS: US Army Corps of Engineers, Waterways Experiment Station.

Brinson, M. M. 1995. "The HGM approach explained." *National Wetlands Newsletter* 17: 7–13.

Brinson, M. M. 1996. "Assessing wetland functions using HGM." *National Wetlands Newsletter* 18:10–16.

Brinson, M. M., A. E. Lugo, and S. Brown. 1981. "Primary productivity, decomposition, and consumer activity in freshwater wetlands." *Annual Review of Ecology and Systematics* 12: 123–161.

Broch, E. S. 1965. "Mechanism of adaptation of the fairy shrimp *Chirocephalus bundyi* Forbes to the temporary pond." *Memoirs of the Cornell University Agricultural Experimental Station* No. 392: 1–48.

Brockelman, W. Y. 1969. "An analysis of density effects and predation in *Bufo americanus* tadpoles." *Ecology* 50: 632–644.

Brodie, E. D., Jr., and D. R. Formanowicz, Jr. 1983. "Prey size preference of predators: Differential vulnerability of larval anurans." *Herpetologica* 39: 67–75.

Brodie, E. D., III. 1989. "Individual variation in antipredator response of *Ambystoma jeffersonianum* to snake predators." *Journal of Herpetology* 23: 307–309.

Brodman, R. 1995. "Annual variation in breeding success of two syntopic species of *Ambystoma* salamanders." *Journal of Herpetology* 29: 111–113.

Brooks, J. L. 1959. "Cladocera." Pp. 587–656 *in*: W. T. Edmondson, (ed.). *Freshwater Biology*. 2nd Edition. New York, NY: J. Wiley and Sons.

Brooks, R. T. 2000. "Annual and seasonal variation and the effects of hydroperiod on benthic macroinvertebrates of seasonal forest ("vernal") ponds in central Massachusetts, USA." *Wetlands* 20: 707–715.

Brooks, R. T., and K. L. Doyle. 2001. "Shrew species richness and abundance in relation to vernal pool habitat in southern New England." *Northeastern Naturalist* 8: 137–148.

Brooks, R. T., and M. Hayashi. 2002. "Depth-area-volume and hydroperiod relationships of ephemeral ("vernal") forest pools in southern New England." *Wetlands* 22: 247–255.

Brooks, R. T., J. Stone, and P. Lyons. 1998. "An inventory of seasonal forest ponds on the Quabbin Reservoir Watershed, Massachusetts." *Northeastern Naturalist* 5: 219–230.

Brown, K. M. 1991. "Mollusca: Gastropoda." Pp. 285–314 *in*: J. H. Thorp and A. P. Covich, (eds.). *Ecology and Classification of North American Freshwater Invertebrates*. New York, NY: Academic Press.

Brown, L. R., and L. H. Carpelan. 1971. "Egg hatching and life history of a fairy shrimp, *Branchinecta mackini* Dexter (Crustacea: Anostraca) in a Mojave Desert playa (Rabbit Dry Lake)." *Ecology* 52: 41–54.

Brtek, J. 1967. "*Eubranchipus (Creaseria) moorei* n. sp." *Annotationes Zoologicae et Botanicae, Bratislava* 36: 1–7.

Burgason, B. N. 1992. "Four-toed salamander." Pp. 42–45 *in*: M. L. Hunter, Jr., J. Albright, and J. Arbuckle, (eds.). *The Amphibians and Reptiles of Maine*. Maine Agricultural Experimental Station Bulletin 838. Orono, ME: University of Maine.

Burian, S. K., and K. E. Gibbs. 1991. *Mayflies of Maine: An Annotated Faunal List*. Maine Agricultural Experiment Station Bulletin 142. Orono, ME: University of Maine.

Burke, D. M., and E. Nol. 1998. "Influence of food abundance, nest-site habitat, and forest fragmentation on breeding ovenbirds." *Auk* 115: 96–104.

Burke, V. J., and J. W. Gibbons. 1995. "Terrestrial buffer zones and wetland conservation: A case study of freshwater turtles in a Carolina Bay." *Conservation Biology* 9: 1365–1369.

Burks, B. D. 1953. *The Mayflies, or Ephemeroptera, of Illinois*. Bulletin 26. Urbana, IL: Illinois Natural History Survey.

Burne, M. R. 2000a. Conservation of vernal pool-breeding amphibian communities: Habitat and landscape associations with community richness. MS Thesis, Forestry and Wildlife, University of Massachusetts, Amherst, MA.

Burne, M. R. 2000b. *Aerial Photo Survey of Potential Vernal Pools, Plymouth and Bristol Counties, Massachusetts*. Westborough, MA: Massachusetts Division of Fisheries and Wildlife, Natural Heritage and Endangered Species Program.

Burne, M. R. 2001. *Massachusetts Aerial Photo Survey of Potential Vernal Pools*. Westborough, MA: Massachusetts Division of Fisheries and Wildlife, Natural Heritage and Endangered Species Program.

Burton, T. M., and G. E. Likens. 1975. "Salamander populations and biomass in Hubbard Brook Experimental Forest, New Hampshire." *Copeia* 1975: 541–546.

Burton, T. M., K. E. Ullrich, and S. K. Haack. 1988. "Community dynamics of bacteria, algae, and insects in a first-order stream in New Hampshire, USA." *Verhandlungen Internationale Vereinigung für Theoretische und Angewandte Limnologie* 23: 170–175.

Bury, B. 1983. "Differences in amphibian populations in logged and old-growth redwood forest." *Northwest Science* 57: 167–178.

Byers, G. W. 1996. "Tipulidae." Pp. 549–570 *in:* R. W. Merritt and K. W. Cummins, (eds.). *An Introduction to the Aquatic Insects of North America*. Third edition. Dubuque, IA: Kendall/Hunt Publishing Company.

Calabrese, D. M. 1977. "The habitats of *Gerris* F. (Hemiptera: Heteroptera: Gerridae) in Connecticut." *Annals of the Entomological Society of America* 70: 977–983.

Calabrese, D. M. 1979. "Pterygomorphism in 10 Nearctic species of *Gerris*." *American Midland Naturalist* 101: 61–68.

Calhoun, A. J. K. 1997. *Maine Citizen's Guide to Locating and Describing Vernal Pools*. Falmouth, ME: Maine Audubon Society.

Calhoun, A. J. K., and M. W. Klemens. 2002. *Best Development Practices for Conservation of Pool-breeding Amphibians in Residential and Commercial Developments in the Northeastern US*. Technical Paper No. 5. Bronx, NY: Metropolitan Conservation Alliance.

Calhoun, A. J. K., and P. K. deMaynadier. 2003. *Forestry Habitat Management Guidelines for Vernal Pool Wildlife*. Boston, MA: US Environmental Protection Agency.

Calhoun, A. J. K., T. E. Walls, S. S. Stockwell, and M. McCollough. 2003. "Evaluating vernal pools as a basis for conservation strategies: A Maine case study." *Wetlands* 32: 70–81.

Camper, J. D. 1986. "*Ambystoma tigrinum tigrinum* (eastern tiger salamander)." *Herpetological Review* 17: 19.

Cannings, R. A., S. G. Cannings, and R. J. Cannings. 1980. "The distribution of the genus *Lestes* in a saline lake series in British Columbia, Canada." *Odonatologica* 9: 19–28.

Caprioli, M., and C. Ricci. 2001. "Recipes for successful anhydrobiosis in bdelloid rotifers." *Hydrobiologia* 446: 13–17.

Carpenter, C. C. 1952. "Comparative ecology of the common garter snake (*Thamnophis s. sirtalis*), the ribbon snake (*Thamnophis s. sauritus*), and Butler's garter snake (*Thamnophis butleri*) in mixed populations." *Ecological Monographs* 22: 235–258.

Carr, A. F. 1952. *Handbook of Turtles of the United States, Canada, and Baja California*. Ithaca, NY: Comstock Publishing Associates.

Casper, G. S. 2002. "Use of ephemeral wetlands by Blanding's turtles, *Emydoidea blandingii*." Presentation at US EPA Region 5 Conference on Midwestern Ephemeral Wetlands: A Vanishing Habitat. Chicago, IL, February 20, 2002. Abstract available at *http://www.epa.gov/region5/water/ephemeralwetlands/pdf/eph_wet_abstracts.pdf*.

Chapin, F. S., and R. A. Kedrowski. 1983. "Seasonal changes in nitrogen and phosphorus fractions and autumn retranslocation in evergreen and deciduous taiga trees." *Ecology* 64: 376–391.

Chengaloth, R. 1987. *Bibliographia Invertebratorum Aquaticum Canadensium. Volume 7. Synopsis Speciorum. Crustacea: Branchiopoda*. Ottawa, ON: National Museum of Canada.

Childs, N., and B. Colburn. 1995. *Vernal Pool Lessons and Activities*. A curriculum companion to *Certified: A Citizen's Step-by-Step Guide to Protecting Vernal Pools*. Lincoln, MA: Massachusetts Audubon Society.

Christie, B. 2002. "Ontario's ephemeral wetlands. The Vernalis Project: Searching Ontario for ephemeral wetland habitat." *Amphibian Voice* 12(2): 1–2. Newsletter of the Adopt-a-Pond Wetland Conservation Programme. Toronto, ON: Toronto Zoo. Available at *http://www.torontozoo.com/adoptapond/newsletter/SUMMER2002.pdf*.

Clark, J. 1979. "Fresh water wetlands: Habitats for aquatic invertebrates, amphibians, reptiles, and fish." Pp. 330–343 *in*: P. E. Greeson, J. R. Clark, and J. E. Clark, (eds.). *Wetland Functions and Values: The State of Our Understanding*. Minneapolis, MN: American Water Resources Association.

Clark, K. L. 1986a. "Distributions of anuran populations in central Ontario relative to habitat acidity." *Water, Air and Soil Pollution* 30: 727–734.

Clark, K. L. 1986b. "Responses of spotted salamander (*Ambystoma maculatum*) populations in central Ontario to habitat acidity." *Canadian Field-Naturalist* 100: 463–469.

Clark, K. L., and R. J. Hall. 1985. "Effects of elevated hydrogen ion and aluminum concentration on the survival of amphibian embryos and larvae." *Canadian Journal of Zoology* 63: 116–123.

Clark, K. L., and B. D. LaZerte. 1985. "A laboratory study of the effects of aluminum and pH on amphibian eggs and tadpoles." *Canadian Journal of Fisheries and Aquatic Sciences* 42: 1544–1551.

Clarke, A. H. 1981. *The Freshwater Molluscs of Canada*. Ottawa, ON: National Museum of Natural Sciences, National Museums of Canada.

Cobb, B. 1956. *A Field Guide to the Ferns*. Boston, MA: Houghton Mifflin.

Clymo, R. S. 1964. "The origin of acidity in *Sphagnum* bogs." *Bryologist* 67: 427–431.

Clymo, R. S. 1983. "Peat." Pp. 159–224 *in*: A. J. P. Gore, (ed.). *Mires: Swamp, Bog, Fen and Moor*. Ecosystems of the World 4A. Amsterdam: Elsevier.

Coffman, W. P., and L. M . Ferrington. 1996. "Chironomidae." Pp. 635–590 *in*: R. W. Merritt and K. W. Cummins, (eds.). *Aquatic Insects of North America*. 3rd edition. Dubuque, IA: Kendall/Hunt Publishing Company.

Colburn, E. A. 1975. Distribution and ecology of aquatic pulmonate snails on a Wisconsin River floodplain. MS Thesis, Zoology, University of Wisconsin, Madison, WI.

Colburn, E. A. 1984. "Diapause in a salt-tolerant desert caddisfly: The life cycle of *Limnephilus assimilis* (Trichoptera) in Death Valley." *American Midland Naturalist* 111: 280–287.

Colburn, E. A., (ed.). 1996. *Certified: A Citizen's Step-by-Step Guide to Protecting Vernal Pools*. 7th Edition. (1st Edition 1988). Lincoln, MA: Massachusetts Audubon Society.

Colburn, E. A., and F. M. Garretson. 1997. "Patterns of biodiversity in Massachusetts vernal pools." Presentation at 45th Annual Meeting, North American Benthological Society, San Marcos, TX, May 26–30, 1997. Abstract, *Bulletin of the North American Benthological Society* 14: 89.

Colburn, E. A., and F. M. Garretson Clapp. In press. "Habitat and life history of a northern caddisfly at the southern extreme of its range." *Northeastern Naturalist.*

Cole, C. A., R. P. Brooks, and D. H. Wardrop. 1997. "Wetland hydrology as a function of hydrogeomorphic (HGM) subclass." *Wetlands* 17: 456–467.

Cole, C. A., and R. P. Brooks. 2000. "Patterns of wetland hydrology in the ridge and valley province, Pennsylvania, USA." *Wetlands* 20: 438–337.

Cole, G. A. 1953. "Notes on copepod encystment." *Ecology* 34: 208–211.

Cole, G. A. 1959. "A summary of our knowledge of Kentucky crustaceans." *Transactions of the Kentucky Academy of Sciences* 20: 66–81.

Cole, J., and S. G. Fisher. 1978. "Annual metabolism of a temporary pond ecosystem." *American Midland Naturalist* 100: 15–22.

Collins, F. H., and R. K. Washino. 1985. "Insect predators." Pp. 25–41 *in*: H. C. Chapman, (ed.). *Biological Control of Mosquitoes*. Bulletin No. 6. Fresno, CA: American Mosquito Control Association.

Collins, J. P. 2003. "Pathogens and amphibian declines." *Froglog* 55. Available at *http://www.open.ac.uk/daptf/froglog/FROGLOG-55-2.html.*

Collins, J. P., and A. Storfer. 2003. "Global amphibian declines: Sorting the hypotheses." *Diversity and Distributions* 9: 89–98.

Collins, J. P., and H. M. Wilbur. 1979. "Breeding habits and habitats of the amphibians of the Edwin S. George Reserve, Michigan, with notes on the local distribution of fishes." *Occasional Papers of the Museum of Zoology of the University of Michigan* 686:1–34.

Conant, R. 1975. *A Field Guide to Reptiles and Amphibians of Eastern and Central North America*. Boston, MA: Houghton Mifflin Company.

Congdon, J. D., D. W. Tinkle, G. L. Breitenbach, and R. C. van Loben Sels. 1983. "Nesting ecology and hatching success in the turtle *Emydoidea blandingii*." *Herpetologica* 39: 417–429.

Congdon, J. D., A. E. Dunham, and R. C. van Loben Sels. 1993. "Delayed sexual maturity and demographics of Blanding's turtles: Implications for conservation and management." *Conservation Biology* 7: 826–833.

Conrad, H. S. 1956. *How to Know the Mosses and Liverworts*. Dubuque, IA: W. C. Brown Company.

Conservation Foundation. *Midwestern Ephemeral Wetlands: A Vanishing Habitat.* Available at *http://herpcenter.ipfw.edu/parcmw.htm.*

Cook, D. R. 1959. "Studies on the Thyasinae of North America (Acarina: Hydryphantidae)." *American Midland Naturalist* 62: 402–438.

Cook, F. R. 1981. "Amphibians and reptiles of the Ottawa district." *Trail and Landscape* 15: 75–109.

Cook, F. R. 1992. "Pitfalls in quantifying amphibian populations in Canada." Pp. 83–86 *in*: C. A. Bishop and K. E. Pettit, (eds.). *Declines in Canadian Amphibian Populations: Designing a National Monitoring Strategy*. Occasional Paper No. 76. Burlington, ON: Canadian Wildlife Service.

Cook, H. H., and C. F. Powers. 1958. "Early biochemical changes in the soils and waters of artificially created marshes in New York." *New York Fish and Game* 5: 9–65.

Cook, R. P. 1978. Effects of acid precipitation on embryonic mortality of spotted salamanders (*Ambystoma maculatum*) and Jefferson salamanders (*Ambystoma jeffersonianum*) in the Connecticut Valley of Massachusetts. PhD Dissertation, Forestry and Wildlife Management, University of Massachusetts, Amherst, MA.

Cook, R. P. 1983. "Effects of acid precipitation on embryonic mortality of *Ambystoma* salamanders in the Connecticut Valley of Massachusetts." *Biological Conservation* 27: 77–88.

Coombs, P. 1982. "The amphibians and reptiles of Maine." *Maine Biologist* 14: 3–17.

Corbet, P. S. 1962. *A Biology of Dragonflies*. London, UK: H. F. and G. Witherby, Ltd.

Corbett, P. S. 1999. *Dragonflies. Behavior and Ecology of Odonata*. Ithaca, NY: Cornell University Press.

Cortwright, S. A. 1988. "Intraguild predation and competition: An analysis of net growth shifts in larval amphibian prey." *Canadian Journal of Zoology* 66: 1813–1821.

Cory, L., and J. L. Manion. 1953. "Predation on eggs of the wood-frog, *Rana sylvatica,* by leeches." *Copeia* 1953: 66.

Cowardin, L. M., V. Carter, F. C. Golet, and E. T. Laroe. 1979. *Classification of Wetlands and Deepwater Habitats of the United States.* Report FWS/OBS-79/31. Washington, DC: US Fish and Wildlife Service, Biological Services Program.

Crosetti, D., and F. G. Margaritora. 1987. "Distribution and life cycles of cladocerans in temporary pools from central Italy." *Freshwater Biology* 18: 165–175.

Crouch, W. B., and P. W. C. Paton. 2000. "Using egg mass counts to monitor wood frog populations." *Wildlife Society Bulletin* 28: 895–901.

Crowe, J. H. 1971. "Anhydrobiosis: An unsolved problem." *American Naturalist* 105: 563–573.

Cummins, K. W., and M. J. Klug. 1979. "Feeding ecology of stream invertebrates." *Annual Review of Ecology and Systematics* 10: 147–172.

Cuppen, J. M. 1983. "On the habitats of three species of *Hygrotus* Stephens (Coleoptera: Dytiscidae)." *Freshwater Biology* 13: 579–588.

Cutko, A. 1997. "A botanical and natural community assessment of selected vernal pools in Maine." Unpublished report submitted to Maine Department of Inland Fisheries and Wildlife, Augusta, ME.

Cyberlizard (UK). 2004. "The family Plethodontidae: Newts and salamanders. Genus *Hemidactylium* — four-toed salamander." Webpage. *http://www.nafcon.dircon.co.uk/plethodontidae_hemidactylium.htm.*

Daborn, G. R. 1974. "Biological features of an aestival pond in western Canada." *Hydrobiologia* 44: 287–299.

Daborn, G. R. 1976a. "Colonization of isolated aquatic habitats." *Canadian Field-Naturalist* 90: 56–57.

Daborn, G. R. 1976b. "The life cycle of *Eubranchipus bundyi* (Forbes) (Crustacea: Anostraca) in a temporary vernal pool of Alberta." *Canadian Journal of Zoology* 54: 193–201.

Daborn, G. R. 1977a. "The life history of *Branchinecta mackini* Dexter (Crustacea: Anostraca) in an argillotrophic lake of Alberta." *Canadian Journal of Zoology* 55: 280–287.

Daborn, G. R. 1977b. "On the distribution and biology of an Arctic fairy shrimp *Artemiopsis stefanssoni* Johansen, 1921 (Crustacea: Anostraca)." *Canadian Journal of Zoology* 55: 280–287.

Dale, J. M., B. Freedman, and J. Kerekes. 1985. "Acidity and associated water chemistry of amphibian habitats in Nova Scotia." *Canadian Journal of Zoology* 63: 97–105.

Darsie, R. F., Jr. 1981. *Identification and Geographical Distribution of the Mosquitoes of North America, North of Mexico.* Mosquito Systematics, Supplement 1. Fresno, CA: American Mosquito Control Association.

Darwin, C. 1859. *On The Origin of Species by Means of Natural Selection: Or, the Preservation of Favoured Races in the Struggle for Life*. London, reprinted with introduction by Ernst Mayr. 1964. Cambridge, MA: Harvard University Press.

Dazsak P., L. Berger, A. A. Cunningham, A. D. Hyatt, D. E. Green, and R. Speare. 1999. "Emerging infectious diseases and amphibian population decline." *Emerging Infectious Diseases* 5: 735–748.

Daussin, G. L. 1979. "Rediscovery of *Hygrotus sylvanus* (Coleoptera: Dytiscidae)." *Entomological News* 90: 207–208.

Davies, R. W. 1991. "Annelida: Leeches, Polychaetes, and Acanthobdellids." Pp. 437–479 *in*: J. H. Thorp and A. P. Covich, (eds.). *Ecology and Classification of North American Freshwater Invertebrates*. New York, NY: Academic Press.

Debrey, L. S., S. P. Schell, and J. A. Lockwood. 1991. "Sympatric occurrence of three species of Eubranchiopoda in a vernal prairie pond in southeastern Wyoming." *American Midland Naturalist* 126: 399–400.

DeGraaf, R. M., and D. D. Rudis. 1983. *Amphibians and Reptiles of New England*. Amherst, MA: University of Massachusetts Press.

DeGraaf, R. M., and D. D. Rudis. 1986. "New England wildlife: Habitat, natural history, and distribution." General Technical Report NE-108. Broomall, PA: U.S. Department of Agriculture, Forest Service, Northeastern Forest Experiment Station.

DeGraaf, R. M., and D. D. Rudis. 1990. "Herpetofaunal species composition and relative abundance among three New England forest types." *Forestry and Ecological Management* 32: 155–165.

Delorme, L. D. 1970a. "Freshwater ostracodes of Canada. Part I. Subfamily Cypridinae." *Canadian Journal of Zoology* 48: 153–168.

Delorme, L. D. 1970b. "Freshwater ostracodes of Canada. Part II. Subfamily Cypridopsinae and Herpetocypridinae, and family Cyclocyprididae." *Canadian Journal of Zoology* 48: 253–266.

Delorme, L. D. 1970c. "Freshwater ostracodes of Canada. Part III. Family Candonidae." *Canadian Journal of Zoology* 48: 1099–1127.

Delorme, L. D. 1970d. "Freshwater ostracodes of Canada. Part IV. Families Ilyocyprididae, Notodromadidae, Darwinulidae, Cytherideidae, and Entocytheridae." *Canadian Journal of Zoology* 48: 1251–1259.

Delorme, L. D. 1971. "Freshwater ostracodes of Canada. Part V. Families Limnocytheridae, Loxoconchidae." *Canadian Journal of Zoology* 49: 43–64.

deMaynadier, P. G. 1996. Patterns of movement and habitat use by amphibians in Maine's managed forests. PhD Dissertation, Wildlife Ecology, University of Maine, Orono, ME.

deMaynadier, P. G. 2000. "Amphibian movement responses to forest management practices surrounding temporary pools." Presentation at Conference on Vernal Pools of the Northeast: Ecology, Conservation and Education, University of Rhode Island, April 1, 2000.

deMaynadier, P. G., and M. L. Hunter, Jr. 1995. "The relationship between forest management and amphibian ecology: A review of the North American literature." *Environmental Review* 3: 230–261.

deMaynadier, P. G., and M. L. Hunter, Jr. 1998. "Effects of silvicultural edges on the distribution and abundance of amphibians in Maine." *Conservation Biology* 12: 340–352.

Department of Pathology, University of Cambridge. 2004. "The basic lifecycle of the major groups of the digeneans: The mesocercaria, metacercaria and juvenile flukes." Webpage: *http://www.path.cam.ac.uk/~schisto/OtherFlukes/Flukes_Gen/Fluke_Life5.html.*

DeWitt, R. M. 1953. Studies on the biology of *Physa gyrina* Say; Ecology and life history. PhD Dissertation, Biology, University of Michigan, Ann Arbor, MI.

DeWitt, R. M. 1954. "Reproductive capacity in a pulmonate snail (*Physa gyrina* Say)." *American Naturalist* 88: 159–164.

Dexter, R. W. 1946. "Further studies on the life history and distribution of *Eubranchipus vernalis* (Verrill)." *Ohio Journal of Science* 46: 31–44.

Dexter, R. W. 1953. "Studies on North American fairy shrimps with the description of two new species." *American Midland Naturalist* 49: 751–771.

Dexter, R. W. 1956. "A new fairy shrimp from western United States, with notes on other North American species." *Journal of the Washington Academy of Science* 46: 159–165.

Dexter, R. W. 1959. "Anostraca." Pp. 558–571 *in*: W. T. Edmonson, (ed.). *Freshwater Biology*. 2nd Edition. New York, NY: J. Wiley and Sons.

Dexter, R. W., and C. H. Kuehnle. 1948. "Fairy shrimp populations of northeastern Ohio in the seasons of 1945 and 1946." *Ohio Journal of Science* 48: 15–26.

Dexter, R. W., and C. H. Kuehnle. 1951. "Further studies on the fairy shrimp populations of northeastern Ohio." *Ohio Journal of Science* 51: 73–86.

Dexter, R. W., and L. E. Sheary. 1943. "Records of anostracan phyllopods in northeastern Ohio." *Ohio Journal of Science* 43: 176–179.

Dickerson, J. E., and J. V. Robinson. 1985 "Microcosms as islands: A test of the MacArthur-Wilson equilibrium theory." *Ecology* 66: 966–980.

Dickerson, M. C. 1906, reprinted 1969. *The Frog Book*. New York, NY: Dover Publications, Inc.

Digiani, M. C. 2002. "Belostomatidae (Insecta: Heteroptera) as intermediate hosts of digenetic trematodes." *Comparative Parasitology* 69: 89–92.

DiMauro, D., and M. L. Hunter, Jr. 2002. "Reproduction of amphibians in natural and anthropogenic temporary pools in managed forests." *Forest Science* 48: 397–406.

Docherty, D. E., and P. G. Slota. 1988. "Use of muscovy duck fibroblasts for the isolation of virusus from wild birds." *Journal of Tisue Culture Methods* 11: 165–170.

Dodd, C. K., Jr. 1991. "Drift-fence associated sampling bias of amphibians at a Florida sandhills temporary pond." *Journal of Herpetology* 25: 296–310.

Dodge, H. R. 1966. "Studies on mosquito larvae. II. The first-stage larvae of North American Culicidae and of world Anophelinae." *Canadian Entomologist* 98: 337–393.

Dodson, S. I., and D. G. Frey. 1991. "Cladocera and other Branchiopoda." Pp. 723–786 *in*: J. H. Thorp and A. P. Covich, (eds.). *Ecology and Classification of North American Freshwater Invertebrates*. New York, NY: Academic Press.

Dodson, S. I., and R. A. Lillie. 2001. "Zooplankton communities of restored depressional wetlands in Wisconsin, USA." *Wetlands* 21: 292–300.

Donald, D. B. 1983. "Erratic occurrence of anostracans in a temporary pond: Colonization and extinction or adaptation to variations in annual weather?" *Canadian Journal of Zoology* 61: 1492–1498.

Doody, J. S. 1995. "A photographic mark-recapture method for patterned amphibians." *Herpetological Review* 26: 19–21.

Douglas, M. E. 1979. "Migration and sexual selection in *Ambystoma jeffersonianum*." *Canadian Journal of Zoology* 57: 2303–2310.

Douglas, M. E., and B. L. Monroe, Jr. 1981. "A comparative study of topographical orientation in *Ambystoma* (Amphibia: Caudata)." *Copeia* 1981: 460–463.

Downer, A. 1988. *Spring Pool: A Guide to the Ecology of Temporary Ponds*. Boston, MA: New England Aquarium.

Downs, F. L. 1978. "Unisexual *Ambystoma* from Bass Islands of Lake Erie." *Occasional Papers of the Museum of Zoology, University of Michigan* 685: 1–36.

Downs, F. L. 1989. "*Ambystoma maculatum* (Shaw), spotted salamander." Pp. 108–124 *in*: R. A. Pfingsten and F. L. Downs, (eds.). *Salamanders of Ohio*. Columbus, OH: Ohio State University.

Ducey, P. K. 1989. "Agonistic behavior and biting during intraspecific encounters in *Ambystoma* salamanders." *Herpetologica* 45: 155–160.

Ducey, P. K., K. Schramm, and N. Cambry. 1994. "Interspecific aggression between the sympatric salamanders *Ambystoma maculatum* and *Plethodon cinereus*." *American Midland Naturalist* 131: 320–329.

Duellman W. E., and L. Trueb. 1986. *Biology of Amphibians*. New York, NY: McGraw-Hill.

Dunning, J. B., R. Borgella, K. Clements, and G. K. Meffe. 1995. "Patch isolation, corridor effects, and colonization by a resident sparrow in a managed pine woodland." *Conservation Biology* 9: 542–550.

Dunning, J. B., B. J. Danielson, and H. R. Pulliam. 1992. "Ecological processes that affect populations in complex landscapes." *Oikos* 65: 169–175.

Dunson, W. A., and J. Connell. 1982. "Specific inhibition of hatching in amphibian embryos." *Journal of Herpetology* 16: 314–316.

Dunson, W. A., R. A. Wyman, and E. S. Corbett. 1992. "A symposium on amphibian declines and habitat acidification." *Journal of Herpetology* 26: 349–352.

Ebert, T. A., and M. L. Balko. 1987. "Temporary pools as islands in space and in time: The biota of vernal pools in San Diego, Southern California, USA." *Archiv für Hydrobiologie* 110: 101–123.

Eckblad, J. W. 1973. "Population studies of three aquatic gastropods in an intermittent backwater." *Hydrobiologia* 41: 889–906.

Eddy, S., and A. C. Hodson. 1969. *Taxonomic Keys to the Common Animals of the North Central States*. Minneapolis, MN: Burgess Publishing Company.

Eddy, S., and A. C. Hodson (revised by J. C. Underhill, W. D. Schmid, and D. E. Gilbertson). 1982. Tax*onomic Keys to the Common Animals of the North Central States, Exclusive of the Parasitic Worms, Terrestrial Insects, and Birds.* 4th Edition. Minneapolis, MN: Burgess Publishing Co.

Eder, E. 2001. "Large branchiopod home page." Webpage: *http://mailbox.univie.ac.at/Erich.Eder/UZK/index2.html.*

Edman, J. D., and J. M. Clark. 1988. "Generic environmental impact report on mosquito control practices in Massachusetts." Discussion Draft. Boston, MA: State Reclamation and Mosquito Control Board.

Edmonson, W. T., (ed.). 1959. *Freshwater Biology*. 2nd Edition. New York, NY: J. Wiley and Sons.

Edwards, J. R., K. L. Koster, and D. L. Swanson. 2000. "Time course for cryoprotectant synthesis in the freeze-tolerant chorus frog, *Pseudacris triseriata*." *Comparative Biochemistry and Physiology A* 125: 367–375.

Ehrlich, P. R., and D. D. Murphy. 1987. "Conservation lessons from long-term studies of checkerspot butterflies." *Conservation Biology* 1: 122–131.

Emlen, S. T. 1968. "Territoriality in the bullfrog, *Rana catesbeiana*." *Copeia* 1968: 240–243.

Emlen, S. T. 1976. "Lek organization and mating strategies in the bullfrog." *Behavioral Ecology and Sociobiology* 1: 283–313.

Enge, K. M., and W. R. Marion. 1986. "Effects of clear cutting and site preparation on herpetofauna of a north Florida flatwoods." *Forest Ecology and Management* 154: 177–192.

Epler, J. H. 2001. *Identification Manual for the Larval Chironomidae (Diptera) of North and South Carolina. A Guide to the Taxonomy of the Midges of the Southeastern United States, Including Florida*. Special Publication SJ2001-SP13. Raleigh, NC: North Carolina Department of Environment and Natural Resources, and Palatka, FL: St. Johns River Water Management District. Available at *http://www.esb.enr.state.nc.us/BAUwww/Chironomid.htm*.

Ernst, C. H. 1970. "Home range of the spotted turtle, *Clemmys guttata*, in southeastern Pennsylvania." *Copeia* 1970: 391–393.

Ernst, C. H. 1976. "Ecology of the spotted turtle, *Clemmys guttata* (Reptilia, Testudines, Testudinidae), in southeastern Pennsylvania." *Journal of Herpetology* 10: 25–33.

Etchberger, C. R. 1992. "Painted turtle." Pp. 102–104 *in*: M. L. Hunter, Jr., J. Albright, and J. Arbuckle, (eds.). *The Amphibians and Reptiles of Maine*. Maine Agricultural Experimental Station Bulletin 838. Orono, ME: University of Maine.

Eyre, M. D., R. Carr, R. P. McBlane, and G. N. Foster. 1992. "The effects of varying site-water duration on the distribution of water beetle assemblages, adults and larvae (Coleoptera: Haliplidae, Dytiscidae, Hydrophilidae)." *Archiv für Hydrobiologie* 124: 281–291.

Fahd, K., L. Serrano, and J. Toja. 2000. "Crustacean and rotifer composition of temporary ponds in the Donana National Park (SW Spain) during floods." *Hydrobiologia* 436: 41–49.

Fahrig, L., J. H. Pedlar, S. E. Pope, P. D. Taylor, and J. F. Wegner. 1995. "Effect of road traffic on amphibian density." *Biological Conservation* 73: 177–182.

Fairchild, G. W., J. Cruz, and A. M. Faulds. 2003. "Microhabitat and landscape influences on aquatic beetle assemblages in a cluster of temporary and permanent ponds." *Journal of the North American Benthological Society* 22: 224–240.

Fairchild, G. W., A. M. Faulds, and J. F. Matta. 2000. "Beetle assemblages in ponds: Effects of habitat and site age." *Freshwater Biology* 44: 523–534.

Fall, H. C. 1922. "The North American species of *Gyrinus*." *Transactions of the Entomological Society of America* 47: 269–306.

Fassett, N. C. 1957. *A Manual of Aquatic Plants*. Madison, WI: The University of Wisconsin Press.

Fellman, B., (ed.). 1998. *Our Hidden Wetlands: The Proceedings of a Symposium on Vernal Pools in Connecticut*. New Haven, CT: Yale University and Connecticut Department of Environmental Protection.

Felton, M., J. J. Cooney, and W. C. Moore. 1967. "A quantitative study of the bacteria of a temporary pond." *Journal of General Microbiology* 47: 25–31.

Ferguson, E., Jr. 1944. "Studies on the seasonal life history of three species of freshwater Ostracoda." *American Midland Naturalist* 32: 713–727.

Ferguson, M. S. 1939. "Observations on *Eubranchipus vernalis* in south-western Ontario and eastern Illinois." *American Midland Naturalist* 22: 466–469.

Fernando, C. H. 1958. "The colonization of small freshwater habitats by aquatic insects. 1. General discussion, methods and colonization in the aquatic Coleoptera." *Ceylon Journal of Science* 1: 117–154.

Fernando, C. H., and D. Galbraith. 1973. "Seasonality and dynamics of aquatic insects colonizing small habitats." *Verhandlungen Internationale Vereinigung für Theoretische und Angewandte Limnologie* 18: 1564–1575.

Findlay, C. S., and J. Houlahan. 1997. "Anthropogenic correlates of species richness in southeastern Ontario wetlands." *Conservation Biology* 11: 1000–1009.

Fischer, Z. 1966. "Food selection and energy transformation in larvae of *Lestes sponsa* (Odonata) in astatic waters." *Verhandlungen Internationale Vereinigung für Theoretische und Angewandte Limnologie* 16: 600–603.

Fisher, S., and G. E. Likens. 1973. "Energy flow in Bear Brook, New Hampshire: An integrative approach to stream ecosystem metabolism." *Ecological Monographs* 43: 421–439.

Flannagan, J. F., and S. R. MacDonald. 1987. "Ephemeroptera and Trichoptera of peatlands and marshes in Canada." *Memoirs of the Entomological Society of Canada* 140: 47–56.

Flannagan, M. 2003. "Spring is here: Dive into a vernal pool near you!" *Great Lakes Habitat News* 11: 1.

Flint, O. S. 1960. "Taxonomy and ecology of Nearctic limnephilid larvae (Trichoptera), with special reference to species in eastern United States." *Entomologica America* 40: 1–120.

Floyd, M. A. 1995. "Larvae of the caddisfly genus *Oecetis* (Trichoptera: Leptoceridae) in North America." *Bulletin of the Ohio Biological Survey New Series* 10: 1–85.

Forbes, S. A. 1876. "List of Illinois Crustacea with descriptions of new species." *Bulletin of the Illinois Museum of Natural History* 1: 3–25.

Fore, S., J. R. Karr, and R. W. Wisseman. 1996. "Assessing invertebrate responses to human activities: Evaluating alternative approaches." *Journal of the North American Benthological Society* 15: 212–231.

Forester, D. C., and D. V. Lykens. 1986. "Significance of satellite males in a population of spring peepers." *Copeia* 1986: 719–724.

Formanowicz, D. R., Jr. 1986. "Anuran tadpole/aquatic insect predator-prey interactions: Tadpole size and predator capture success." *Herpetologica* 42: 367–373.

Formanowicz, D. R., Jr., and E. D. Brodie, Jr. 1981. "Prepupation behavior and a description of pupation in the predaceous diving beetle *Dytiscus verticalis* (Coleoptera: Dytiscidae)." *Journal of the New York Entomological Society* 89: 152–157.

Foster, D. R., G. Motzkin, D. Orwig, J. Aber, J. Melillo, R. Bowden, and F. Bazzaz. 1998. "Case studies in forest history and ecology from the Harvard Forest." *Arnoldia* 58: 32–44.

Fowler, J. A. 1991. "The marbled salamander." *The Jack-Pine Warbler* 68: 9–10.

Freda, J. 1983. "Diet of larval *Ambystoma maculatum* in New Jersey." *Journal of Herpetology* 17: 177–179.

Freda, J., and W. A. Dunson. 1984. "Sodium balance of amphibian larvae exposed to low environmental pH." *Physiological Zoology* 57: 435–443.

Freda, J., and W. A. Dunson. 1985a. "Field and laboratory studies of ion balance and growth rates of ranid tadpoles chronically exposed to low pH." *Copeia* 1985: 415–423.

Freda, J., and W. A. Dunson. 1985b. *The Effect of Acidic Precipitation on Amphibian Breeding in Temporary Ponds in Pennsylvania*. Biological Report 80(40.22). Washington, DC: US Fish and Wildlife Service, Eastern Energy and Land Use Team.

Freda, J., and W. A. Dunson. 1985c. "The influence of external cation concentration on the hatching of amphibian embryos in water of low pH." *Canadian Journal of Zoology* 63: 2649–2656.

Freda, J., and W. A. Dunson. 1986. "Effects of low pH and other chemical variables on the local distribution of amphibians." *Copeia* 1986: 454–466.

Freda, J., and P. J. Morin. 1984. *Adult Home Range of the Pine Barrens Treefrog (Hyla andersoni) and the Physical, Chemical, and Ecological Characteristics of its Preferred Breeding Ponds*. Final Report to the NJ Endangered and Nongame Species Project. New Brunswick, NJ: Division of Pinelands Research, Center for Coastal and Environmental Studies at Rutgers — The State University of New Jersey.

Friday, L. E. 1987. "The diversity of macroinvertebrate and macrophyte communities in ponds." *Freshwater Biology* 18: 87–104.

Froom, B. 1982. *Amphibians of Canada.* Toronto, ON: McClelland and Stewart.

Fuchs, C., and R. Mannesmann. 2003. "Indigenous Trematodes." Webpage. *http://www.uni-bielefeld.de/biologie/Didaktik/Zoologie/html_eng/trematoden_eng.html.*

Fuentes, L., and O. Mistretta. 1995. "Vernal pools in CPC California gardens." *Plant Conservation* 9: 1–2.

Galewski, K. 1971. "A study on morphobiotic adaptations of European species of the Dytiscidae (Coleoptera)." *Polski Pismo Entomologiczne* 41: 488–702.

Garcia, R., and K. S. Hagen. 1987. "Summer dormancy in adult *Agabus disintegratus* (Crotch) (Coleoptera: Dytiscidae) in dried ponds in California." *Annals of the Entomological Society of America* 80: 267–271.

Gascon, C., and D. Planas. 1986. "Spring water chemistry and the reproduction of the wood frog, *Rana sylvatica*." *Canadian Journal of Zoology* 64: 543–550.

Gates, J. E., and E. L. Thompson. 1982. "Small pool habitat selection by red-spotted newts in western Maryland." *Journal of Herpetology* 16: 7–15.

Gee, J. H., and R. C. Waldick. 1995. "Ontogenetic buoyancy changes and hydrostatic control in larval anurans." *Copeia* 1995: 861–870.

George, S. A., and M. R. Lennartz. 1980. "Methods for determining ploidy in amphibians: Nucleolar number and erythrocyte size." *Experientia* 36: 687–688.

Gernes, M. C., and J. C. Helgen. 1999. *Indexes of Biotic Integrity (IBI) for Wetlands: Vegetation and Invertebrate IBI's*. Final Report to US EPA, Assistance Number CD995525-01, April 1999. St. Paul, MN: Minnesota Pollution Control Agency, Environmental Outcomes Division.

Gibbs, J. P. 1993. "Importance of small wetlands for the persistence of local populations of wetland-associated animals." *Wetlands* 13: 25–31.

Gilbert, P. W. 1942. "Observations on the eggs of *Ambystoma maculatum* with especial reference to the green algae found within the egg envelopes." *Ecology* 23: 216–227.

Gilbert, P. W. 1944. "The alga-egg relationship in *Ambystoma maculatum*, a case of symbiosis." *Ecology* 25: 366–369.

Gilhen, J. 1984. *Amphibians and Reptiles of Nova Scotia*. Halifax, NS: Nova Scotia Museum.

Gillilland, M. G., III, and P. M. Muzzall. 2002. "Amphibians, trematodes, and deformities: An overview from southern Michigan." *Comparative Parasitology* 69: 81–85.

Glover, J. B. 1996. *Larvae of the Genus Triaenodes and Ylodes (Trichoptera: Leptoceridae) in North America*. New Series Bulletin 11(2). Columbus, OH: Ohio Biological Survey.

Glowa, J. 1992. "Spotted turtle." Pp. 105–107 *in*: M. L. Hunter, Jr., J. Albright, and J. Arbuckle, (eds.). *The Amphibians and Reptiles of Maine*. Maine Agricultural Experimental Station Bulletin 838. Orono, ME: University of Maine.

Goater, C. P. 1994. "Growth and survival of post-metamorphic toads: Interactions among larval history, density, and parasitism." *Ecology* 75: 2264–2274.

Goater, T. M., and C. P. Goater. 2001. "Ecological monitoring and assessment network (EMAN) protocols for measuring biodiversity: Parasites of amphibians and reptiles." Webpage: *http://eqb-dqe.cciw.ca/eman/ecotools/protocols/terrestrial/herp_parasites/intro.html.*

Gorbushin, A. 2002. "Trematode life cycles." Webpage. *http://www.angelfire.com/sc/gorbushin/cycles.html.*

Gosner, K. L., and I. H. Black. 1957. "The effects of acidity on the development and hatching of New Jersey frogs." *Ecology* 38: 256–262.

Gosselink, J. G., and R. E. Turner. 1978. "The role of hydrology in freshwater wetland ecosystems." Pp. 63–78 *in*: R. E. Good, D. F. Whigham, and R. L. Simpson, (eds.). *Freshwater Wetlands: Ecological Processes and Management Potential*. New York, NY: Academic Press.

Gosz, J. R., G. E. Likens, and F. H. Bormann. 1972. "Nutrient content of litter fall on the Hubbard Brook Experimental Forest, New Hampshire." *Ecology* 53: 769–784.

Gosz, J. R., G. E. Likens, and F. H. Bormann. 1973. "Nutrient release from decomposing leaf and branch litter in the Hubbard Brook Forest, New Hampshire." *Ecological Monographs* 43: 173–191.

Gosz, J. R., G. E. Likens, and F. H. Bormann. 1976. "Organic matter and nutrient dynamics of the forest floor in the Hubbard Brook forest." *Oecologia* 22: 305–320.

Gower, J. L., and E. J. Kormondy. 1963. "Life history of the damselfly *Lestes rectangularis* with special reference to seasonal regulation." *Ecology* 44: 398–402.

Graham, T. E. 1992. "Blanding's turtle." Pp. 112–115 *in*: M. L. Hunter, Jr., J. Albright, and J. Arbuckle, (eds.). *The Amphibians and Reptiles of Maine*. Maine Agricultural Experimental Station Bulletin 838. Orono, ME: University of Maine.

Graham, T. E. 1995. "Habitat use and populations parameters of the spotted turtle *Clemmys guttata*, a species of special concern in Massachusetts." *Chelonian Conservation and Biology* 1: 207–214.

Graham, T. E., and T. S. Doyle. 1977. "Growth and population characteristics of Blanding's turtles, *Emydoidea blandingii* in Massachusetts." *Herpetologica* 33: 410–414.

Graham, T. E., and T. S. Doyle. 1979. "Dimorphism, courtship, eggs, and hatchlings of the Blanding's turtle *Emydoidea blandingii* in Massachusetts." *Journal of Herpetology* 13: 125–127.

Graham, T. H. 1995. "Sympatric occurrence of Eubranchiopoda in ephemeral pools: A comment." *American Midland Naturalist* 133: 371–372.

Grant, K. P., and L. E. Licht. 1993. "Acid tolerance of anuran embryos and larvae from central Ontario." *Journal of Herpetology* 27: 1–6.

Green, A. J., J. Figuerola, and M. I. Sánchez. 2002. "Implications of waterbird ecology for the dispersal of aquatic organisms." *Acta Oecologica* 23: 177–189.

Gromko, M. H., F. S. Mason, and S. J. Smith-Gill. 1973. "Analysis of the crowding effect in *Rana pipiens* tadpoles." *Journal of Experimental Zoology* 186: 63–72.

Groves, J. D. 1980. "Mass predation on a population of the American toad, *Bufo americanus*." *American Midland Naturalist* 103: 202–203.

Gulve, P. S. 1994. "Distribution and extinction patterns within a northern metapopulation of the pool frog, *Rana lessonae*." *Ecology* 75: 1357–1367.

Hairston, N. G., Jr., and B. T. DeStasio, Jr. 1988. "Rate of evolution slowed by a dormant propagule pool." *Nature* 336: 239–242.

Hairston, N. G., Jr., and W. R. Munns, Jr. 1984. "The timing of copepod diapause as an evolutionarily stable strategy." *American Naturalist* 123: 733–751.

Hairston, N. G., Sr. 1987. *Community Ecology and Salamander Guilds*. Cambridge, UK: Cambridge University Press.

Hall, R. J., and E. Kolbe. 1980. "Bioconcentration of organophosphorus pesticides to hazardous levels by amphibians." *Journal of Toxicology and Environmental Health* 6: 853.

Hamilton, W. J., Jr. 1940. "The feeding habits of larval newts with reference to availability and predilection of food items." *Ecology* 21: 351–356.

Hammer, D. A. 1992. *Creating Freshwater Wetlands*. Chelsea, MI: Lewis Publishers.

Harding, J. H. 1997. *Amphibians and Reptiles of the Great Lakes Region*. Ann Arbor, MI: The University of Michigan Press.

Harris, R. N., R. A. Alford, and H. M. Wilbur. 1988. "Density and phenology of *Notophthalmus viridescens dorsalis* in a natural pond." *Herpetologica* 22: 234–242.

Hartland-Rowe, R. 1965. "The Anostraca and Notostraca of Canada with some new distribution records." *Canadian Field-Naturalist* 79: 185–189.

Hartland-Rowe, R. 1967. "*Eubranchipus intricatus*, n. sp., a widely distributed North American fairy shrimp, with a note on its ecology." *Canadian Zoologist* 45: 663–666.

Hasler, A. D., and A. T. Scholz. 1983. *Olfactory Imprinting and Homing in Salmon*. Berlin: Springer-Verlag.

Hassinger, D. D. 1970. "Notes on the thermal properties of frog egg." *Herpetologica* 26: 49–51.

Hassinger, D. D., J. D. Anderson, and G. H. Dalrymple. 1970. "The early life history and ecology of *Ambystoma tigrinum* and *Ambystoma opacum* in New Jersey." *American Midland Naturalist* 84: 474–495.

Hauer, F. R., and G. A. Lamberti, (eds.). 1996. *Methods in Stream Ecology*. Boston, MA: Academic Press.

Hausmann, P. S. 2002. "Regulation of isolated (non-federal) wetlands in Wisconsin — the Wisconsin response to SWANCC." Presentation at US EPA Region 5 Conference on Midwestern Ephemeral Wetlands: A Vanishing Habitat. Chicago, IL, February 21, 2002. Abstract available at *http://www.epa.gov/region5/water/ephemeralwetlands/pdf/eph_wet_abstracts.pdf*.

Healy, W. R. 1974. "Population consequences of alternative life histories in *Notophthalmus v. viridescens*." *Copeia* 1974: 221–229.

Heard, W. H. 1977. "Reproduction of fingernail clams (Sphaeriidae: *Sphaerium* and *Musculium*)." *Malacologia* 16: 421–455.

Heatwole, H. 1961. "Habitat selection and activity of the wood frog, *Rana sylvatica* LeConte." *American Midland Naturalist* 66: 197–200.

Heatwole, H., and L. L. Getz. 1960. "Studies of the amphibians and reptiles of Mud Lake Bog in southern Michigan." *The Jack-Pine Warbler* 38: 107–112.

Hecnar, S. J., and R. T. M'Closkey. 1996. "Amphibian species richness and distribution in relation to pond water chemistry in south-western Ontario, Canada." *Freshwater Biology* 36: 7–15.

Heinen, J. T. 1994. "Antipredator behavior of newly metamorphosed American toads (*Bufo a. americanus*), and mechanisms of hunting by eastern garter snakes (*Thamnophis s. sirtalis*)." *Herpetologica* 50: 137–145.

Helgen, J., R. G. McKinnell, and M. C. Gernes. 1998. "Investigation of malformed northern leopard frogs in Minnesota." Pp. 288–297 *in*: M. J. Lannoo, (ed.). *Status and Conservation of Midwestern Amphibians*. Iowa City, IA: University of Iowa Press.

Helgen, J., C. Prudhomme, and C. Miller. 2002. "Approaches for biological assessments of ephemeral wetlands in Minnesota." Presentation at US EPA Region 5 Conference on Midwestern Ephemeral Wetlands: A Vanishing Habitat. Chicago, IL, February 20, 2002. Abstract available at *http://www.epa.gov/region5/water/ephemeralwetlands/pdf/eph_wet_abstracts.pdf*.

Hemond, H. F. 1990. "Acid neutralizing capacity, alkalinity, and acid-base status of natural waters containing organic acids." *Environmental Science and Technology* 24: 1486–1489.

Henderson, M. T., G. Merriam, and J. Wegner. 1985. "Patchy environments and species survival: Chipmunks in an agricultural mosaic." *Biological Conservation* 31: 95–105.

Henson, E. B. 1988. "Macro-invertebrate associations in a marsh ecosystem." *Verhandlungen Internationale Vereinigung für Theoretische und Angewandte Limnologie* 23: 1049–1056.

Herreid, C. F., II, and S. Kinney. 1966. "Survival of Alaskan woodfrog (*Rana sylvatica*) larvae." *Ecology* 47: 1039–1041.

Herreid, C. F., II, and S. Kinney. 1967. "Temperature and development of the woodfrog, *Rana sylvatica*, in Alaska." *Ecology* 48: 579–590.

Higgins, M. J., and R. W. Merritt. 1997. "What is the role of leaf litter in temporary woodland pond foodchains?" Poster presented at 45th Annual Meeting, North American Benthological Society, San Marcos, TX, May 26–30, 1997. Abstract, *Bulletin of the North American Benthological Society* 14: 189.

Higgins, M. J., and R. W. Merritt. 1998. "Predator-prey patterns in temporary woodland ponds." Poster presented at 46th Annual Meeting, North American Benthological Society, Charlottetown, PE, June 1–3, 1998. Abstract, *Bulletin of the North American Benthological Society* 15: 212–213.

Higgins, M. J., and R. W. Merritt. 2001. "The influence of predation on cladoceran succession in a temporary woodland pond in Michigan." Poster presented at the 49th Annual Meeting, North American Benthological Society, La Crosse, WI, 2001. Abstract, *Bulletin of the North American Benthological Society* 18: 248.

Higley, R. 1918. *Morphology and Biology of some Turbellaria from the Mississippi Basin*. Illinois Biological Monographs 4(3). Urbana, IL: University of Illinois.

Hildrew, A. G. 1985. "A quantitative study of the life history of a fairy shrimp (Branchiopoda: Anostraca) in relation to the temporary nature of its habitat, a Kenyan rainpool." *Journal of Animal Ecology* 54: 99–110.

Hill, B. H., and W. T. Perrotte, Jr. 1995. "Microbial colonization, respiration, and breakdown of maple leaves along a stream-marsh continuum." *Hydrobiologia* 312: 11–16.

Hilsenhoff, W. L. 1975. "Notes on Nearctic *Acilius* (Dytiscidae) with the description of a new species." *Annals of the Entomological Society of America* 68: 271–274.

Hilsenhoff, W. L. 1991. "Diversity and classification of Insects and Collembola." Pp. 593–663 *in*: J. H. Thorp and A. P. Covich, (eds.). *Ecology and Classification of North American Freshwater Invertebrates*. New York, NY: Academic Press.

Hilsenhoff, W. L. 1994. "Dytiscidae and Noteridae of Wisconsin (Coleoptera). VI. Distribution, habitat, life cycle, and identification of species of Hydroporinae, except *Hydroporus* Clairville *sensu lato* (Hydroporinae)." *Great Lakes Entomologist* 28: 1–23.

Hinshaw, S. H., and B. K. Sullivan. 1990. "Predation on *Hyla versicolor* and *Pseudacris crucifer* during reproduction." *Journal of Herpetology* 24: 196–197.

Hoff, C. C. 1942. *The Ostracods of Illinois, their Biology and Taxonomy*. Illinois Biological Monographs 19(1–2). Urbana, IL: University of Illinois.

Holland, R., and S. Jain. 1981. "Insular biogeography of vernal pools in the Central Valley of California." *American Naturalist* 117: 24–37.

Holomuzki, J. R., and N. Hemphill. 1996. "Snail-tadpole interactions in streamside pools." *American Midland Naturalist* 136: 315–327.

Homan, R. N., J. M. Reed, and B. S. Windmiller. 2003. "Analysis of spotted salamander (*Ambystoma maculatum*) growth rates based on long-bone growth rings." *Journal of Herpetology* 37: 617–621.

Homan, R. N., J. V. Regosin, D. M. Rodrigues, J. M. Reed, B. S. Windmiller, and L. M. Romero. 2003. "Impacts of varying habitat quality on the physiological stress of spotted salamanders (*Ambystoma maculatum*)." *Animal Conservation* 6: 11–18.

Hopey, M. E., and J. W. Petranka. 1994. "Restriction of wood frogs to fish-free habitats: How important is adult choice?" *Copeia* 1994: 1023–1025.

Hoppe, D. M., and D. Pettus. 1984. "Developmental features influencing color polymorphism in chorus frogs." *Journal of Herpetology* 18: 113–120.

Hornbach, D. J., C. M. Way, and A. J. Burky. 1980. "Reproductive strategies in the freshwater sphaeriid clam, *Musculium partumeium* (Say), from a permanent and a temporary pond." *Oecologia* 44: 164–170.

Horne, M. T., and W. A. Dunson. 1994a. "Behavioral and physiological responses of the terrestrial life stages of the Jefferson salamander, *Ambystoma jeffersonianum*, to low soil pH." *Archives of Environmental Contamination and Toxicology* 27: 232–238.

Horne, M. T., and W. A. Dunson. 1994b. "Exclusion of the Jefferson salamander, *Ambystoma jeffersonianum*, from some potential breeding ponds in Pennsylvania: Effects of pH, temperature, and metals on embryonic development." *Archives of Environmental Contamination and Toxicology* 27: 323–330.

Horsfall, W. R. 1963. "Eggs of floodwater mosquitoes (Diptera: Culicidae). IX. Local distribution." *Annals of the Entomological Society of America* 56: 426–441.

Hoskins, D. 1998. "Connecticut's regulation of vernal pools." Pp. 49–50 *in:* B. Fellman, (ed.). *Our Hidden Wetlands: The Proceedings of a Symposium on Vernal Pools in Connecticut.* New Haven, CT: Yale University and Connecticut Department of Environmental Protection.

Hosseinie, S. O. 1976. " Comparative life histories of three species of *Tropisternus* in the laboratory (Coleoptera: Hydrophilidae)." *Internationale Revue der Gesamten Hydrobiologie* 61: 261–268.

Howard, R. D. 1978. "The influence of male-defended oviposition sites on early embryo mortality in bullfrogs." *Ecology* 59: 789–798.

Huckery, P. 1998. "The Massachusetts perspective." Pp. 51–53 *in:* B. Fellman, (ed.). *Our Hidden Wetlands: The Proceedings of a Symposium on Vernal Pools in Connecticut.* New Haven, CT: Yale University and Connecticut Department of Environmental Protection.

Hungerford, H. B. 1948. *The Corixidae of the Western Hemisphere (Hemiptera).* Science Bulletin 32. Lawrence, KS: University of Kansas.

Hungerford, H. B. 1959. "Hemiptera." Pp. 958–972 *in*: W. T. Edmonson, (ed.). *Freshwater Biology*. 2nd Edition. New York, NY: J. Wiley and Sons.

Hunter, M. L., Jr., J. Albright, and J. Arbuckle, (eds.). 1992. *The Amphibians and Reptiles of Maine*. Maine Agricultural Experimental Station Bulletin 838. Orono, ME: University of Maine.

Hunter, R. D. 1975. "Variation in populations of *Lymnaea palustris* in upstate New York." *American Midland Naturalist* 94: 401–420.

Husting, E. L. 1965. "Survival and breeding structure in a population of *Ambystoma maculatum*." *Copeia* 1965: 352–362.

Hynes, H. B. N. 1970. *The Ecology of Running Waters*. Toronto, ON: University of Toronto Press.

Hynes, H. B. N. 1975. "The stream and its valley." *Verhandlungen Internationale Vereinigung für Theoretische und Angewandte Limnologie* 19: 1–15.

Ingwerson, S., and B. Johnson. 2002. "An ephemeral WHAT land?: Toronto Zoo's Ephemeral Wetland Education Project." Presentation at US EPA Region 5 Conference on Midwestern Ephemeral Wetlands: A Vanishing Habitat. Chicago, IL, February 21, 2002. Abstract available at *http://www.epa.gov/region5/water/ephemeralwetlands/pdf/eph_wet_abstracts.pdf.*

Innes, D. J. 1997. "Sexual reproduction of *Daphnia pulex* in a temporary habitat." *Oecologia* 111: 53–60.

Ireland, P. H. 1989. "Larval survivorship in two populations of *Ambystoma maculatum*." *Journal of Herpetology* 23: 209–215.

Jackson, D. J. 1958. "Egg-laying and egg-hatching in *Agabus bipustulatus* L., with notes on oviposition in other species of *Agabus*." *Transactions of the Royal Entomological Society of London* 110: 58–80.

Jackson, M. E., D. E. Scott, and R. A. Estes. 1989. "Determinants of nest success in the marbled salamander (*Ambystoma opacum*)." *Canadian Journal of Zoology* 67: 2277–2281.

Jackson, S. D. 1988. "Occurrence of listed herptiles in a large Massachusetts wetland." Unpublished report to Massachusetts Natural Heritage and Endangered Species Program, Massachusetts Division of Fisheries and Wildlife, Westboro, MA.

Jackson, S. D., and C. R. Griffin. 1991. "Effects of pond chemistry on two syntopic mole salamanders, *Ambystoma jeffersonianum* and *A. maculatum*, in the Connecticut Valley of Massachusetts." Water Resources Research Center Publication 163. Amherst, MA: University of Massachusetts.

James, H. G. 1957. "*Mochlonyx velutinus* (Ruthe) (Diptera: Culicidae), an occasional predator of mosquito larvae." *Canadian Entomologist* 89: 472–480.

James, H. G. 1961. "Some predators of *Aedes stimulans* (Walk) and *Aedes trichurus* (Dyar) (Diptera: Culicidae) in woodland pools." *Canadian Journal of Zoology* 39: 533–540.

James, H. G. 1966a. "Insect predators of univoltine mosquitoes in woodland pools of the Precambrian shield in Ontario." *Canadian Entomologist* 98: 550–555.

James, H. G. 1966b. "Location of univoltine *Aedes* eggs in woodland pool areas and experimental exposure to predators." *Mosquito News* 26: 59–63.

James, H. G. 1967. "Seasonal activity of mosquito predators in woodland pools in Ontario." *Mosquito News* 27: 453–457.

James, H. G. 1969. "Immature stages of five diving beetles (Coleoptera: Dytiscidae), notes on their habits and life history, and a key to aquatic beetles of vernal woodland pools in southern Ontario." *Proceedings of the Entomological Society of Ontario* 100: 52–97.

James, H. G., and R. Redner. 1965. "An aquatic trap for sampling mosquito predators." *Mosquito News* 25: 35–37.

Jass, J. 1976. "Wisconsin record of *Eubranchipus serratus* Forbes 1897 questioned." *Crustaceana* 30: 313.

Jeffries, Michael. 1996. "Effects of *Notonecta glauca* predation on *Cyphon* larvae (Coleoptera: Scirtidae) in small seasonal ponds." *Archiv für Hydrobiologie* 136: 413–420.

Johnson, C. M., and L. B Johnson. 2002. "Patterns of ephemeral wetland use by wood frogs (*Rana sylvatica*)." Presentation at US EPA Region 5 Conference on Midwestern Ephemeral Wetlands: A Vanishing Habitat. Chicago, IL, February 20, 2002. Abstract available at *http://www.epa.gov/region5/water/ephemeralwetlands/pdf/eph_wet_abstracts.pdf.*

Johnson, D. M., and P. H. Crowley. 1980. "Habitat and seasonal segregation among coexisting odonate naiads." *Odonatologica* 9: 297–308.

Johnson, P. T. J., K. B. Lunde, E. G. Ritchie, and A. E. Launer. 1999. "The effect of trematode infection on amphibian limb development and survivorship." *Science* 294: 802–804.

Johnson, P. T. J., and D. R. Sutherland. 2003. "Amphibian deformities and *Ribeiroia* infection: An emerging helminthiasis." *Trends in Parasitology* 19: 332–336.

Jokinen, E. H. 1978. "The aestivation pattern of a population of *Lymnaea elodes* (Say) (Gastropoda: Lymnaeidae)." *American Midland Naturalist* 100: 43–53.

Jokinen, E. H. 1985. "Comparative life history patterns within a littoral zone snail community." *Verhandlungen Internationale Vereinigung für Theoretische und Angewandte Limnologie* 22: 3292–3299.

Jokinen, E. H. 1991. "The malacofauna of the acid and non-acid lakes and rivers of the Adirondack Mountains and surrounding lowlands, New York State, USA." *Verhandlungen Internationale Vereinigung für Theoretische und Angewandte Limnologie* 24: 2940–2946.

Jokinen, E. H. 1992. *The Freshwater Snails of New York State*. Bulletin 482. Albany, NY: New York State Museum.

Jokinen, E. H., and J. Morrison. 1998. "Invertebrate community structure and physico-chemical parameters of temporary ponds in Connecticut, USA." Poster presented at 46th Annual Meeting, North American Benthological Society, Charlottetown, PE, June 1–5, 1999. Abstract, *Bulletin of the North American Benthological Society* 15: 212.

Joyal, L. A. 1996. Ecology of Blanding's (*Emydoidea blandingii*) and spotted (*Clemmys guttata*) turtles in southern Maine: Population structure, habitat use, movements, and reproductive biology. MS Thesis, Wildlife Ecology, University of Maine, Orono.

Joyal, L. A., M. McCollough, and M. J. Hunter, Jr. 2001. "Landscape ecology approaches to wetland species conservation: A case study of two turtle species in southern Maine." *Conservation Biology* 15: 1755–1762.

Judd, W. W. 1957. "The food of Jefferson's salamander, *Ambystoma jeffersonianum*, in Rondeau Park, Ontario." *Ecology* 38: 77–81.

Kaplan, R. H. 1980. "The implications of ovum size variability for offspring fitness and clutch size within several populations of salamanders (*Ambystoma*)." *Evolution* 34: 51–64.

Kappel-Smith, D. 1996. "Pools of life." *Country Journal* May/June 1996: 23–25.

Karns, D. R. 1992. "Effects of acidic bog habitats on amphibian reproduction in a northern Minnesota peatland." *Journal of Herpetology* 26: 401–412.

Kats, L. B., and A. Sih. 1992. "Oviposition site selection and avoidance of fish by streamside salamanders (*Ambystoma barbouri*)." *Copeia* 1992: 468–473.

Kats, L. B., J. W. Petranka, and A. Sih. 1988. "Antipredator defenses and the persistence of amphibian larvae with fishes." *Ecology* 69: 1865–1870.

Kaushik, N. K., and H. B. N. Hynes. 1971. "The fate of the dead leaves that fall into streams." *Archiv für Hydrobiologie* 68: 465–515.

Keeler-Wolf, T., D. R. Elam, K. Lewis, and S. A. Flint. 2004. "California Vernal Pool Assessment Preliminary Report. Appendix C. Plant and Animal Taxa Known to be Associated with Vernal Pools." Modified 2 March 2004. Sacramento, CA: California Department of Fish and Game. Available at *http://maphost.dfg.ca.gov/wetlands/vp_asses_rept/app_c_anml.htm#Ostracoda*.

Keeley, J. E., and P. R. Zedler. 1998. "Characterization and global distribution of vernal pools." Pp. 1–14 *in*: C. W. Witham, E. T. Bauder, D. Belk, W. R. Ferren, Jr., and R. Ornduff, (eds.). *Ecology, Conservation, and Management of Vernal Pool Ecosystems — Proceedings from a 1996 Conference*. Sacramento, CA: California Native Plant Society.

Kenk, R. 1944. *The Freshwater Triclads of Michigan*. Miscellaneous Publications of the Museum of Zoology No. 60. Ann Arbor, MI: University of Michigan.

Kenk, R. 1949. *The Animal Life of Temporary and Permanent Ponds in Southern Michigan*. Miscellaneous Publications of the Museum of Zoology No. 71. Ann Arbor, MI: University of Michigan.

Kenk, R. 1972. *Freshwater Planarians (Turbellaria) of North America*. Biota of Freshwater Systems, Identification Manual No. 1. Washington, DC: US Environmental Protection Agency.

Kenk, R. 1989. *Revised List of the North American Freshwater Planarians (Platyhelminthes: Tricladida: Paludicola)*. Smithsonian Contributions to Zoology No. 476. Washington, DC: National Museum of Natural History, Smithsonian Institution.

Kenney, L. P. 1991. Vernal pools in a suburban community. MLA Thesis, Biology, Harvard University Extension, Cambridge, MA.

Kenney, L. P. 1994. *Wicked Big Puddles*. Reading, MA: Vernal Pool Association, Reading Memorial High School.

Kenney, L. P. 1996a. "Student certification of Massachusetts vernal pools." *Wetland Journal* 8: 4–7.

Kenney, L. P., (ed.). 1996b. *Diving into Wicked Big Puddles: An Educator's Resource Kit for the Study and Teaching of Vernal Pools*. Reading, MA: Vernal Pool Association, Reading Memorial High School.

Kenney, L. P., and M. R. Burne. 2000. *A Field Guide to the Animals of Vernal Pools*. Westboro, MA: Massachusetts Division of Fisheries and Wildlife, Natural Heritage and Endangered Species Program, and Reading, MA: Vernal Pool Association, Reading Memorial High School.

Kiesecker, J. M. 2002. "Synergism between trematode infection and pesticide exposure: A link to amphibian limb deformities in nature?" *Proceedings of the National Academy of Sciences* 99: 9900–9904.

Kiesecker, J. M., and D. K. Skelley. 2000. "Choice of oviposition site by gray treefrogs: The role of potential parasitic infection." *Ecology* 81: 2939–2943.

Kiesecker, J. M., and D. K. Skelley. 2001. "Effects of disease and pond drying on gray tree frog growth, development, and survival." *Ecology* 82: 1956–1963.

King, R. S., K. T. Nunnery, and C. J. Richardson. 2000. "Macroinvertebrate assemblage response to highway crossings in forested wetlands: Implications for biological assessment." *Wetlands Ecology and Management* 8: 243–256.

King, W. 1935. "Ecological observations on *Ambystoma opacum*." *Ohio Journal of Science* 35: 4–15.

Kingsbury, B., and J. Gibson. 2002. *Habitat Management Guidelines for Amphibians and Reptiles of the Midwest.* Partners in Reptile and Amphibian Conservation. Fort Wayne, IN: Center for Reptile and Amphibian Conservation and Management, Indiana-Purdue University.

Kirkman, L. K., S. W. Golladay, L. Laclaire, and R. Sutter. 1999. "Biodiversity in southeastern seasonally ponded, isolated wetlands: Management and policy perspectives for research and conservation." *Journal of the North American Benthological Society* 18: 553–562.

Kiviat, E., P. M. Groffman, G. Stevens, S. Nyman, and G. C. Hanson. 1994. "Reference wetlands in western New York." Report to the US Environmental Protection Agency, Region 2. Annandale, NY: Hudsonia, Ltd.

Kiviat, E., and K. MacDonald. 2002. "Hackensack Meadowlands, New Jersey, biodiversity: A review and synthesis." Report to the Hackensack Meadowlands Partnership. Annandale, NY: Hudsonia, Ltd.

Kleeberger, S. R., and J. K. Werner. 1983. "Post-breeding migration and summer movement of *Ambystoma maculatum*." *Journal of Herpetology* 17: 176–177.

Klemens, M. W. 1990. The herpetofauna of southwestern New England. PhD Dissertation, Conservation Biology and Ecology, University of Kent, Canterbury, UK.

Klemens, M. W. 1993. *Amphibians and Reptiles of Connecticut and Adjacent Regions*. Bulletin 112. Hartford, CT: State Geological and Natural History Survey of Connecticut.

Klemens, M. W. 1998. "Ephemeral wetlands — ephemeral protection?" Pp. 15–17 *in*: B. Fellman, (ed.). *Our Hidden Wetlands: The Proceedings of a Symposium on Vernal Pools in Connecticut*. New Haven, CT: Yale University and Connecticut Department of Environmental Protection.

Klemm, D. J. 1972. *Freshwater Leeches (Annelida: Hirudinea) of North America*. Biota of Freshwater Ecosystems, Identification Manual No. 8. Washington, DC: US Environmental Protection Agency.

Klemm, D. J. 1976. "Leeches (Annelida: Hirudinea) found in North American molluscs." *Malacological Review* 9: 63–76.

Knox, C. B. 1992. "Wood frog." Pp. 86–91 *in*: M. L. Hunter, Jr., J. Albright, and J. Arbuckle, (eds.). *The Amphibians and Reptiles of Maine*. Maine Agricultural Experimental Station Bulletin 838. Orono, ME: University of Maine.

Knutson, M. G., J. R. Sauer, D. A. Olsen, M. J. Mossman, L. M. Hemesath, and M. J. Lannoo. 1999. "Effects of landscape composition and wetland fragmentation on frog and toad abundance and species richness in Iowa and Wisconsin, USA." *Conservation Biology* 13: 1437–1446.

Kolasa, J. 1991. "Flatworms: Turbellaria and Nemertea." Pp. 145–186 *in*: J. H. Thorp and A. P. Covich, (eds.). *Ecology and Classification of North American Freshwater Invertebrates*. New York, NY: Academic Press.

Kraus, F., P. K. Ducey, P. Moler, and M. M. Miyamoto. 1991. "Two new triparental unisexual *Ambystoma* from Ohio and Michigan." *Herpetologica* 47: 429–439.

Kraus, M. L. 1989. "Bioaccumulation of heavy metals in pre-fledgling tree swallows, *Tachycineta bicolor*." *Bulletin of Environmental Contamination and Toxicology* 43: 407–414.

Krenz, J. D., and D. E. Scott. 1994. "Terrestrial courtship affects mating locations in *Ambystoma opacum*." *Herpetologica* 50: 46–50.

Krenz, J. D., and D. M. Sever. 1995. "Mating and oviposition in paedomorphic *Ambystoma talpoideum* precedes the arrival of terrestrial males." *Herpetologica* 51: 387–393.

Kumpf, K. F., and S. C. Yeaton, Jr. 1932. "Observations on the courtship behavior of *Ambystoma jeffersonianum.*" Novitates No. 546. New York, NY: American Museum of Natural History.

Kusler, J. A., and M. E. Kentula, (eds.). 1990. *Wetland Creation and Restoration — The Status of the Science*. Washington, DC: Island Press.

Lagler, K. F. 1943. "Food habits and economic relations of turtles of Michigan with special reference to fish management." *American Midland Naturalist* 29: 257–312.

Lahr, J., A. O. Diallo, B. Gadji, P. S. Diouf, J. J. M. Bedaux, A. Badji, K. B. Ndour, J. E. Andreasen, and N. M. Van Straalen. 2000. "Ecological effects of experimental insecticide applications on invertebrates in Sahelian temporary ponds." *Environmental Toxicology and Chemistry* 19: 1278–1289.

Laird, M. 1988. *The Natural History of Larval Mosquito Habitats*. New York, NY: Academic Press.

Lake, P. S., I. A. E. Bayly, and D. W. Morton. 1989. "The phenology of a temporary pond in western Victoria, Australia, with special reference to invertebrate succession." *Archiv für Hydrobiologie* 115: 171–202.

Lanciani, C. A. 1970. "Resource partitioning in species of the water mite genus *Eylais*." *Ecology* 51: 338–342.

Lanciani, C. A. 1971. "Host-related size of parasitic water mites of the genus *Eylais*." *American Midland Naturalist* 85: 242–247.

Lanciani, C. A. 1975. "Parasite-induced alterations in host reproduction and survival." *Ecology* 56: 689–695.

Lanciani, C. A. 1976. "Intraspecific competition in the parasitic water mite, *Hydryphantes tenuabilis*." *American Midland Naturalist* 96: 210–214.

Lanciani, C. A., and J. M. Boyett. 1980. "Demonstrating parasitic water mite-induced mortality in natural host populations." *Parasitology* 81: 465–475.

Landy, M. J. 1967. A study of the life histories of two sympatric species of ambystomatid salamanders, *Ambystoma jeffersonianum* (Green) and *Ambystoma maculatum* (Shaw), from Franklin and Hampshire counties in Massachusetts. MS Thesis, Zoology, University of Massachusetts, Amherst, MA.

Lannoo, M. J., (ed.). 1998. *Status and Conservation of Midwestern Amphibians*. Iowa City, IA: University of Iowa Press.

Larson, D. J. 1985. "Structure in temperate predaceous diving beetle communities (Coleoptera: Dytiscidae)." *Holarctic Ecology* 8: 18–32.

Larson, D. J. 1987a. "Aquatic Coleoptera of peatlands and marshes in Canada." *Memoirs of the Entomological Society of Canada* 140: 99–132.

Larson, D. J. 1987b. "Revision of North American species of *Ilybius* Erichson (Coleoptera: Dytiscidae) with systematic notes on Palearctic species." *Journal of the New York Entomological Society* 95: 341–413.

Larson, D. J. 1996. "Revision of North American *Agabus* Leach (Coleoptera: Dytiscidae): The *opacus* group." *Canadian Entomologist* 128: 613–665.

Larson, D. J., Y. Alarie, and R. E. Roughley. 2000. *Predaceous Diving Beetles (Coleoptera: Dytiscidae) of the Nearctic Region, With an Emphasis on the Fauna of Canada and Alaska*. Ottawa, ON: National Research Council of Canada Research Press.

Layne, J. R. 1991. "External ice triggers freezing in freeze-tolerant frogs at temperatures above their supercooling point." *Journal of Herpetology* 25: 129–130.

LeBlanc, A., A. Maire, and A. Aubin. 1981. "Écologie et dynamique des populations de Copépodes (Cyclopoida) des principaux types de milieux astatiques temporaries de la zone tempérée du Québec méridional." *Canadian Journal of Zoology* 59: 722–732.

Leech, H. B. 1939. "On some Nearctic species of *Agabus,* with the description of a new species (Coleoptera: Dytiscidae)." *Canadian Entomologist* 71: 217–221.

Lehmkuhl, D. M. 1971. "Stoneflies (Plecoptera: Nemouridae) from temporary lentic habitats in Oregon." *American Midland Naturalist* 85: 514–515.

Leibowitz, S. G. 2003. "Isolated wetlands and their functions: An ecological perspective." *Wetlands* 23: 517–531.

Leibowitz, S. G., and T.-L. Nadeau. 2003. "Isolated wetlands: State-of-the-science and future directions." *Wetlands* 23: 663–684.

Leonard, M. D. 1928. *A List of the Insects of New York, with a List of the Spiders and Certain Other Allied Groups.* Ithaca, NY: Cornell University Press.

Licht, L. E. 1974. "Survival of embryos, tadpoles, and adults of the frogs *Rana aurora aurora* and *Rana pretiosa pretiosa* sympatric in southwestern British Columbia." *Canadian Journal of Zoology* 52: 613–627.

Licht, L. E., and J. P. Bogart. 1987. "Comparative size of epidermal cell nuclei from shed skin of diploid, triploid, and tetraploid salamanders (Genus *Ambystoma*)." *Copeia* 1987: 284–290.

Licht, L. E., and J. P. Bogart. 1989. "Embryonic development and temperature tolerance in diploid and polyploid salamanders (Genus *Ambystoma*)." *American Midland Naturalist* 122: 401–407.

Likens, G. E., F. H. Bormann, R. S. Pierce, J. S. Eaton, and N. M. Johnson. 1977. *Biogeochemistry of a Forested Ecosystem*. New York, NY: Springer-Verlag.

Ling, R. W., J. W. van Amberg, and J. K. Werner. 1986. "Pond acidity and its relationship to larval development of *Ambystoma maculatum* and *Rana sylvatica* in upper Michigan." *Journal of Herpetology* 20: 230–236.

Lipps, K. R. 1998. "Decline of a tropical montane amphibian fauna." *Conservation Biology* 12: 106–117.

Litzgus, J. D., and R. J. Brooks. 1998. "Reproduction in a northern population of *Clemmys guttata*." *Journal of Herpetology* 32: 252–273.

Litzgus, J. D., J. P. Costanzo, R. J. Brooks, and R. E. Lee, Jr. 1999. "Phenology and ecology of hibernation in spotted turtles (*Clemmys guttata*) near the northern limit of their range." *Canadian Journal of Zoology* 77: 1348–1357.

Loafman, P. 1991. "Identifying individual spotted salamanders by spot pattern." *Herpetological Review* 22: 91–92.

Lockyer, A. E., C. S. Jones, L. R. Noble, and D. Rollinson. 2004. "Trematodes and snails: An intimate association." *Canadian Journal of Zoology* 82: 251–269.

Longcore, J. E., A. P. Pessier, and D. K. Nichols. 1999. "*Batrachochytrium dendrobatidis* gen. et sp. Nov., a chytrid pathogenic to amphibians." *Mycologia* 9: 219–227.

Lowcock, L. A., H. Griffin, and R. W. Murphy. 1992. "Size in relation to sex, hybridity, ploidy, and breeding dynamics in Central Ontario populations of the *Ambystoma laterale-jeffersonianum* complex." *Journal of Herpetology* 26: 46–53.

Lowe, R. L., and G. D. Laliberte. 1996. "Biomass and pigments of benthic algae." Pp. 295–314 *in*: F. R. Hauer and G. A. Lamberti, (eds.). *Methods in Stream Ecology*. Boston, MA: Academic Press.

Lykens, D. V., and D. C. Forester. 1987. "Age structure in the spring peeper: Do males advertise longevity?" *Herpetologica* 43: 216–223.

MacArthur, R. H., and E. O. Wilson. 1967. *The Theory of Island Biogeography*. Princeton, NJ: Princeton University Press.

MacGregor, H. C., and T. M. Uzzell, Jr. 1964. "Gynogenesis in salamanders related to *Ambystoma jeffersonianum*." *Science* 143: 1043–1045.

Mackie, G. L. 1979. "Growth dynamics in natural populations of Sphaeriidae clams *(Sphaerium, Musculium, Pisidium)*." *Canadian Journal of Zoology* 57: 441–456.

Mackie, G. L., and S. U. Qadri. 1974. "Calyculism in *Musculium securis* (Pelecypoda: Sphaeriidae) and its significance." *Canadian Journal of Zoology* 52: 977–980.

Mackie, G. L., S. U. Qadri, and A. H. Clarke. 1976. "Reproductive habits of four populations of *Musculium securis* (Bivalvia: Sphaeriidae) near Ottawa, Canada." *Nautilus* 90: 76–86.

Mackie, G. L., S. U. Qadri, and R. M. Reed. 1978. "Significance of litter size in *Musculium securis* (Bivalvia: Sphaeriidae)." *Ecology* 59: 1069–1074.

Madill, J. 1985. *Bibliographia Invertebratorum Aquaticum Canadensium. Volume 5. Synopsis Speciorum. Annelida: Hirudinea*. Ottawa, ON: National Museum of Canada.

Magee, D. W. 1981. *Freshwater Wetlands: A Guide to Common Indicator Plants of the Northeast.* Amherst, MA: University of Massachusetts Press.

Maguire, B. J. 1963. "The passive dispersal of small aquatic organisms and their colonization of isolated bodies of water." *Ecological Monographs* 33: 161–185.

Mahoney, D. L., M. A. Mort, and B. E. Taylor. 1990. "Species richness of calanoid copepods, cladocerans, and other branchiopods in Carolina Bay temporary ponds." *American Midland Naturalist* 123: 244–258.

Maier, G. 1989. "The seasonal cycle of *Metacyclops gracilis* (Lilljeborg) in a shallow pond." *Archiv für Hydrobiologie* 115: 97–110.

Maier, G. 1992. "Development, reproduction, and growth pattern of two coexisting, pond dwelling cladocerans." *Internationale Revue der Gesamten Hydrobiologie* 77: 621–632.

Maier, G. 1993. "The life histories of two temporarily coexisting pond dwelling cladocerans." *Internationale Revue der Gesamten Hydrobiologie* 78: 83–93.

Maier, G., J. Hössler, and U. Tessenow. 1998. "Succession of physical and chemical conditions and of crustacean communities in some small, man made water bodies." *International Review of Hydrobiology* 83: 405–418.

Malcolm, S. E. 1971. *The Water Beetles of Maine: Including the Families Gryrinidae, Haliplidae, Dytiscidae, Noteridae, and Hydrophilidae*. Maine Agricultural Experimental Station Bulletin 48. Orono, ME: University of Maine.

Mantilacci, L., M. Mearelli, and O. Tiberi. 1976. "Halophilic Odonata in the lakes of Portonovo (Marche-Italy)." *Rivista di Idrobiologia* 15: 169–172. Abstract, *Biological Abstracts* 67: 16151.

Marcus, N. H., R. Lutz, W. Burnett, and P. Cable. 1994. "Age, viability, and vertical distribution of zooplankton resting eggs from an anoxic basin: Evidence of an egg bank." *Limnology and Oceanography* 39: 154–158.

Mark, R. 1999. "A Guide to the Wildflowers of Twin Swamps Nature Preserve in Posey County, Indiana." Webpage: *http://www.usi.edu/science/biology/TwinSwamps/Hottonia_inflata.htm*.

Marshall, W. H., and M. F. Buell. 1955. "A study of the occurrence of amphibians in relation to a bog succession, Itasca State Park, Minnesota." *Ecology* 36: 381–387.

Marten, G. G. 1999. "Cyclopoids, mosquitoes, and dengue hemmorhagic fever." *Monoculus* 38: 19–25. Available at *http://www.copepoda.uconn.edu/Monoculus_38.pdf*.

Martin, J. W., and D. Belk. 1988. "Review of the clam shrimp family Lynceidae Stebbing, 1902 (Branchiopoda: Conchostraca) in the Americas." *Journal of Crustacean Biology* 8: 451–482.

Martin, J. W., B. E. Felgenhaure, and L. G. Abele. 1986. "Redescription of the clam shrimp *Lynceus gracilicornis* (Packard) (Branchiopoda, Conchostraca, Lynceidae) from Florida, with notes on its biology." *Zoologica Scripta* 15: 221–232.

Martof, B. 1953. "Home range and movements of the green frog, *Rana clamitans*." *Ecology* 34: 529–543.

Martof, B., and R. L. Humphries. 1959. "Geographic variation in the wood frog, *Rana sylvatica*." *American Midland Naturalist* 61:350–389.

Massachusetts Audubon Society. 1995. *PondWatchers: Guide to Ponds and Vernal Pools of Eastern North America*. Laminated field identification guide and information sheet. Lincoln, MA: Center for Biological Conservation.

Massachusetts Audubon Society. 1999. "Amphibians of the Cape Cod National Seashore: Interim Report to the National Park Service." Lincoln, MA: Center for Biological Conservation.

Massachusetts Department of Environmental Management. 1995. "Massachusetts forestry best management practices manual." Boston, MA.

Massachusetts Department of Environmental Protection. 1995. "Interpretations of 310 CMR 10.57(2)(b) Definition of Isolated Land Subject to Flooding." Wetlands Program Policy 85–2, revised 1995. Boston, MA: Wetlands Division.

Massachusetts Department of Public Health. 1940. *Survey of Mosquitoes of Massachusetts*. Boston, MA.

Massachusetts Division of Fisheries and Wildlife. 1988. "Guidelines for certification of vernal pool habitat." Westboro, MA: Natural Heritage and Endangered Species Program.

Massey, A. 1990. "Notes on the reproductive ecology of red-spotted newts (*Notophthalmus viridescens*)." *Journal of Herpetology* 24: 106–107.

Masters, C. O. 1968. *Pond Life: A Field Guide to the Inhabitants of Temporary Ponds*. Jersey City, NJ: TFH Publications, Inc.

Mastrantuono, L. 1994. "Cyclopoids (Crustacea: Copepoda) of temporary pools of Latium (Italy): Composition, distribution, and comparison with other data of central and southern Italy." Abstract, *Biological Abstracts* 97: 39310.

Matheson, R. 1914. "Notes on *Hydrophilus triangularis* Say." *Canadian Entomologist* 46: 337–343.

Matta, J. F., and A. G. Michael. 1976. "A new subspecies of *Acilius* (Coleoptera: Dytiscidae) from the southeastern United States." *Entomological News* 87: 11–16.

Mattox, N. T. 1937. "Studies on the life history of a new species of fairy shrimp, *Eulimnadia diversa*." *Transactions of the American Microscopical Society* 56: 249–255.

Mattox, N. T. 1939. "Descriptions of two new species of the genus *Eulimnadia* and notes on the other Phyllopoda of Illinois." *American Midland Naturalist* 22: 642–653.

Mauger, D. 2002. "A summary of conservation efforts of forested ephemeral wetlands undertaken by the Forest Preserve District of Will County." Poster presentation at US EPA Region 5 Conference on Midwestern Ephemeral Wetlands: A Vanishing Habitat. Chicago, IL, February 20–21, 2002. Abstract available at *http://www.epa.gov/region5/water/ephemeralwetlands/pdf/eph_wet_abstracts.pdf.*

Mayer, M. S., and G. E. Likens. 1987. "The importance of algae in a shaded headwater stream as food for an abundant caddisfly (Trichoptera)." *Journal of the North American Benthological Society* 6: 262–269.

McCraw, B. M. 1959. "The ecology of the snail *Lymnaea humilis* Say." *Transactions of the American Microscopical Society* 78: 101–121.

McCraw, B. M. 1960. "Life history and growth of the snail *Lymnaea humilis* Say." *Transactions of the American Microscopical Society* 80: 16–27.

McDiffett, W., and T. E. Jordan. 1978. "The effects of an aquatic detritivore on the release of inorganic N and P from decomposing leaf litter." *American Midland Naturalist* 99: 36–44.

McGee, E., E. O. Mill, and D. Mauger. 1989. "Baseline survey of a spotted turtle (*Clemmys guttata*) population at Romeoville Prairie Nature Reserve, Will Co., Illinois." Unpublished report. Charleston, IL: Forest Preserve District of Will County.

McKee, P. M., and G. L. Mackie. 1979. "Incidence of *Marvinmeyeria lucida* (Hirudinea: Glossiphoniidae) in the fingernail clam, *Sphaerium occidentale*." *Canadian Journal of Zoology* 57: 499–503.

McKee, P. M., and G. L. Mackie. 1980. "Desiccation resistance in *Sphaerium occidentale* and *Musculium securis* (Bivalvia: Sphaeriidae) from a temporary pond." *Canadian Journal of Zoology* 58: 1693–1696.

McKee, P. M., and G. L. Mackie. 1981. "Life history adaptations of the fingernail clams *Sphaerium occidentale* and *Musculium securis* to ephemeral habitats." *Canadian Journal of Zoology* 59: 2219–2229.

McKee, P. M., and G. L. Mackie. 1983. "Respiratory adaptations of the fingernail clams *Sphaerium occidentale* and *Musculium securis* to ephemeral habitats." *Canadian Journal of Fisheries and Aquatic Sciences* 40: 783–791.

McLeod, R. F., and J. E. Gates. 1998. "Response of herpetofaunal communities to forest cutting and burning at Chesapeake Farms, Maryland." *The American Midland Naturalist* 139: 164–177.

McMahon, R. F. 1991. "Mollusca: Bivalvia." Pp. 315–399 *in*: J. H. Thorp and A. P. Covich, (eds.). *Ecology and Classification of North American Freshwater Invertebrates*. New York, NY: Academic Press.

McPeek, M. A. 1990a. "Determination of species composition in the *Enallagma* damselfly assemblages of permanent lakes." *Ecology* 71: 83–98.

McPeek, M. A. 1990b. "Behavioral differences between *Enallagma* species (Odonata) influencing differential vulnerability to predators." *Ecology* 71: 1714–1726.

McWilliams, S. R., and M. D. Bachmann. 1989. "Predatory behavior of larval small-mouthed salamanders (*Ambystoma texanum*)." *Herpetologica* 45: 459–467.

Means, D. B., J. G. Palis, and M. Baggett. 1996. "Effects of slash pine silviculture on a Florida population of flatwoods salamander." *Conservation Biology* 10: 426–437.

Medland, V. L., and B. E. Taylor. 2001. "Strategies of emergence from diapause for cyclopoid copepods in a temporary pond." *Archives of Hydrobiology* 150: 329–349.

Meeks, D. E., and J. W. Nagel. 1973. "Reproduction and development of the wood frog, *Rana sylvatica*, in Tennessee." *Herpetologica* 29: 188–191.

Merritt, R. W., and K. W. Cummins. 1996a. "Trophic relations of macroinvertebrates." Pp. 453–474 *in*: J. H. Thorp and A. P. Covich, (eds.). *Ecology and Classification of North American Freshwater Invertebrates*. New York, NY: Academic Press.

Merritt, R. W., and K. W. Cummins, (eds.). 1996b. *An Introduction to the Aquatic Insects of North America*. 3rd Edition. Dubuque, IA: Kendall/Hunt Publishing Company.

Meyer, J. L., and C. Johnson. 1983. "The influence of elevated nitrogen concentration on the rate of leaf decomposition in a stream." *Freshwater Biology* 13: 177–183.

Middleton, J., and G. Merriam. 1983. "Distribution of woodland species in farmland woods." *Journal of Applied Ecology* 20: 625–244.

Miersma, K. 2003. "Western Chorus Frog, *Pseudacris triseriata*." Webpage: *http://kmier.net/ecology/chorus.html*.

Milam, J. C. 1997. Home range, habitat use, and conservation of spotted turtles (*Clemmys guttata*) in central Massachusetts. MS Thesis, Forestry and Wildlife Management, University of Massachusetts, Amherst, MA.

Milam, J. C., and S. M. Melvin. 1995. "Spotted turtles habitat use and seasonal movements in central Massachusetts." Presentation at Society of Wetland Scientists 16th Annual Meeting, June 1, 1995, Boston, MA.

Milam, J. C., and S. M. Melvin. 2001. "Density, habitat use, movements, and conservation of spotted turtles (*Clemmys guttata*) in Massachusetts." *Journal of Herpetology* 35: 418–427.

Milius, S. 1998. "Fatal skin fungus found in US frogs." *Science News* 154: 7.

Milliger, L. E., K. W. Stewart, and J. K. G. Silvey. 1971. "The passive dispersal of viable algae, protozoans, and fungi by aquatic and terrestrial Coleoptera." *Annals of the Entomological Society of America* 64: 36–45.

Milne, L., and M. Milne. 1980. *The Audubon Society Field Guide to North American Insects and Spiders*. New York, NY: Alfred A. Knopf.

Minshall, G. W. 1967. "Role of allochthonous detritus in the trophic structure of a woodland springbrook community." *Ecology* 48: 139–149.

Minshall, G. W. 1996. "Organic matter budgets." Pp. 591–606 *in*: F. R. Hauer and G. A. Lamberti, (eds.). *Methods in Stream Ecology*. Boston, MA: Academic Press.

Minton, S. A., Jr. 1954. "Salamanders of the *Ambystoma jeffersonianum* complex in Indiana." *Herpetologica* 10: 173–179.

Mitchell, R. 1954. "Checklist of North American water-mites." *Fieldiana: Zoology* 35: 29–70.

Mitsch, W. J., and J. G. Gosselink. 2000. *Wetlands*. 3rd Edition. New York, NY: J. Wiley and Sons.

Modlin, R. F. 1982. "A comparison of two *Eubranchipus* species (Crustacea: Anostraca)." *American Midland Naturalist* 107: 107–113.

Moore, J. P. 1949. "*Dina bucera*, n. sp." P. 38 *in*: R. Kenk. *The Animal Life of Temporary and Permanent Ponds in Southern Michigan*. Miscellaneous Publications of the Museum of Zoology No. 71. Ann Arbor, MI: University of Michigan.

Moore, P. D., and D. J. Bellamy. 1974. *Peatlands*. New York, NY: Springer-Verlag.

Moore, W. G. 1959. "Observations on the biology of the fairy shrimp, *Eubranchipus holmani*." *Ecology* 40: 398–403.

Moore, W. G. 1963. "Some interspecies relationships in Anostraca populations in certain Louisiana ponds." *Ecology* 44: 131–139.

Moore, W. G. 1979. "The effect of prolonged wet storage on the hatchability of eggs of *Streptocephalus seali* Ryder (Branchiopoda: Anostraca)." *Southwest Naturalist* 24: 201–203.

Moore, W. G., and B. F. Faust. 1972. "Crayfish as possible agents of dissemination of fairy shrimp into temporary ponds." *Ecology* 53: 314–316.

Morey, S. R. 1998. "Pool duration influences age and body mass at metamorphosis in the western spadefoot toad: Implications for vernal pool conservation." Pp. 86–91 *in*: C. W. Witham, E. T. Bauder, D. Belk, W. R. Ferren, Jr., and R. Ornduff, (eds.). *Ecology, Conservation, and Management of Vernal Pool Ecosystems — Proceedings from a 1996 Conference*. Sacramento, CA: California Native Plant Society.

Morgan, A. H. 1930. *Field Book of Ponds and Streams*. New York, NY: G. P. Putnams Sons.

Morin, P. J. 1981. "Predatory salamanders reverse the outcome of competition among three species of anuran tadpoles." *Science* 212: 1284–1286.

Morin, P. J. 1983. "Competitive and predatory interactions in natural and experimental populations of *Notophthalmus viridescens dorsalis* and *Ambystoma tigrinum*." *Copeia* 1983: 628–639.

Morris, M. A. 1974. "An Illinois record for a triploid species of the *Ambystoma jeffersonianum* complex." *Journal of Herpetology* 8: 255–256.

Morris, M. A. 1985. "A hybrid *Ambystoma platineum* x *Ambystoma tigrinum* from Indiana." *Herpetologica* 41: 267–272.

Morse, J. C. (ed.). 2001. "Trichoptera world checklist." Webpage. Version 1: 22 May 1999. Version 2: 28 July 2000. Version 3: 8 January 2001. *http://entweb.clemson.edu/database/trichopt/index/htm.*

Morse, M. 1904. "Batrachians and reptiles of Ohio." *Proceedings of the Ohio State Academy of Science* 4: 95–144.

Morton, D. W., and I. A. E. Bayly. 1977. "Studies on the ecology of some temporary freshwater pools in Victoria with special reference to microcrustaceans." *Australian Journal of Marine and Freshwater Research* 28: 439–454.

Mozley, A. 1928. "Notes on some fresh water Mollusca inhabiting temporary ponds in western Canada." *Nautilus* 42: 19–20.

Mozley, A. 1932. "A biological study of a temporary pond in western Canada." *American Midland Naturalist* 66: 235–249.

Mui, J. M., and R. E. Szafoni. 2002. "Ephemeral pond restoration for amphibian conservation: A case history from Vermillion County, Illinois." Presentation at US EPA Region 5 Conference on Midwestern Ephemeral Wetlands: A Vanishing Habitat. Chicago, IL, February 21, 2002. Abstract available at *http://www.epa.gov/region5/water/ephemeralwetlands/pdf/eph_wet_abstracts.pdf.*

Munz, P. A. 1920. "A study of the food habits of the Ithacan species of Anura during transformation." *Journal of Entomology and Zoology* 12: 33–56.

Mura, G. 1993. "Italian Anostraca: Distribution and status." *Anostracan News* 1: 3.

Mura, G., G. Fancello, and S. Di Giuseppe. 2003. "Adaptive strategies in populations of *Chirocephalus diaphanus* (Crustacea: Anostraca) from temporary waters in the Reatine Apennines (central Italy)." *Journal of Limnology* 62: 35–40.

Musgrove, K. L., and E. A. Colburn. 2003. "Using hydroperiod and historical vegetation data to explain patterns in the distribution of malacostracan crustaceans in Massachusetts vernal pools." Poster presented at 51st Annual Meeting, North American Benthological Society. Abstract, *Bulletin of the North American Benthological Society* 20: 324–325.

Mushinsky, H. R., and E. D. Brodie, Jr. 1975. "Selection of substrate pH by salamanders." *American Midland Naturalist* 93: 440–443.

NatureServe. 2003. *NatureServe Explorer: An Online Encyclopedia of Life [web application]*. Version 1.8. Arlington, VA: NatureServe. *http://www.natureserve.org/explorer*.

Nayar, J. K., and A. Ali. 2003. "A review of monomolecular surface films as larvicides and pupicides of mosquitoes." *Journal of Vector Ecology* 28: 190–199.

Needham, J. G. 1903. "Aquatic Insects in New York State. Part 3. Life histories of Odonata, sub-order Zygoptera." *Bulletin of the New York State Museum* 68: 218–279.

Needham, J. G., and M. J. Westfall. 1954. *A Manual of the Dragonflies of North America (Anisoptera)*. Berkeley, CA: University of California Press.

Needham, J. G., M. J. Westfall, and M. L. May. 2000. *Dragonflies of North America.* Gainesville, FL: Scientific Publishers.

Nelson, C. E., and R. R. Humphrey. 1972. "Artificial interspecific hybridization among *Ambystoma*." *Herpetologica* 28: 27–32.

Newcomb, L. 1977. *Newcomb's Wildflower Guide.* Toronto, ON: Little, Brown and Company.

Newell, I. M. 1959. "Acari." Pp. 1080–1116 *in*: W. T. Edmonson, (ed.). *Freshwater Biology*. 2nd Edition. New York, NY: J. Wiley and Sons.

New England Aquarium. 1985. *Spring Pool.* Video and film on salamander migration and vernal pool ecology. Boston, MA.

New England Interstate Water Pollution Control Commission. 1998. *Field Indicators for Identifying Hydric Soils in New England.* Version 2, July 1998. Lowell, MA: NEIWPCC.

New Jersey Agricultural Experiment Station. 1960. "The story of the mosquito." Circular 585. New Brunswick, NJ: Rutgers, The State University.

New Jersey Division of Fish and Wildlife. 2003. "2003 vernal pool training seminars." Webpage: *http://www.njfishandwildlife.com/ensp/vpoltrn03.htm.*

New York State Department of Environmental Conservation. 2003. "Eastern tiger salamander fact sheet." Webpage: *http://www.dec.state.ny.us/website/dfwmr/wildlife/endspec/tisafs.html.*

Niering, W. A. 1985. *Wetlands*. Audubon Society Nature Guides. New York, NY: Alfred A. Knopf.

Nilsson, A. N. 1985. "Life cycle and larval instars of the Holarctic *Agabus opacus* Aubé (Coleoptera: Dytiscidae)." *Entomologica Scandinavia* 15: 492–496.

Nilsson, A. N., and O. Soderstrom. 1988. "Larval consumption rates, interspecific predation, and local guild composition of egg-overwintering *Agabus* (Coleoptera, Dytiscidae) species in vernal ponds." *Oecologia* 76: 131–137.

Nimmo, A. P. 1971. "The adult Rhyacophilidae and Limnephilidae (Trichoptera) of Alberta and eastern British Columbia and their post glacial origin." *Quaestiones Entomologicae* 7: 3–234.

Nisbet, I. C. T. 1972. "Mosquitoes and man." Lincoln, MA: Massachusetts Audubon Society.

Nix, M. H., and D. G. Jenkins. 2000. "Life history comparisons of *Daphnia obtusa* from temporary ponds, cultured with a low-quality food." *Aquatic Ecology* 34: 19–27.

Noble, G. K., and M. K. Brady. 1933. "Observations on the life history of the marbled salamander, *Ambystoma opacum* Gravenhorst." *Zoologica* 11: 79–87.

Nolon, J. 1998. "Can land use be used to protect vernal pools?" Pp. 33–38 *in:* B. Fellman, (ed.). *Our Hidden Wetlands: The Proceedings of a Symposium on Vernal Pools in Connecticut.* New Haven, CT: Yale University and Connecticut Department of Environmental Protection.

Northern Prairie Research Center. 2002. "Chorus frogs." Jamestown, ND: US Geological Survey. Webpage: *http://www.npwrc.usgs.gov/narcam/idguide/chorus.htm.*

Novak, K., and F. Sehnal. 1963. "The development cycle of some species of the genus *Limnephilus* (Trichoptera)." *Casopis Ceskoslovenske Spolecnosti Entomologicke* 60: 68–80.

Nyman, S. 1991. "Ecological aspects of syntopic larvae of *Ambystoma maculatum* and the *Ambystoma laterale-jeffersonianum* complex in two New Jersey ponds." *Journal of Herpetology* 25: 505–509.

Nyman, S., M. J. Ryan, and J. D. Anderson. 1988. "The distribution of the *Ambystoma jeffersonianum* complex in New Jersey." *Journal of Herpetology* 22: 224–228.

O'Brien, W. J., and F. deNoyelles, Jr. 1972. "Photosynthetically elevated pH as a factor in zooplankton mortality in nutrient enriched ponds." *Ecology* 53: 605–614.

O'Donnell, J. D. 1937. "Natural history of the ambystomatid salamanders of Illinois." *American Midland Naturalist* 18: 1063–1071.

Ohio Environmental Council. 2003. "Vernal pool monitoring program." Webpage: *http://www.theoec.org/cwater_vernal.html.*

Oldham, R. S. 1966. "Spring movements in the American Toad, *Bufo americanus*." *Canadian Journal of Zoology* 44: 63–100.

Ostrofsky, M. L. 1997. "Relationship between chemical characteristics of autumn-shed leaves and aquatic processing rates." *Journal of the North American Benthological Society* 16: 750–759.

Packard, A. S., Jr. 1883. "A monograph on the phyllopod Crustacea of North America, with remarks on the Order Phyllocarida." Pp. 295–592 *in*: F. V. Hayden, (ed.). 12th Annual Report of the United States Geological and Geographic Survey of the Territories: A report on the progress of the exploration of Wyoming and Idaho for the year 1878. Part 1, Section 2. Washington, DC: US Geological Survey.

Palik, B., D. P. Batzer, R. Buech, L. Egeland, and D. Streblow. 2002. "Landscape variation in seasonal forest pond characteristics in northern forests." Presentation at US EPA Region 5 Conference on Midwestern Ephemeral Wetlands: A Vanishing Habitat. Chicago, IL, February 20, 2002. Abstract available at *http://www.epa.gov/region5/water/ephemeralwetlands/pdf/eph_wet_abstracts.pdf.*

Palik, B. J., R. Buech, and L. Egeland. 2003. "Using an ecological land hierarchy to predict seasonal-wetland abundance in upland forests." *Ecological Applications* 13: 1153–1163.

Parsons, J. K., and R. A. Matthews. 1995. "Analysis of the associations between macroinvertebrates and macrophytes in a freshwater pond." *Northwest Science* 69: 265–275.

Paton, P. 2000. "Movement chronology and population sizes of pond-breeding amphibians in Rhode Island." Presentation at Conference on Vernal Pools of the Northeast: Vernal Pool Ecology, Conservation, and Education. University of Rhode Island, Kingston, RI, March 31–April 1, 2000.

Paton, P. W. C., and S. Egan. 2001. "Effects of roads on amphibian community structure at breeding ponds in Rhode Island." Unpublished Final Report to the Transportation Environmental Research Program (TERP), Federal Highway Administration. Cited in *http://www.uri.edu/cels/nrs/paton/bib.html#paton2001*.

Pearman, P. B., A. M. Velasco, and A. Lopez. 1995. "Tropical amphibian monitoring: A comparison of methods for detecting inter-site variation in species composition." *Herpetologica* 51: 325–337.

Pecharsky, B. L., P. R. Fraissinet, M. A. Penton, and D. J. Conklin, Jr. 1990. *Freshwater Macroinvertebrates of Northeastern North America*. Ithaca, NY: Comstock Publishing Associates.

Pechmann, J. H. K. 1995. "Use of large field enclosures to study the terrestrial ecology of pond-breeding amphibians." *Herpetologica* 51: 434–450.

Pechmann, J. H. K., D. E. Scott, R. D. Semlitsch, J. P. Caldwell, L. J.Vitt, and J. W. Gibbons. 1991. "Declining amphibian populations: The problem of separating human impacts from natural fluctuations." *Science* 253: 892–895.

Pennak, R. W. 1978. *Fresh-water Invertebrates of the United States*. New York, NY: J. Wiley and Sons.

Pennak, R. W. 1989. *Fresh-water Invertebrates of the United States*. 3rd Edition. New York, NY: J. Wiley and Sons.

Pesce, G. L. 2003. "Copepod web portal." Webpage: *http://copepods.interfree.it/arpa/atthe.htm.*

Peterson, C. L., R. F. Wilkinson, D. Mott, and T. Holder. 1991. "Premetamorphic survival of *Ambystoma annulatum*." *Herpetologica* 47: 96–100.

Peterson, J. A., and A. R. Blaustein. 1992. "Relative palatabilities of anuran larvae to natural aquatic insect predators." *Copeia* 1992: 577–584.

Petranka, J. W. 1982. "Geographic variation in the mode of reproduction and larval characteristics of the small-mouthed salamander (*Ambystoma texanum*) in the east-central United States." *Herpetologica* 38: 475–485.

Petranka, J. W. 1984. "Breeding migrations, breeding season, clutch size, and oviposition of stream-breeding *Ambystoma texanum*." *Journal of Herpetology* 18: 106–112.

Petranka, J. W. 1989a. "Density-dependent growth and survival of larval *Ambystoma*: Evidence from whole-pond manipulations." *Ecology* 70: 1752–1767.

Petranka, J. W. 1989b. "Response of toad tadpoles to conflicting chemical stimuli: Predator avoidance versus 'optimal' foraging." *Herpetologica* 45: 283–293.

Petranka, J. W. 1990. "Observations of nest site selection, nest desertion, and embryonic survival in marbled salamanders." *Journal of Herpetology* 24: 229–234.

Petranka, J. W. 1994. "Response to impact of timber harvesting on salamanders." *Conservation Biology* 8: 302–304.

Petranka, J. W. 1998. *Salamanders of the United States and Canada*. Washington, DC: Smithsonian Institution Press.

Petranka, J. W., M. E. Eldridge, and K. E. Haley. 1993. "Effects of timber harvesting on southern Appalachian salamanders." *Conservation Biology* 7: 363–370.

Petranka, J. W., M. E. Hopey, B. T. Jennings, S. D. Baird, and S. J. Boone. 1994. "Breeding habitat segregation of wood frogs and American toads: The role of interspecific tadpole predation and adult choice." *Copeia* 1994: 691–697.

Petranka, J. W., J. J. Just, and E. C. Crawford. 1982. "Hatching of amphibian embryos: The physiological trigger." *Science* 217: 257–259.

Petranka, J. W., and J. G. Petranka. 1980. "Selected aspects of the larval ecology of the marbled salamander *Ambystoma opacum* in the southern portion of its range." *American Midland Naturalist* 104: 352–363.

Petranka, J. W., and J. G. Petranka. 1981. "On the evolution of nest site selection in the marbled salamander, *Ambystoma opacum*." *Copeia* 1981: 387–391.

Petrides, G. A. 1986. *A Field Guide to Trees and Shrubs*. Boston, MA: Houghton Mifflin.

Pfingsten, R. A., and F. L. Downs, (eds.). 1989. *Salamanders of Ohio*. Columbus, OH: Ohio State University.

Phillips, C. A. 1992. "Variation in metamorphosis in spotted salamanders *Ambystoma maculatum* from eastern Missouri." *American Midland Naturalist* 128: 276–280.

Pierce, B. A. 1985. "Acid tolerance in amphibians." *Bioscience* 35: 239–243.

Pierce, B. A. 1987. "The effects of acid rain on amphibians." *American Biology Teacher* 49: 342–347.

Pierce, B. A., and J. B. Harvey. 1987. "Geographic variation in acid tolerance of Connecticut wood frogs." *Copeia* 1987: 94–103.

Pierce, B. A., J. B. Hoskins, and E. Epstein. 1984. "Acid tolerance in Connecticut wood frogs (*Rana sylvatica*)." *Journal of Herpetology* 18: 159–168.

Pierce, B. A., M. A. Margolis, and L. J. Nirtaut. 1987. "The relationship between egg size and acid tolerance in *Rana sylvatica*." *Journal of Herpetology* 21: 174-184.

Pierce, B. A., and J. R. Shayevitz. 1982. "Within and among population variation in spot number of *Ambystoma maculatum*." *Journal of Herpetology* 16: 402–405.

Pierce, B. A., and D. K. Wooten. 1992. "Genetic variation in tolerance of amphibians to low pH." *Journal of Herpetology* 26: 422–429.

Piersol, W. H. 1910. "Spawn and larva of *Ambystoma jeffersonianum*." *American Naturalist* 44: 732–738.

Platt, T. M., D. M. Sever, and V. L. Gonzalez. 1993. "First report of the predaceous leech *Helobdella stagnalis* (Rhynchobdellida: Glossiphoniidae) as a parasite of an amphibian *Ambystoma tigrinum* (Amphibia: Caudata)." *American Midland Naturalist* 129: 208–210.

Pollan, M. 1998. "Dream pond: Just add water. Then add more." *The New York Times*, January 22, 1998, Pp. F1, F4.

Poole, R. W. 1997. *Nomina Insecta Nearctica: A Check List of the Insects of North America. Vol. 4: Non-Holometabolous Orders.* Rockville, MD: Entomological Information Services.

Poole, R. W., and P. Gentili. 1996a. *Nomina Insecta Nearctica: A Check List of the Insects of North America. Vol. 1: Coleoptera, Strepsiptera.* Rockville, MD: Entomological Information Services.

Poole, R. W., and P. Gentili. 1996b. *Nomina Insecta Nearctica: A Check List of the Insects of North America. Vol. 2: Hymenoptera, Mecoptera, Megaloptera, Neuroptera, Raphidoptera, Trichoptera*. Rockville, MD: Entomological Information Services.

Poole, R. W., and R. E. Lewis. 1996. *Nomina Insecta Nearctica: A Check List of the Insects of North America. Vol. 3: Diptera, Lepidoptera, Siphonaptera.* Rockville, MD: Entomological Information Services.

Pope, S. E., L. Fahrig, and H. G. Merriam. 2000. "Landscape complementation and metapopulation effects on leopard frog populations." *Ecology* 81: 2498–2508.

Portnoy, J. W. 1987. "Vernal ponds of the Cape Cod National Seashore: Location, water chemistry, and *Ambystoma* breeding biology." Unpublished report. Wellfleet, MA: US National Park Service, Cape Cod National Seashore.

Portnoy, J. W. 1990. "Breeding biology of the spotted salamander *Ambystoma maculatum* (Shaw) in acidic temporary ponds at Cape Cod, USA." *Biological Conservation* 53: 61–75.

Portnoy, J. W., C. T. Roman, R. Sobczak, and T. Cambareri. 1997. "Progress report: Potential groundwater withdrawal effects on plant distributions, soils, and water chemistry of seasonally-flooded wetlands and kettle ponds of Cape Cod National Seashore." Submitted to National Park Service, Water Resources Division-Water Rights Branch, Fort Collins, CO, June 17, 1997.

Pough, F. H. 1976. "Acid precipitation and embryonic mortality of spotted salamanders, *Ambystoma maculatum*." *Science* 192: 68–70.

Pough, F. H., and R. E. Wilson. 1976. "Acid precipitation and reproductive success of *Ambystoma* salamanders." *Proceedings of the International Symposium on Acid Rain and the Forest Ecosystem*. USDA Forest Service General Technical Report 23: 531–543.

Pough, F. Harvey, and R. E. Wilson. 1977. "Acid precipitation and reproductive success of *Ambystoma* salamanders." *Water Air and Soil Pollution* 7: 307–316.

Pough, F. H., E. M. Smith, D. H. Rhodes, and A. Collazo. 1987. "The abundance of salamanders in forest stands with different histories of disturbance." *Forest Ecology and Management* 20: 1–9.

Prescott, G. W. 1980. *How to Know the Aquatic Plants*. Dubuque, IA: W. C. Brown Company.

Proctor, H., and G. Pritchard. 1989. "Neglected predators: Water mites (Acari: Parasitengona: Hydrachnellae) in freshwater communities." *Journal of the North American Benthological Society* 8: 100–111.

Proctor, V. W. 1964. "Viability of crustacean eggs recovered from ducks." *Ecology* 45: 656–658.

Proctor, V. W., and C. R. Malone. 1965. "Further evidence of the passive dispersal of small aquatic organisms via the intestinal tract of birds." *Ecology* 46: 728–729.

Proctor, V. W., C. R. Malone, and V. L. DeVlaming. 1967. "Dispersal of aquatic organisms: Viability of disseminules recovered from the intestinal tract of captive killdeer." *Ecology* 48: 672–676.

Prosser, C. L. 1973. *Comparative Animal Physiology*. 3rd Edition. Philadelphia, PA: W. B. Saunders Company.

Ptacek, M. B., H. C. Gerhardt, and R. D. Sage. 1994. "Speciation by polyploidy in treefrogs: multiple origins of the tetraploid, *Hyla versicolor*." *Evolution* 48: 898–908.

Rafi, F. 1985. *Bibliographia Invertebratorum Aquaticum Canadensium. Volume 4. Synopsis Speciorum. Crustacea: Isopoda et Tanaidacea*. Ottawa, ON: National Museum of Canada.

Rawinski, T. J. 1997. "Vegetation ecology of the Grafton Ponds, York County, Virginia, with notes on waterfowl use." Natural Heritage Technical Report 97-10. Richmond, VA: Virginia Department of Conservation and Recreation, Division of Natural Heritage.

Raymond, L. R., and L. M. Hardy. 1991. "Effects of a clearcut on a population of the mole salamander, *Ambystoma talpoideum*, in an adjacent unaltered forest." *Journal of Herpetology* 25: 509–512.

Redington, C. B. 1994. *Plants in Wetlands, Redington Field Guides to Biological Interactions*. Dubuque, IA: Kendall/Hunt Publishing Company.

Regosin, J. V., B. S. Windmiller, and J. M. Reed. 2003a. "Influence of abundance of small-mammal burrows and conspecifics on the density and distribution of spotted salamanders (*Ambystoma maculatum*) in terrestrial habitats." *Canadian Journal of Zoology* 81: 596–605. Available at *http://www.hyla-ecological.com/publications.html*.

Regosin, J. V., B. S. Windmiller, and J. M. Reed. 2003b. "Terrestrial habitat use and winter densities of the wood frog (*Rana sylvatica*)." *Journal of Herpetology* 37: 390–394. Available at *http://www.hyla-ecological.com/publications.html*.

Reh, W., and A. Seitz. 1990. "The influence of land use on the genetic structure of populations of the common frog, *Rana temporaria*." *Biological Conservation* 54: 239–249.

Reid, J. W. 2000. "Copepod crustaceans of Great Smoky Mountains National Park: First collections, preliminary impressions, and results." Report to the US National Park Service and Discover Life In America. Washington, DC: Smithsonian Institution. Available at *http://www.discoverlife.org/pa/ev/me/copepod.html*.

Reilly, P., and T. K. McCarthy. 1993. "Attachment site selection of *Hydrachna* and *Eylais* (Acari, Hydrachnellidae) water mite larvae infecting Corixidae (Hemiptera, Heteroptera)." *Journal of Natural History* 27: 599–607.

Reinert, H. K. 1991. "Translocation as a conservation strategy for amphibians and reptiles: Some comments, concerns, and observations." *Herpetologica* 43: 347–363.

Ribera, I., and J. Isart. 1994. "Classification of the communities of Hydradephaga (Coleoptera) from the Spanish Pyrenees." *Verhandlungen Internationale Vereinigung für Theoretische und Angewandte Limnologie* 25: 2475–2477.

Ricci, C. 2001. "Dormancy patterns in rotifers." *Hydrobiologia* 446: 1–11.

Ricci, C., and M. Caprioli. 1998. "Stress during dormancy: Effect on recovery rates and life-history traits of anhydrobiotic animals." *Aquatic Ecology* 32: 353–359.

Rich, P. H., and R. G. Wetzel. 1978. "Detritus in the lake ecosystem." *American Naturalist* 112: 57–71.

Richards, S. J., and C. M. Bull. 1990. "Non-visual detection of anuran tadpoles by odonate larvae." *Journal of Herpetology* 24: 311–313.

Richardson, J. S., and R. J. Mackay. 1984. "A comparison of the life history and growth of *Limnephilus indivisus* (Trichoptera: Limnephilidae) in three temporary pools." *Archives of Hydrobiology* 99: 515–528.

Richter, K. O. 1995. "A simple aquatic funnel trap and its application to wetland amphibian monitoring." *Herpetological Review* 26: 90–91.

Richter, K. O., and A. L. Azous. 1995. "Amphibian occurrence and wetland characteristics in the Puget Sound Basin." *Wetlands* 15: 305–312.

Ritke, M. E., J. G. Babb, and M. K. Ritke. 1991. "Breeding-site specificity in the gray treefrog *Hyla chrysoscelis*." *Journal of Herpetology* 25: 123–125.

Roble, S. 1988. "Life in fleeting waters." *Massachusetts Wildlife* Spring: 22–27.

Roble, S. M. 1998. "A zoological inventory of the Grafton Ponds sinkhole complex, York County, Virginia." Natural Heritage Technical Report 98-3. Richmond, VA: Virginia Department of Conservation and Recreation, Division of Natural Heritage. Unpublished report submitted to the US Environmental Protection Agency.

Ross, H. H. 1944. *The Caddisflies or Trichoptera of Illinois*. Bulletin 23. Urbana, IL: Illinois Natural History Survey.

Ross, H. H. 1956. *Evolution and Classification of the Mountain Caddisflies*. Urbana, IL: University of Illinois Press.

Ross, H. H., and D. R. Merkley. 1952. "An annotated key to the Nearctic males of *Limnephilus* (Trichoptera: Limnephilidae)." *American Midland Naturalist* 47: 435–455.

Ross, T. E. 1987. "A comprehensive bibliography of the Carolina Bays literature." *The Journal of the Elisha Mitchell Scientific Society* 103: 28–42.

Roth, A. H., and J. F. Jackson. 1987. "The effect of pool size on recruitment of predatory insects and on mortality in a larval anuran." *Herpetologica* 43: 224–232.

Rouen, K., (ed.). 2001. *European Temporary Ponds*. Contributions from a conference held in Birmingham, UK, February 2001. Ambleside, UK: Freshwater Biological Association.

Roughley, R. E. 2004. "Hierarchical Classification of Dytiscidae Leach (Coleoptera)." Webpage: *http://www.inhs.uiuc.edu/cbd/waterbeetles/thayer/dytiscid/wp5*.

Roughley, R. E., and D. J. Larson. 1991. "Aquatic Coleoptera of springs in Canada." *Memoirs of the Entomological Society of Canada* 155: 125–140.

Roughley, R. E., and A. N. Nilsson. 1994. "Taxonomy and distribution of the Holarctic diving beetle *Laccophilus biguttatus* Kirby (Coleoptera: Dytiscidae)." *Journal of the New York Entomological Society* 102: 91–101.

Rowe, C. L., and W. A. Dunson. 1993. "Relationships among abiotic parameters and breeding effort by three amphibians in temporary wetlands of central Pennsylvania." *Wetlands* 13: 237–246.

Rowe, C. L., and P. D. N. Hebert. 1999. "Cladoceran Web Site." Webpage: *http://www.cladocera.uoguelph.ca*.

Rowe, C. L., W. J. Sadinski, and W. A. Dunson. 1994. "Predation on larval and embryonic amphibians by acid-tolerant caddisfly larvae (*Ptilostomis postica*)." *Journal of Herpetology* 28: 357–364.

Rowe, J. M., S. K. Meegan, E. S. Engstrom, S. A. Perry, and W. B. Perry. 1996. "Comparison of leaf processing rates under different temperature regimes in three headwater streams." *Freshwater Biology* 36: 277–288.

Rowe, J. W. 1992. "Dietary habits of the Blanding's turtle (*Emydoidea blandingi*) in northeastern Illinois." *Journal of Herpetology* 26: 111–114.

Ruiter, D. E. 1995. *The Genus Limnephilus Leach (Trichoptera: Limnephilidae) of the New World.* New Series Bulletin 11(1). Columbus, OH: Ohio Biological Survey.

Ruth, B. C., W. A. Dunson, C. L. Rowe, and S. B. Hedges. 1993. "A molecular and functional evaluation of the egg mass color polymorphism of the spotted salamander, *Ambystoma maculatum*." *Journal of Herpetology* 27: 306–314.

Ruttner, F. 1963. *Fundamentals of Limnology*. Toronto, ON: University of Toronto Press.

Saber, P. A., and W. A. Dunson. 1978. "Toxicity of bog water to embryonic and larval anuran amphibians." *Journal of Experimental Zoology* 204: 33–42.

Sadinski, W. J., and W. A. Dunson. 1992. "A multilevel study of effects of low pH on amphibians of temporary ponds." *Journal of Herpetology* 26: 413–422.

Salett, M. C. 1992. "Temperature and precipitation effects on breeding abundance of two species of syntopic mole salamanders: *Ambystoma maculatum* and *Ambystoma laterale*." Unpublished report to Broadmoor Wildlife Sanctuary, Sherborn, MA. Boston, MA: Bioscience 307, Simmons College.

Sawchyn, W. W., and C. Gillott. 1974a. "The life history of *Lestes congener* (Odonata: Zygoptera) on the Canadian prairies." *Canadian Entomologist* 106: 367–376.

Sawchyn, W. W., and C. Gillott. 1974b. "The life history of three species of *Lestes* (Odonata: Zygoptera) in Saskatchewan." *Canadian Entomologist* 106: 1283–1293.

Sawchyn, W. W., and C. Gillott. 1975. "The biology of two related species of coenagrionid dragonflies (Odonata: Zygoptera) in western Canada." *Canadian Entomologist* 107: 119–128.

Schilling, M. R. 1996. "Vernal pools: An overview." *Wetland Journal* 8: 6–7.

Schlichter, L. C. 1981. "Low pH affects the fertilization and development of *Rana pipiens* eggs." *Canadian Journal of Zoology* 59: 1693–1699.

Schmid, F. 1980. *Genera des Trichoptères du Canada et des États Adjacents*. Part 7, *The Insects and Arachnids of Canada*. Ottawa, ON: Biosystematics Research Institute.

Schneider, D. W. 1990. Habitat duration and the community ecology of temporary ponds. PhD Dissertation, Zoology, University of Wisconsin, Madison, WI.

Schneider, D. W. 1997. "Predation and food web structure along a habitat duration gradient." *Oecologia* 110: 567–575.

Schneider, D. W. 1999. "Snow-melt ponds in Wisconsin. Influence of hydroperiod on invertebrate community structure." Pp. 299–318 *in*: D. P. Batzer, R. B. Rader, and S. A. Wissinger, (eds.). *Invertebrates in Freshwater Wetlands of North America: Ecology and Management*. New York, NY: J. Wiley and Sons.

Schneider, D. W., and T. M. Frost. 1996. "Habitat duration and community structure in temporary ponds." *Journal of the North American Benthological Society* 15: 64–86.

Schroeder, T. 2002. "Mate recognition in *Epiphanes senta* (Monogononta, Rotifera)." Presentation at American Society of Limnology and Oceanography, Victoria, BC, June 10, 2002. Abstract available at *http://aslo.org/meetings/victoria2002/archive/866.html*.

Schwartz, S. S., and P. D. N. Hebert. 1985. "*Daphniopsis ephemeralis* sp. n. (Cladocera, Daphniidae): A new genus for North America." *Canadian Journal of Zoology* 63: 2689–2693.

Schwartz, S. S., and P. D. N. Hebert. 1987. "Breeding system of *Daphniopsis ephemeralis*: Adaptations to a transient environment." *Hydrobiologia* 145: 195–200.

Schwartz, S. S., and D. J. Jenkins. 2000. "Temporary aquatic habitats: Constraints and opportunities." *Aquatic Ecology* 34: 3–8.

Scott, D. E. 1994. "The effect of larval density on adult demographic traits in *Ambystoma opacum*." *Ecology* 75: 1397–1405.

Scott, D. E., and M. R. Fore. 1995. "The effect of food limitation on lipid levels, growth, and reproduction in the marbled salamander *Ambystoma opacum*." *Herpetologica* 51: 462–471.

Seale, D. B. 1980. "Influence of amphibian larvae on primary production, nutrient flux, and competition in a pond ecosystem." *Ecology* 61: 1531–1550.

Seale, D. B. 1982. "Physical factors influencing oviposition by the woodfrog, *Rana sylvatica*, in Pennsylvania." *Copeia* 1982: 627–635.

Sedell, J. R., F. J. Triska, and N. S. Triska. 1975. "The processing of conifer and hardwood leaves in two coniferous forest streams: I. Weight loss and associated invertebrates." *Verhandlungen Internationale Vereinigung für Theoretische und Angewandte Limnologie* 19: 1617–1627.

Seibert , H. C. 1989. "*Ambystoma opacum.*" Pp. 125–140 *in:*: R. A. Pfingsten and F. L. Downs, (eds.). *Salamanders of Ohio*. Columbus, OH: Ohio State University.

Seigel, R. A. 1983. "Natural survival of eggs and tadpoles of the wood frog, *Rana sylvatica*." *Copeia* 1983: 1096–1098.

Semlitsch, R. D. 1985. "Reproductive strategy of a facultatively paedomorphic salamander *Ambystoma talpoideum*." *Oecologia* 65: 305–313.

Semlitsch, R. D. 1987. "Relationship of pond drying to the reproductive success of the salamander *Ambystoma talpoideum*." *Copeia* 1987: 61–69.

Semlitsch, R. D. 1998. "Biological delineation of terrestrial buffer zones for pond-breeding salamanders." *Conservation Biology* 12: 1113–1119.

Semlitsch, R. D., (ed.). 2003. *Amphibian Conservation*. Washington, DC: Smithsonian Books.

Semlitsch, R. D., and J. R. Bodie. 1998. "Are small, isolated wetlands expendable?" *Conservation Biology* 12: 1129–1133.

Semlitsch, R. D., and J. W. Gibbons. 1985. "Phenotypic variation in metamorphosis and paedomorphosis in the salamander *Ambystoma talpoideum*." *Ecology* 66: 1123–1130.

Semlitsch, R. D., D. E. Scott, J. H. K. Pechmann, and J. W. Gibbons. 1996. "Structure and dynamics of an amphibian community: Evidence from a 16-year study of a natural pond." Pp. 217–248 *in*: M. L. Cody and J. A. Smallwood, (eds.). *Long-term Studies of Vertebrate Communities*. San Diego, CA: Academic Press.

Semlitsch, R. D., and S. C. Walls. 1993. "Competition in two species of larval salamanders: A test of geographic variation in competitive ability." *Copeia* 1993: 587–595.

Serrano, L., and J. Toja. 1995. "Limnological description of four temporary ponds in the Doñano National Park (Spain)." *Archiv für Hydrobiologie* 113: 497–516.

Sessions, S. K. 1982. "Cytogenetics of diploid and triploid salamanders of the *Ambystoma jeffersonianum* complex." *Chromosoma* 84: 599–621.

Sessions, S. 1998. "Amphibians and trematodes." Froglog 26: 1. Available at *http://www.open.ac.uk/daptf/froglog/FROGLOG-26-1.html.*

Sessions, S. K., and S. B. Ruth.1990. "Explanation for naturally occurring supernumerary limbs in amphibians." *Journal of Experimental Zoology* 254: 38–47.

Sessions, S. K., G. Stopper, L. Hecker, V. Horner, and A. Franssen. 1999a. "Deformed amphibian research at Hartwick College." Webpage: *http://info.hartwick.edu/biology/def_frogs/Introduction/Introduction.html.*

Sessions, S. K., G. Stopper, L. Hecker, V. Horner, and A. Franssen. 1999b. "Update on our research." Webpage: *http://info.hartwick.edu/biology/def_frogs/Introduction/Update.html.*

Sever, D. M., and C. F. Dineen. 1978. "Reproductive ecology of the tiger salamander, *Ambystoma tigrinum*, in northern Indiana." *Proceedings of the Indiana Academy of Science* 87: 189–203.

Sexton, O. J. 1959. "Spatial and temporal movements of a population of the painted turtle, *Chrysemys picta marginata* (Agassiz)." *Ecological Monographs* 29: 113–140.

Shaffer, H. B. 1983. "Biosystematics of *Ambystoma rosaceum* and *A. tigrinum* in northwestern Mexico." *Copeia* 1983: 67–78.

Shoop, C. R. 1965. "Orientation of *Ambystoma maculatum*: Movements to and from breeding ponds." *Science* 149: 558–559.

Shoop, C. R. 1968. "Migratory orientation of *Ambystoma maculatum*: Movements near breeding ponds and displacements of migrating individuals." *Biological Bulletin* 135: 230–238.

Shoop, C. R. 1974. "Yearly variation in larval survival of *Ambystoma maculatum*." *Ecology* 55: 440–444.

Shoop, C. R., and T. L. Doty. 1972. "Migratory orientation by marbled salamanders (*Ambystoma opacum*) near a breeding area." *Behavioral Biology* 7: 131–136.

Simovich, M. A. 1998. "Crustacean biodiversity and endemism in California's ephemeral wetlands." Pp. 107–118 *in*: C. W. Witham, E. T. Bauder, D. Belk, W. R. Ferren, Jr., and R. Ornduff, (eds.). *Ecology, Conservation, and Management of Vernal Pool Ecosystems — Proceedings from a 1996 Conference*. Sacramento, CA: California Native Plant Society.

Simpson, M. 1997. "Salamander populations in two vernal pools." Unpublished report to Broadmoor Wildlife Sanctuary, Sherborn, MA. Keene, NH: Amphibian Biology class, Antioch New England Graduate School.

Skelly, D. 1998. "Landscape change and amphibian populations." Pp. 12–14 *in*: B. Fellman, (ed.). *Our Hidden Wetlands: The Proceedings of a Symposium on Vernal Pools in Connecticut*. New Haven, CT: Yale University and Connecticut Department of Environmental Protection.

Skelly, D. K., L. K. Freidenburg, and J. M. Kiesecker. 2002. "Forest canopy and the performance of larval amphibians." *Ecology* 83: 983–992.

Skelly, D. K., and J. Golon. 2003. "Assimilation of natural benthic substrates by two species of tadpoles." *Herpetologica* 59; 37–42.

Skelly, D. K., E. E. Werner, and S. A. Cortwright. 1999. "Long-term distributional dynamics of a Michigan amphibian assemblage." *Ecology* 80: 2326–2337.

Smetana, A. 1988. "Review of the family Hydrophilidae of Canada and Alaska (Coleoptera)." *Memoirs of the Entomological Society of Canada* 142: 3–316.

Smith, D. C. 1983. "Factors controlling tadpole populations of the chorus frog (*Pseudacris triseriata*) in Isle Royale, Michigan." *Ecology* 64: 501–510.

Smith, D. G. 1992. *Keys to the Freshwater Macroinvertebrates of Massachusetts*. 2nd Edition. Sunderland, MA: D. G. Smith.

Smith, D. G. 1995. "Notes on the status and natural history of limnadiid clam shrimp in southern New England." *Anostracan News* 3(2): 3–4.

Smith, D. G. 2001. *Pennak's Freshwater Invertebrates of the United States: Porifera to Crustacea*. 4th Edition. New York, NY: J. Wiley and Sons.

Smith, D. G., and A. A. Gola. 2001. "The discovery of *Caenestheriella gynecia* Mattox 1950 (Branchiopoda, Cyzicidae) in New England, with ecological and systematic notes." *Northeastern Naturalist* 8: 443–454.

Smith, G. R., and J. E. Rettig. 1996. "Effectiveness of aquatic funnel traps for sampling amphibian larvae." *Herpetological Review* 27: 190–191.

Smith, I. M. 1987. "Water mites of peatlands and marshes in Canada." *Memoirs of the Entomological Society of Canada* 140: 31–46.

Smith, I. M., (ed.). 1997. *Assessment of Species Diversity in the Mixedwood Plains Ecozone*. Ecological Monitoring and Assessment Network. Printed summary. Available on CD-ROM in hybrid format. Burlington, ON: Ecological Monitoring Coordinating Office, Environment Canada. Available at *http://www.cciw/ca/eman-temp/reports/publications/Mixedwood/intro.html.*

Smith, I. M., and D. R. Cook. 1991. "Water mites." Pp. 523–592 *in*: J. H. Thorp and A. P. Covich, (eds.). *Ecology and Classification of North American Freshwater Invertebrates*. New York, NY: Academic Press.

Smith, I. M., D. R. Cook, and B. P. Smith. 2001. "Water mites (Hydrachnida) and other arachnids." Pp. 551–659 *in*: J. H. Thorp and A. P. Covich, (eds.). *Ecology and Classification of North American Freshwater Invertebrates*. 2nd Edition. New York, NY: Academic Press.

Smith, R. L. 1973. "Aspects of the biology of three species of the genus *Rhantus* (Coleoptera: Dytiscidae) with special reference to the acoustical behavior of two." *Canadian Entomologist* 105: 909–919.

Snodgrass, J., M. J. Komoroski, A. L. Bryan, Jr., and J. Burger. 2000. "Relationships among isolated wetland size, hydroperiod, and amphibian species richness: Implications for wetland regulation." *Conservation Biology* 14: 414–419.

Snure, P. 1957. "Thallose liverworts, Ricciaceae." P. 42 *in*: N. C. Fassett. *A Manual of Aquatic Plants*. Madison, WI: The University of Wisconsin Press.

Sobczak, R. 1999. Unpublished data, hydrologic studies on vernal pools in Eastham, MA. Barnstable, MA: Cape Cod Commission.

Sobczak, R. 2000. "Surface Water Fluctuations in Depressional Wetlands, Cape Cod National Seashore, Eastham, MA, in Relation to Precipitation and Groundwater, 1982–1997. A Pilot Study." Powerpoint presentation illustrating hydrologic model developed for 14 Cape Cod vernal pools. Barnstable, MA: Water Resources Office, Cape Cod Commission.

Sobczak, R. V., T. C. Cambareri, and J. W. Portnoy. 2003. "Physical hydrology of selected vernal pools and kettle hole ponds in the Cape Cod National Seashore, Massachusetts: Ground and surface water interactions." Report to the US National Park Service, Water Rights Branch, Water Resources Division, Fort Collins, CO, and Cape Cod National Seashore. Water Resources Office, Cape Cod Commission, Barnstable, MA.

Spandl, H. 1925. "Euphyllopoda." Pp. 2–22 *in*: P. Schulze, (ed.). *Biologie der Tiere Deutschlands* (Berlin), 14(14).

Speare, R., and L. Berger. 2000. "Global distribution of chytridiomycosis in amphibians." Webpage: *http://www.jcu.edu.au/school/phtm/PHTM/frogs/chyglob.htm*.

Spencer, R. W. 1967. "What the people of Essex County need to know about mosquitoes and their control." Circular No. 1. Gloucester, MA: Essex County Mosquito Control Project.

Spielman, A. 1964. "Swamp mosquito, *Culiseta melanura*: Occurrence in an urban habitat." *Science* 143: 361–362.

Sredl, M. J., and J. P. Collins. 1992. "The interaction of predation, competition, and habitat complexity in structuring an amphibian community." *Copeia* 1992: 607–614.

Stangel, P. W. 1988. "Premetamorphic survival of the salamander *Ambystoma maculatum* in eastern Massachusetts." *Journal of Herpetology* 22: 345–347.

Stearns, S. C. 1976. "Life-history tactics: A review of the ideas." *Quarterly Review of Biology* 51: 3–47.

Stearns, S. C. 1992. *The Evolution of Life Histories.* Oxford, UK; New York, NY: Oxford University Press.

Stearns, S. C. 2000. "Life history evolution: Successes, limitations, and prospects." *Naturwissenschaften* (2000) 87: 476–486. Available at *http://folk.uio.no/avollest/Stearns.pdf.*

Stenhouse, S. L. 1985. "Migration, orientation, and homing in *Ambystoma maculatum* and *Ambystoma opacum.*" *Copeia* 1985: 631–637.

Stenhouse, S. L. 1987. "Embryo mortality and recruitment of juveniles of *Ambystoma maculatum* and *Ambystoma opacum* in North Carolina." *Herpetologica* 43: 496–501.

Stenhouse, S. L., N. G. Hairston, and E. E. Cobey. 1983. "Predation and competition in *Ambystoma* larvae: Field and laboratory experiments." *Journal of Herpetology* 17: 210–220.

Stevens, P. H., and D. G. Jenkins. 2000. "Analyzing species distributions among temporary ponds with a permutation test approach to the joint-count statistic." *Aquatic Ecology* 34: 91–99.

Stewart, B. A. 1992. "The effect of invertebrates on leaf decomposition rates in two small woodland streams in southern Africa." *Archiv für Hydrobiologie* 124: 19–33.

Stickel, W. H. 1967. "Wildlife, pesticides, and mosquito control." Lincoln, MA: Massachusetts Audubon Society.

Stickel, W. H. 1968. "Mosquito control practices from a conservation viewpoint." Amherst, MA: Northeastern Pesticide Coordinators of the Cooperative Extension Service.

Stille, W. T. 1954. "Eggs of the salamander *Ambystoma jeffersonianum* in the Chicago area." *Copeia* 1954: 300.

Stine, C. J. 1984. "The life history and status of the Eastern Tiger Salamander, *Ambystoma tigrinum.*" *Bulletin of the Maryland Herpetological Society* 20(3).

Stine, C. J., Jr., J. A. Fowler, and R. S. Simmons. 1954. "Occurrence of the eastern tiger salamander, *Ambystoma tigrinum tigrinum* (Green) in Maryland, with notes on its life history." *Annals of the Carnegie Museum* 33: 145–148.

Stockwell, S. S., and M. L. Hunter, Jr. 1989. "Relative abundance of herpetofauna among eight types of Maine peatland vegetation." *Journal of Herpetology* 23: 409–414.

Stone, J. S. 1992. Vernal pools in Massachusetts: Aerial photographic identification, biological and physiographic characteristics, and state certification criteria. MS Thesis, Forestry and Wildlife, University of Massachusetts, Amherst, MA.

Storey, K. B., and J. M. Storey. 1986a. "Freeze tolerance and intolerance as strategies of winter survival in terrestrially-hibernating amphibians." *Comparative Biochemistry and Physiology: A: Comparative Physiology* 83: 613–617.

Storey, K. B., and J. M. Storey. 1986b. "Freeze-tolerant frogs: Cryoprotectants and tissue metabolism during freeze-thaw cycles." *Canadian Journal of Zoology* 64: 49–56.

Storey, K. B., and J. M. Storey. 1987. "Persistence of freeze tolerance in terrestrially hibernating frogs after spring emergence." *Copeia* 1987: 720–726.

Stout, B. M., III, and K. K. Stout. 1992. "Predation by the caddisfly *Banksiola dossuaria* on egg masses of the spotted salamander *Ambystoma maculatum*." *American Midland Naturalist* 127: 368–372.

Stratman, D. 2000. "Using micro and macrotopography in wetland restoration." Indiana Biology Technical Note 1. Indianapolis, IN: USDA Natural Resources Conservation Service. Available at *http://www.parcplace.org/publications/PDFfiles/IndianaTechNote1.pdf* and *http://www.in.nrcs.usda.gov/PlanningandTechnology/biology/Indiana_Tech_note_1.pdf.*

Strayer, D. 1990. "Freshwater Mollusca." Pp. 333–372 *in*: B. Pecharsky, P. Fraissinet, M. Penton, and D. Conklin, (eds.). *Freshwater Macroinvertebrates of Northeastern North America.* Ithaca, NY: Comstock Publishing Associates.

Sutcliffe, D. W. 1960. "Observations on the salinity tolerance and habits of a euryhaline caddis larva, *Limnephilus affinis* (Trichoptera: Limnephilidae)." *Proceedings of the Royal Society of London* A35: 156–162.

Szafoni, R. E., C. A. Phillips, S. R. Ballard, R. A. Brandon, and G. Kruse. 2002. *Illinois Landowner's Guide to Amphibian Conservation.* Special Publication 22. Champaign, IL: Illinois Natural History Survey.

Tabacchi, E., and P. Marmonier. 1994. "Dynamics of the interstitial ostracod assemblage of a pond in the Ardour alluvial plain." *Archiv für Hydrobiologie* 131: 321–340.

Takahashi, R. M., W. H. Wilder, and T. Muira. 1984. "Field evaluations of ISA-20E for mosquito control and effects on aquatic nontarget arthropods in experimental plots." *Mosquito News* 44: 363.

Talentino, K., and E. Landre. 1991. "Comparative development of two species of sympatric *Ambystoma* salamanders." *Journal of Freshwater Ecology* 6: 395–401.

Tappan, A. 1997. *Identification and Documentation of Vernal Pools*. Concord, NH: New Hampshire Association of Wetlands Scientists and New Hampshire Department of Fish and Game.

Taylor, B. E., and D. E. Scott. 1997. "Effects of larval density dependence on population dynamics of *Ambystoma opacum*." *Herpetologica* 53: 132–145.

Taylor, B. E., G. A. Wyngaard, and D. L. Mahoney. 1990. "Hatching of *Diaptomus stagnalis* eggs from a temporary pond after a prolonged dry period." *Archiv für Hydrobiologie* 117: 271–278.

Taylor, P., L. Fahrig, K. Heinen, and G. Merriam. 1993. "Connectivity is a vital element of landscape structure." *Oikos* 68: 571–573.

Tejedo, M. 1993. "Size-dependent vulnerability and behavioral responses of tadpoles of two anuran species to beetle larvae predators" *Herpetologica* 49: 287–294.

Tesauro, J. 2001. "New Jersey's vernal pools." Webpage: *http://www.state.nj.us/dep/fgw/vpoolart.htm.*

Tessier, A. J., and P. Woodruff. 2002. "Trading off the ability to exploit rich versus poor food quality." *Ecology Letters* 5: 685–692.

Thomas, G. L. 1963. "Study of a population of sphaeriid clams in a temporary pond." *Nautilus* 77: 37–43.

Thompson, E. L., and J. E. Gates. 1982. "Breeding pool segregation by the mole salamanders, *Ambystoma jeffersonianum* and *A. maculatum*, in a region of sympatry." *Oikos* 38: 273–279.

Thompson, E. L., J. E. Gates, and G. J. Taylor. 1980. "Distribution and breeding habitat selection of the Jefferson salamander, *Ambystoma jeffersonianum*, in Maryland." *Journal of Herpetology* 14: 113–120.

Thorp, J. H., and A. P. Covich, (eds.). 1991. *Ecology and Classification of North American Freshwater Invertebrates*. New York, NY: Academic Press.

Thorp, J. H., and A. P. Covich, (eds.). 2001. *Ecology and Classification of North American Freshwater Invertebrates.* 2nd Edition. New York, NY: Academic Press.

Tiner, R. W. 2003a. "Geographically isolated wetlands of the United States." *Wetlands* 23: 494–516.

Tiner, R. W. 2003b. "Estimated extent of geographically isolated wetlands in selected areas of the United States." *Wetlands* 23: 636–652.

Tiner, R. W., H. C. Bergquist, G. P. DeAlessio, and M. J. Starr. 2002. *Geographically Isolated Wetlands: A Preliminary Assessment of their Characteristics and Status in Selected Areas of the United States*. Hadley, MA: US Fish and Wildlife Service, Northeast Region. Available at *http://wetlands.fws.gov/Pubs_Reports.isolated.geoisolated.htm.*

Tome, M. A., and F. H. Pough. 1982. "Responses of amphibians to acid precipitation." Pp. 245–254 *in:* R. E. Johnson, (ed.). *Acid Rain/Fisheries*. Proceedings of an International Symposium on Acid Precipitation and Fisheries Impacts in Northeastern North America, Cornell University, Ithaca, NY, August 2–5, 1981. Bethesda, MD: American Fisheries Society.

Tordoff, W. 1965. Some aspects of the biology of Fowler's Toad, *Bufo woodhousei fowleri*, Hinckley, on Monomoy Island, Chatham, Massachusetts. Senior Honor's Thesis, Zoology, University of Massachusetts, Amherst, MA.

Torrey, T. W. 1967. *Morphogenesis of the Vertebrates*. New York, NY: J. Wiley and Sons.

Tracy, C. R., K. A. Christian, M. P. O'Connor, and C. R. Tracy. 1993. "Behavioral thermoregulation by *Bufo americanus*: The importance of the hydric environment." *Herpetologica* 49: 375–382.

Travis, J. 1983. "Variation in development patterns of larval anurans in temporary ponds. I. Persistent variation within a *Hyla gratiosa* population." *Evolution* 37:496–512.

Tressler, W. L. 1959. "Ostracoda." Pp. 657–734 *in*: W. T. Edmonson, (ed.). *Freshwater Biology*. 2nd Edition. New York, NY: J. Wiley and Sons.

Trombulak, S. C., and C. A. Frissell. 2000. "Review of ecological effects of roads on terrestrial and aquatic communities." *Conservation Biology* 14: 18–30.

Trottier, R. 1966. "The emergence and sex ratio of *Anax junius* Drury (Odonata: Aeshnidae) in Canada." *Canadian Entomologist* 98: 794–798.

Turgeon, D. D., J. F. Quinn, Jr., A. E. Bogan, E. V. Coan, F. G. Hochberg, W. G. Lyons, et al. 1998. *Common and Scientific Names of Aquatic Invertebrates from the United States and Canada: Mollusks*. 2nd Edition. Special Publication 26. Bethesda, MD: American Fisheries Society.

Tyning, T. F. 1990. *A Guide to Amphibians and Reptiles*. Stokes Nature Guides. Boston, MA: Little, Brown, and Co.

US Department of Agriculture. 2002. "Integrated Taxonomic Information System (ITIS)." Webpage: *http://www.itis.usda.gov/*. Updated 20 August 2002.

US Geological Survey. 2002. "North American reporting center for amphibian malformations." Webpage: *http://www.npwrc.usgs.gov/narcam/info/BiblioAU.htm*.

US Soil Conservation Service. 1985. *Hydrology, Section 4, National Engineering Handbook*. Washington, DC: US Department of Agriculture.

US Soil Conservation Service. 1986. *Urban Hydrology for Small Watersheds*. Technical Release 55. 2nd Edition. PB87-101580. Washington, DC: US Department of Agriculture.

University of Rhode Island. 2001. "Rhode Island vernal pools." Webpage: *http://www.uri.edu/cels/nrs/paton/*. Updated 26 November 2001.

Uzzell, T. 1969. "Notes on spermatophore production by salamanders of the *Ambystoma jeffersonianum* complex." *Copeia* 1969: 602–612.

Uzzell, T. M., Jr. 1963. "Natural triploidy in salamanders related to *Ambystoma jeffersonianum*." *Science* 139: 113–115.

Uzzell, T. M., Jr. 1964. "Relations of the diploid and triploid species of the *Ambystoma jeffersonianum* complex (Amphibia, Caudata)." *Copeia* 1964: 257–300.

Uzzell, T. M., Jr. 1970. "Meiotic mechanisms of naturally occurring unisexual vertebrates." *American Naturalist* 104: 433–445.

Uzzell, T. M., Jr., and S. M. Goldblatt. 1967. "Serum proteins of salamanders of the *Ambystoma jeffersonianum* complex, and the origin of the triploid species of this group." *Evolution* 21: 345–354.

Vallières, L., and A. Aubin. 1986. "Effet de certains paramètres physico-chimiques sur les relations prédateurs-proies dans les populations de copépodes cylcopoides des mares temporaires d'eau douce." *Canadian Zoologist* 64: 8–11.

Vannote, R. L., G. W. Minshall, K. W. Cummins, J. R. Sedell, and C. E. Cushing. 1980. "The river continuum concept." *Canadian Journal of Fisheries and Aquatic Sciences* 37: 130–137.

Vekhoff, N. V., and T. P. Vekhova. 1995. "Anostraca of Russia and adjacent lands in relation to conservation policy." *Anostracan News* 3: 3–4.

Voss, S. R. 1993. "Effect of temperature on body size, developmental stage, and timing of hatching in *Ambystoma maculatum*." *Journal of Herpetology* 27: 329–333.

Wacasey, J. W. 1961. An ecological study of two sympatric species of salamanders, *Ambystoma maculatum* and *Ambystoma jeffersonianum* in southern Michigan. PhD Dissertation, Zoology, Michigan State University, East Lansing, MI.

Waldman, B., and M. J. Ryan. 1983. "Thermal advantages of communal egg mass deposition in wood frogs (*Rana sylvatica*)." *Journal of Herpetology* 17: 70-72.

Walker, E. M. 1953. *The Odonata of Canada and Alaska. Vol. 1, Part I: General, and Part II: The Zygoptera*. Toronto, ON: University of Toronto Press.

Walker, E. M. 1958. *The Odonata of Canada and Alaska. Vol. 2, Part III: The Anisoptera, Four Families*. Toronto, ON: University of Toronto Press.

Walker, E. M., and P. S. Corbet. 1975. *The Odonata of Canada and Alaska. Vol. 3, Part III: The Anisoptera, Three Families.* Toronto, ON: University of Toronto Press.

Wallace, R. L., and T. W. Snell. 1999. "Rotifera." Pp. 187–248 *in*: J. H. Thorp and A. P. Covich, (eds.). *Ecology and Classification of North American Freshwater Invertebrates*. New York, NY: Academic Press.

Wallis, J. B. 1939. "The genus *Ilybius* Er. in North America (Coleoptera: Dytiscidae)." *Canadian Entomologist* 71: 192–199.

Walls, S. C., and R. Altig. 1986. "Female reproductive biology and larval life history of *Ambystoma* salamanders: A comparison of egg size, hatchling size, and larval growth." *Herpetologica* 42: 334–345.

Waltz, R. D., and W. P. McCafferty. 1979. *Freshwater Springtails (Hexapoda: Collembola) of North America*. Agricultural Experiment Station Research Bulletin 960. West Lafayette, IN: Purdue University.

Warren, C. E. 1971. *Biology and Water Pollution Control*. Philadelphia, PA: W. B. Saunders Company.

Way, C. M., D. J. Hornbach, and A. J. Burky. 1980. "Comparative life history tactics of the sphaeriid clam, *Musculium partumeium* (Say), from a permanent and a temporary pond." *American Midland Naturalist* 104: 319–327.

Weaver, C. R. 1943. "Observations on the life cycle of the fairy shrimp, *Eubranchipus vernalis*." *Ecology* 24: 500–502.

Webster, J., and E. F. Benfield. 1986. "Vascular plant breakdown in freshwater ecosystems." *Annual Review of Ecology and Systematics* 17: 567–594.

Weeks, S. C. 1997. "Branchiopod images." Webpage: *http://www3.uakron.edu/biology/brnchmgs.html*.

Weeks, S. C. 2002a. "Large branchiopod bibliography." Webpage: *http://www3.uakron.edu/biology/bibintro.html*.

Weeks, S. C. 2002b. "'Large' branchiopods of the Midwest." Presentation at US EPA Region 5 Conference: Midwestern Ephemeral Wetlands: A Vanishing Habitat. Chicago, IL, February 20, 2002. Abstract available at *http://www.epa.gov/region5/water/ephemeralwetlands/pdf/eph_wet_abstracts.pdf*.

Weeks, S. C., and V. Marcus. 1997. "A survey of the branchiopod crustaceans of Ohio." *Ohio Journal of Science* 97: 86–89.

Weir, J. S. 1974. "Odonata collected in and near seasonal pools in Wankie National Park, Rhodesia, with notes on the physical-chemical environments in which nymphs were found." *Journal of the Entomological Society of South Africa* 37: 135–145.

Welch, P. S. 1952. *Limnology*. New York, NY: McGraw-Hill.

Wellborn, G. A., D. K. Skelly, and E. E. Werner. 1996. "Mechanisms creating community structure across a freshwater habitat gradient." *Annual Review of Ecology and Systematics* 27: 337–363.

Weller, W. F., C. A. Campbell, J. Lovisek, B. MacKenzie, D. Servage, and T. N. Tobias. 1979. "Additional records of salamanders of the *Ambystoma jeffersonianum* complex from Ontario, Canada." *Herpetological Review* 10: 61–62.

Weller, W. F., and R. V. Palermo. 1976. "A northern range extension for the western chorus frog, *Pseudacris triseriata triseriata* (Wied), in Ontario." *Canadian Field-Naturalist* 90: 163–166.

Weller, W. F., W. G. Sprules, and T. P. Lamarre. 1978. "Distribution of salamanders of the *Ambystoma jeffersonianum* complex in Ontario." *Canadian Field-Naturalist* 92: 174–181.

Weller, W. F., and B. W. Menzel. 1979. "Occurrence of the salamander *Ambystoma platineum* (Cope) in southern Ontario." *Journal of Herpetology* 13: 193–197.

Wells, K. D., T. L. Taigen, S. W. Rusch, and C. C. Robb. 1995. "Seasonal and nightly variation in glycogen reserves of calling gray treefrogs (*Hyla versicolor*)." *Herpetologica* 51: 359–368.

Welsh, H. H., Jr. 1990. "Relictual amphibians and old-growth forests." *Conservation Biology* 4: 309–319.

Werner, E. E. 1992. "Competitive interactions between wood frog and northern leopard frog larvae: The influence of size and activity." *Copeia* 1992: 26–35.

Werner, E. E., and M. A. McPeek. 1994. "The roles of direct and indirect effects on the distributions of two frog species along an environmental gradient." *Ecology* 75: 1368–1382.

Werner, J. K., J. Weaselhead, and T. Plummer. 1999. "The accuracy of estimating eggs in anuran egg masses using weight or volume measurements." *Herpetological Review* 30: 30–31.

Westfall, M. J., and M. L. May. 1996. *Damselflies of North America.* Gainesville, FL: Scientific Publishers.

Wetzel, R. G. 1975. *Limnology*. Philadelphia, PA: W. B. Saunders Company.

Whigham, D. F., and T. E. Jordan. 2003. "Isolated wetlands and water quality." *Wetlands* 23: 541–549.

Whitford, W. G., and A. Vinegar. 1966. "Homing, survivorship, and overwintering larvae in spotted salamanders, *Ambystoma maculatum*." *Copeia* 1966: 515–519.

Whitlock, A. L., N. M. Jarman, and J. S. Larson. 1994. *WEThings: Wetland Habitat Indicators for Nongame Species. Wetland-dependent Amphibians, Reptiles, and Mammals of New England.* Vol. II. The Environmental Institute Publication 94-2. Amherst, MA: University of Massachusetts.

Wiggins, G. B. 1973. *A Contribution to the Biology of Caddisflies (Trichoptera) in Temporary Pools*. Life Sciences Contributions No. 88. Toronto, ON: Royal Ontario Museum.

Wiggins, G. B. 1977. *Larvae of the North American Caddisfly Genera (Trichoptera)*. Toronto, ON: University of Toronto Press.

Wiggins, G. B. 1996. *Larvae of the North American Caddisfly Genera (Trichoptera)*. 2nd Edition. Toronto, ON: University of Toronto Press.

Wiggins, G. B. 1998. *The Caddisfly Family Phryganeidae (Trichoptera)*. Toronto, ON: University of Toronto Press.

Wiggins, G. B., R. J. Mackay, and I. M. Smith. 1980. "Evolutionary and ecological strategies of animals in annual temporary pools." *Archiv für Hydrobiologie* (Supplement) 38: 97–206.

Wilbur, H. M. 1971. "The ecological relationship of the salamander *Ambystoma laterale* to its all-female, gynogenetic associate." *Evolution* 25: 168–179.

Wilbur, H. M. 1972. "Competition, predation, and structure of the *Ambystoma-Rana sylvatica* community." *Ecology* 53: 3–12.

Wilbur, H. M. 1976. "Density-dependent aspects of metamorphosis in *Ambystoma* and *Rana sylvatica*." *Ecology* 57: 1289–1296.

Wilbur, H. M. 1977a. "Density-dependent aspects of growth and metamorphosis in *Bufo americanus*." *Ecology* 58: 196–200.

Wilbur, H. M. 1977b. "Interactions of food level and population density in *Rana sylvatica*." *Ecology* 58: 206–209.

Wilbur, H. M. 1977c. "Propagule size, number, and dispersion pattern in *Ambystoma* and *Asclepias*." *The American Naturalist* 111: 43–68.

Wilbur, H. M., and J. P. Collins. 1973. "Ecological aspects of amphibian metamorphosis." *Science* 182: 1305–1314.

Williams, D. D. 1983. "The natural history of a Nearctic temporary pond in Ontario with remarks on continental variation in such habitats." *Internationale Revue der Gesamten Hydrobiologie* 68: 239–253.

Williams, D. D. 1987. *The Ecology of Temporary Waters*. Portland, OR: Timber Press.

Williams, D. D. 1991. "Life history traits of aquatic arthropods in springs." *Memoirs of the Entomological Society of Canada* 155: 63–87.

Williams, D. D. 1996. "Environmental constraints in temporary waters and their consequences for the insect fauna." *Journal of the North American Benthological Society* 15: 634–650.

Williams, D. D. 1997. "Temporary ponds and their invertebrate communities." *Aquatic Conservation* 7: 105–117.

Williams, D. D. 1998. "The role of dormancy in the evolution and structure of temporary water invertebrate communities." *Archiv für Hydrobiologie – Advances in Limnology* 52: 109–124.

Williams, D. D., and H. B. N. Hynes. 1976. "The ecology of temporary streams. I. The faunas of two Canadian streams." *Internationale Revue der Gesamten Hydrobiologie* 61: 761–787.

Williams, P. K. 1973. Seasonal movements and population dynamics of four sympatric mole salamanders, genus *Ambystoma*. PhD Dissertation, Zoology, Indiana University, Bloomingdale, IN.

Williamson C. E., and J. W. Reid. 2001. "Copepoda." Pp. 915–954 *in*: J. H. Thorp and A. P. Covich, (eds.). *Ecology and Classification of North American Freshwater Invertebrates*. New York, NY: Academic Press.

Wilson, M. S. 1959. "Calanoida." Pp. 738–794 *in*: W. T. Edmonson, (ed.). *Freshwater Biology*. 2nd Edition. New York, NY: J. Wiley and Sons.

Wilson, R. E. 1976. An ecological study of *Ambystoma maculatum* and *Ambystoma jeffersonianum*. PhD Dissertation, Biology, Cornell University, Ithaca, NY.

Windmiller, B. S. 1990. The limitations of Massachusetts regulatory protection for temporary pool breeding amphibians. MA Thesis, Biology, Tufts University, Medford, MA. Available at *http://www.hyla-ecological.com/publications.html*.

Windmiller, B. S. 1995a. "Habitat use by vernal pool amphibians." Seminar presented to Environmental Research Group, Massachusetts Audubon Society, Lincoln, MA. April 7, 1995.

Windmiller, B. S. 1995b. "Research findings relevant to vernal pool discussion." Memorandum submitted to Vernal Pool Work Group, Massachusetts Natural Heritage and Endangered Species Program, Massachusetts Division of Fisheries and Wildlife, Westboro, MA. March 24, 1995. Concord, MA: Hyla Ecological Services, Inc.

Windmiller, B. S. 1996. The pond, the forest, and the city: Spotted salamander ecology and conservation in a human-dominated landscape. PhD Dissertation, Biology, Tufts University, Medford, MA. Available at *http://www.hylaecological.com/publications.html*.

Windmiller, B. S. 1997. "Decline and extirpation of a wood frog population following large-scale construction in an adjacent upland forest, Shoppers World Vernal Pool, Framingham, Massachusetts." Report to Framingham Conservation Commission, October 20, 1997. Concord, MA: Hyla Ecological Services, Inc.

Windmiller, B. S., and L. P. Kenney. 1995. "Vernal pools and their protection." Presentation at Massachusetts Association of Conservation Commissions Annual Meeting, Holy Cross College, Worcester, MA. March 4, 1995.

Wipfli, M. S., and R. W. Merritt. 1994a. "Disturbance to a stream food web by a bacterial larvicide specific to black flies: Feeding responses of predatory macroinvertebrates." *Freshwater Biology* 32: 91–103.

Wipfli, M. S., and R. W. Merritt. 1994b. "Effects of *Baccillus thuringiensis* var. *israelensis* on nontarget benthic organisms through direct and indirect exposure." *Journal of the North American Benthological Society* 13: 190–205.

Wissinger, S. A., H. H. Whiteman, G. B. Sparks, G. L. Rouse, and W. S. Brown. 1999. "Foraging trade-offs along a predator–permanence gradient in subalpine wetlands." *Ecology* 80: 2102–2116.

Witham, C. W., E. T. Bauder, D. Belk, W. R. Ferren, Jr., and R. Ornduff, (eds.). 1998. *Ecology, Conservation, and Management of Vernal Pool Ecosystems — Proceedings from a 1996 Conference*. Sacramento, CA: California Native Plant Society.

Wolfe, G. W. 1980. "The larva and pupa of *Acilius fraternus fraternus* (Coleoptera: Dytiscidae) from the Great Smoky Mountains, Tennessee." *The Coleopterists Bulletin* 34: 121–126.

Wood, D. M., P. T. Dang, and R. A. Ellis. 1979. *The Mosquitoes of Canada (Diptera: Culicidae)*. Part 6, *The Insects and Arachnids of Canada*. Ottawa, ON: Biosystematics Research Institute.

Woodward, B. D. 1982a. "Local intraspecific variation in clutch parameters in the spotted salamander (*Ambystoma maculatum*)." *Copeia* 1982: 157–160.

Woodward, B. D. 1982b. "Tadpole competition in a desert anuran community." *Oecologia* 54: 96–100.

Woodward, B. D. 1983. "Predator-prey interactions and breeding-pond use of temporary-pond species in a desert anuran community." *Ecology* 64: 1549–1555.

Worth, C. B. 1972. *Of Mosquitoes, Moths, and Mice*. New York, NY: W. W. Norton.

Worthington, R. D. 1968. "Observations on the relative sizes of three species of salamander larvae in a Maryland pond." *Herpetologica* 24: 242–246.

Worthington, R. D. 1969. "Additional observations on sympatric species of salamander larvae in a Maryland pond." *Herpetologica* 25: 227–229.

Wright, A. H., and A. A. Wright. 1949. *Handbook of Frogs and Toads of the United States and Canada*. Ithaca, NY: Comstock Publishing Associates.

Wyman, R. L., and J. Jancola. 1992. "Degree and scale of terrestrial acidification and amphibian community structure." *Journal of Herpetology* 26: 392–401.

Yeatman, H. C. 1959. "Cyclopoida." Pp. 796–814 *in:* W. T. Edmondson, (ed.) *Freshwater Biology*. New York, NY: J. Wiley and Sons.

Young, F. N. 1960. "The water beetles of a temporary pond in southern Indiana." *Proceedings of the Indiana Academy of Science* 69: 154–164.

Zappalorti, R. T. 1993. "Habitat requirements document, chapter on vernal pool breeders." Forked River, NJ: Herpetological Associates, Plant and Wildlife Consultants, Inc.

Zedler, P. H. 1987. *The Ecology of Southern California Vernal Pools*. Biological Report 85(7.11). Washington, DC: US Fish and Wildlife Service.

Zedler, P. H. 2003. "Vernal pools and the concept of 'isolated wetlands.'" *Wetlands* 23: 597–607.

Zimmerman, J. R. 1960. "Seasonal population changes and habitat preferences in the genus *Laccophilus* (Coleoptera: Dytiscidae)." *Ecology* 41: 141–152.

Zimmerman, J. R. 1970. "A taxonomic revision of the aquatic beetle genus *Laccophilus* (Dytiscidae) of North America." *Memoirs of the American Entomological Society (Philadelphia)* 26: 1–275.

Zinn, D. J., and R. W. Dexter. 1962. "Reappearance of *Eulimnadia agassizii* with notes on its biology and life history." *Science* 137: 676–677.

Zweifel, R. G. 1989. "Calling by the frog, *Rana sylvatica*, outside the breeding season." *Journal of Herpetology* 23: 185–186.

Websites with General Information about Vernal Pools and Links to Other Sites

In addition to specific webpages noted above, the following sites will be useful to those who wish to learn more about temporary waters in the glaciated northeast and elsewhere.

Northeastern Vernal Pools

http://www.vernalpool.org. The Vernal Pool Association, an environmental club at Reading Memorial High School, Reading, MA, maintains a website with photos, information, and links to other sources of information on vernal pools, especially in New England.

California Vernal Pools

http://www.vernalpools.org. This website provides information on California vernal pools, including documents related to specific projects, some excellent bibliographies, and links to other websites with information about North American vernal pools.

Amphibian Conservation

http://www.parcplace.org/default.htm. The Partners in Amphibian and Reptile Conservation (PARC) maintains a website that provides broad information on amphibian and reptile conservation, including links to publications, agencies, programs, and projects.

Index